Monoamine Oxidase and its Inhibition

MARY L. C. BERNHEIM

The Ciba Foundation for the promotion of international cooperation in medical and chemical research is a scientific and educational charity established by CIBA Limited – now CIBA-GEIGY Limited – of Basle. The Foundation operates independently in London under English trust law.

Ciba Foundation Symposia are published in collaboration with Elsevier Scientific Publishing Company, Excerpta Medica, North-Holland Publishing Company, in Amsterdam.

Elsevier/Excerpta Medica/North-Holland, P.O. Box 211, Amsterdam

Monoamine Oxidase and its Inhibition

Ciba Foundation Symposium 39 (new series)

In honour of Mary L. C. Bernheim

1976

Elsevier · Excerpta Medica · North-Holland

Amsterdam · Oxford · New York

ISBN Excerpta Medica 90 219 4044 2
ISBN American Elsevier 0444-15205-9

Published in May 1976 by Elsevier/Excerpta Medica/North-Holland, P.O. Box 211, Amsterdam, and American Elsevier, 52 Vanderbilt Avenue, New York, N.Y. 10017.

Suggested series entry for library catalogues: Ciba Foundation Symposia
Suggested publisher's entry for library catalogues: Elsevier/Excerpta Medica/North-Holland

Ciba Foundation Symposium 39 (new series)

Library of Congress Cataloging in Publication Data

Symposium on Monoamine Oxidase and its Inhibition,
Monoamine oxidase and its inhibition.

(Ciba Foundation symposium ; 39 (new ser.))
Includes bibliographical references and indexes.
CONTENTS: Kety, S. S. Introduction.--Tipton, K. F., Houslay, M. D., and Mantle, T. J. The nature and locations of the multiple forms of monamine oxidase.--Maycock, A. L. et al. The action of acetylenic inhibitors on mitochondrial monoamine oxidase: structure of the flavin site in the inhibited enzyme. [etc.]
1. Monoamine oxidase--Congresses. 2. Monoamine oxidase inhibitors--Congresses. I. Bernheim, Molly. II. Title. III. Series: Ciba Foundation. Symposium ; new ser., 39. [DNLM: 1. Monoamine oxidase--Congresses. 2. Monoamine oxidase inhibitors--Congresses. W3 C161F v. 39 / QU135 M7503]
QP603.M6S95 1975 612'.0151'8 76-10396
ISBN 0-444-15205-9 (Amer. Elsevier)

Printed in The Netherlands by Van Gorcum, Assen

Contents

Participants

Symposium on Monoamine Oxidase and its Inhibition, held in honour of Dr Mary L. C. Bernheim, at the Ciba Foundation, London, 7th–9th May 1975

Chairman: S. S. KETY Department of Psychiatry, Harvard Medical School, Massachusetts General Hospital, Fruit Street, Boston, Massachusetts 02114, USA

H. K. F. BLASCHKO University Department of Pharmacology, South Parks Road, Oxford OX1 3QT, UK

A. J. COPPEN MRC Neuropsychiatry Laboratory, West Park Hospital, Epsom, Surrey, UK

T. J. CROW MRC Division of Psychiatry, Clinical Research Centre, Northwick Park Hospital, Watford Road, Harrow, Middlesex HA1 3UJ, UK

R. W. FULLER The Lilly Research Laboratories, Eli Lilly and Co., Indianapolis, Indiana 46206, USA

V. Z. GORKIN Institute of Biological and Medical Chemistry, Academy of Medical Sciences, Pogodinskaya Street 10, Moscow 119117, USSR

A. R. GREEN MRC Clinical Pharmacology Unit, Radcliffe Infirmary, Woodstock Road, Oxford OX2 6HE, UK

L. L. IVERSEN MRC Neurochemical Pharmacology Unit, Department of Pharmacology, Medical School, Hills Road, Cambridge CB2 2QD, UK

J. KNOLL Department of Pharmacology, Semmelweis University of Medicine, 1085 Budapest 8, Hungary

L. MAÎTRE CIBA-GEIGY Ltd, CH-4002, Basle, Switzerland

D. L. MURPHY Section on Clinical Neuropharmacology, Laboratory of Clinical Science, Building 10, 3S 229, National Institutes of Health, Bethesda, Maryland 20014, USA

N. H. NEFF Section on Biochemical Pharmacology, National Institute of Mental Health, Division of Special Mental Health Research, Saint Elizabeths Hospital, WAW Building, Washington D.C. 20032, USA

L. ORELAND Department of Pharmacology, University of Umeå, S-901 87 Umeå, Sweden

C. M. B. PARE Department of Psychological Medicine, St Bartholomew's Hospital, London EC1A 7EB, UK

A. PLETSCHER Research Department, F. Hoffmann-La Roche & Co. Ltd, CH-4002 Basle, Switzerland

H. M. VAN PRAAG Department of Biological Psychiatry, Psychiatric Clinic, State University of Groningen, Groningen, The Netherlands

O. J. RAFAELSEN Psychochemistry Institute, Rigshospitalet, 9 Blegdamsvej, DK-2100 Copenhagen Ø, Denmark

SIR MARTIN ROTH Department of Psychological Medicine, The Royal Victoria Infirmary, Queen Victoria Road, Newcastle upon Tyne NE1 4LP, UK

M. SANDLER Department of Chemical Pathology, Queen Charlotte's Maternity Hospital, Goldhawk Road, London W6 OXG, UK

D. F. SHARMAN ARC Institute of Animal Physiology, Babraham, Cambridge CB2 4AT, UK

T. P. SINGER Veterans Administration Hospital, 4150 Clement Street, San Francisco, California 94121, USA

T. L. SOURKES Department of Psychiatry, McGill University, 1025 Pine Avenue West, Montreal H3A 1A1, Quebec, Canada

K. F. TIPTON Department of Biochemistry, University of Cambridge, Tennis Court Road, Cambridge CB2 1QW, UK

U. TRENDELENBURG Institut für Pharmakologie und Toxikologie der Universität Würzburg, 8700 Würzburg, Germany

M. B. H. YOUDIM MRC Clinical Pharmacology Unit, Radcliffe Infirmary, Woodstock Road, Oxford OX2 6HE, UK

Editors: G. E. W. WOLSTENHOLME and JULIE KNIGHT

Introduction

S. S. KETY

Department of Psychiatry, Harvard Medical School, Boston, Massachusetts

This symposium will be concerned with an important enzyme that was isolated from the liver of rabbits nearly fifty years ago. Its relevance to psychiatry was recognized thirty years later by a serendipitous discovery—undesirable behavioural side-effects in a drug that was useful in the treatment of tuberculosis. Monoamine oxidase has an interesting history, and a history that includes many of the most important names in biochemistry in the past and present generations. I am indebted to Dr Blaschko for the details of this history, which is well outlined in his paper in *Pharmacological Reviews* (1952).

We can trace the history of amine oxidases back to Schmiedeberg, who in 1877 showed that benzylamine was excreted by the dog as hippuric acid, from which he recognized that there was a necessary step in that conversion representing an oxidation of benzylamine and a deamination to benzoic acid.

In 1883 Minkowski demonstrated *in vitro* the conversion of benzylamine to benzoic acid. In 1910 Ewins and Laidlaw perfused the liver of the cat and the rabbit with tyramine, which they found to be converted to *p*-hydroxyphenylacetic acid. They also found that tryptamine was oxidized to indoleacetic acid. They made the interesting observation that tyramine could be oxidatively deaminated on perfusion through muscle, but they pointed out that it is only in muscles with a rich sympathetic nerve supply that this conversion takes place. Muscles that were poor in sympathetic nerve endings would not make this conversion.

The characterization of the enzyme responsible for this conversion waited until 1928, when Mary Hare, then 26 years old and a postgraduate student at Newnham College in Cambridge, described a new enzyme system in the liver—tyramine oxidase. In the same year (1928) in which she won her doctorate she also married another biochemist in Cambridge who had come from the United States, Frederick Bernheim. With him she returned to the States, first to Johns Hopkins and then to join the faculty of the new school of medicine at Duke University. Since that time, Dr Bernheim has been very active, both in biochemistry and in many other pursuits.

In that classic paper (Hare 1928) she showed that the oxidation of tyramine by liver homogenates and particulate fractions of the liver was not inhibited by cyanide and therefore was an exception to Warburg's dictum that all oxidation in cells is dependent upon iron. From that she concluded that this oxidation was not produced by tyrosinase, which is blocked by cyanide, and she concluded that she was working with a new and undescribed enzyme. Although she found that her enzyme preparation did not act on catecholamines, in 1937 Dr Blaschko and his co-workers were able to demonstrate this important action, after excluding the autoxidation of adrenaline in aqueous solution. It was at that time that the concept of an amine oxidase with generalized oxidatively deaminating properties was developed. Zeller (1938) then pointed out the existence of histamine deamination and enzymes responsible for that reaction, and proposed monoamine oxidase as the name for the class of enzymes responsible for the oxidative deamination of the monoamines, and diamine oxidase for histaminase. The presence of monoamine oxidase in brain was first demonstrated by Pugh & Quastel in 1937. As the catecholamines became increasingly important, with the recognition of chemical nervous transmission and the growth of evidence that noradrenaline was the transmitter at peripheral adrenergic endings, attention focused on MAO as the counterpart of acetylcholinesterase for the sympathetic nervous system. This was rather controversial; there were many biochemical reasons why it did not seem to be likely. In 1957 Axelrod described catechol *O*-methyltransferase, and later the reuptake of noradrenaline by the presynaptic ending, as the main mechanisms for the synaptic inactivation of catecholamines. (See Axelrod 1959.)

The relationship of monoamine oxidase to psychiatry is an interesting historical development. Iproniazid was being used successfully in the treatment of tuberculosis in the fifties, but it had one serious disadvantage. In some patients it produced an undesirable behavioural activation. One of the early papers reporting the use of this substance in the treatment of tuberculosis in a sanatorium described the patients 'dancing in the halls'—and it wasn't because they had seen their X-ray pictures! It was soon realized that iproniazid was a nervous system activator of some kind. In 1955 Zeller *et al.* showed it to be a monoamine oxidase inhibitor and the idea came, first, I believe, to George Crane (1956), that the undesirable side-effect in the treatment of tuberculosis might be a beneficial effect in the treatment of depression. Early trials supported this idea and several other monoamine oxidase inhibitors, less toxic than iproniazid, were then developed, by several pharmaceutical houses, with the knowledge that the inhibition of monoamine oxidase was the prerequisite for that activity. We shall have some discussion at this symposium on the therapeutic effectiveness of monoamine oxidase inhibitors in depression and their comparison with other antidepressant drugs.

The amount of monoamine oxidase in platelets was found by Murphy & Wyatt (1972) to be reduced in patients with schizophrenia. This was the first definitive biochemical finding in schizophrenia—made by competent scientists and with a unique control of schizophrenia-associated artifacts. It has been followed by a number of confirmatory reports, but also two which failed to find a difference. In the course of this symposium we shall have an opportunity to discuss that controversy and perhaps find some explanation of the lack of ability to confirm this result by some workers, and to decide whether this is an artifact or really characteristic of some forms of schizophrenia.

The history of monoamine oxidase and its implications in psychiatry and the reason why there is now a Ciba Foundation symposium on this topic is of interest in more than one way. I think it provides a salutary description of how scientific discoveries are made and how one moves from the accumulation of fundamental knowledge to knowledge of immediate social value. It may also have a sobering effect on the current tendencies to target research. One wonders whether any committee charged with programming research on mental illness would have supported Miss Hare in her studies of an enzyme in the liver, the implications of which for the nervous system were obscure. Apart from the interesting observations of Ewins & Laidlaw that the sympathetic innervation of muscle seemed to have something to do with this oxidase activity, one could not have imagined that the enzyme would have had important implications for the nervous system. Miss Hare in the discussion in her 1928 paper on the characteristics of the enzyme commented on its possible physiological role; she suspected that its important function was to detoxify tyramine and other amines coming into the body from bacterial putrefaction in the gut, by way of the liver. Even she did not foresee implications for behaviour and for psychiatric therapy. It was not until the serendipitous discovery of the behavioural activation produced by monoamine oxidase inhibitors as a side-effect in the treatment of tuberculosis that a committee concerned with its relevance to psychiatry could have taken up the matter. And, by that time, there were enough interested scientists able to recognize the implications for the nervous system and for psychiatry to take it up without the necessity of directing their efforts. In any case, our conference will recapitulate that history, in more modern terms; it is going to consider some of the biochemical aspects of the subject, then the pharmacological aspects, and finally the psychiatric implications of this interesting enzyme and its inhibitors.

References

Axelrod, J. (1959) The metabolism of catecholamines *in vivo* and *in vitro*. *Pharmacol. Rev. 11*, 402–408

Crane, G. E. (1956) Psychiatric side effects of iproniazid. *Am. J. Psychiatr. 112*, 494

Blaschko, H., Richter, D. & Schlossmann, H. (1937) The inactivation of adrenaline. *J. Physiol. (Lond.) 90*, 1-19

Ewins, A. J. & Laidlaw, P. P. (1910) The fate of *p*-hydroxyphenylethylamine in the organism. *J. Physiol. (Lond.) 41*, 78-87

Hare, M. L. C. (1928) Tyramine oxidase. I. A new enzyme system in liver. *Biochem. J. 22*, 968-979

Minkowski, O. (1883) Über Spaltungen im Thierkörper. *Arch. Exp. Pathol. Pharmakol. 17*, 445-465

Murphy, D. L. & Wyatt, R. J. (1972) Reduced monoamine oxidase activity in blood platelets from schizophrenic patients. *Nature (Lond.) 238*, 225-226

Pugh, C. E. M. & Quastel, J. H. (1937) Oxidation of aliphatic amines by brain and other tissues. *Biochem. J. 31*, 286-291

Schmiedeberg, O. (1877) Über das Verhältnis des Ammoniaks und der primären Monoaminbasen zur Harnstoffbildung im Thierkörper. *Naunyn-Schmiedebergs Arch. Pharmakol. Exp. Pathol. 8*, 1-14

Zeller, E. A. (1938) Über den enzymatischen Abbau von Histamin und Diaminen. *Helv. Chim. Acta 21*, 881-890

Zeller, E. A., Barsky, J. & Berman, E. R. (1955) Amine oxidases. XI. Inhibition of monoamine oxidase by 1-isonicotinyl-2-isopropylhydrazine. *J. Biol. Chem. 214*, 267-274

The nature and locations of the multiple forms of monoamine oxidase

K. F. TIPTON, M. D. HOUSLAY and T. J. MANTLE

Department of Biochemistry, University of Cambridge

Abstract The apparent multiplicity of monoamine oxidase preparations from several sources may be abolished by treatment of soluble preparations with chaotropic agents such as sodium perchlorate. This treatment, which causes no loss of enzyme activity, results in the release of lipid material from the enzyme. Abolition of the multiple forms may also be effected by elution of the enzyme from DEAE-cellulose with the detergent Triton X-100, although in this case there is some loss of activity. Solubilization of the enzyme by prolonged sonication and detergent treatment causes a decrease in its sensitivity to selective inhibitors and a change in the kinetic reaction mechanism obeyed. Treatment of rat liver mitochondrial outer membranes with a chaotropic agent under mild conditions liberates a soluble form of the enzyme that is similar to the membrane-bound enzyme in its sensitivity to selective inhibitors and more vigorous treatment of this preparation with the chaotropic agent results in the loss of multiple forms without loss in activity. The use of the selective inhibitors clorgyline and deprenyl indicates that the majority of organs in the rat contain two enzyme species but some rat organs, such as spleen and testis, contain predominantly only a single species of the A type. The substrate specificities of the two species from rat liver are delineated.

The concept of a multiplicity of monoamine oxidases arose from studies of the temperature stabilities and inhibitor sensitivities of the activities of monoamine oxidase (amine: oxygen oxidoreductase [deaminating] [flavin-containing]; EC 1.4.3.4) towards different amine substrates (see Squires 1968; Tipton 1975 and Youdim 1973 for reviews). Use of the selective inhibitor clorgyline enabled Johnston (1968) to classify two major types of activity in mitochondrial preparations from a number of sources. These two species were designated the A and B fractions, the former being active towards 5-hydroxytryptamine and tyramine and being more sensitive to inhibition by clorgyline than the B fraction, which was active towards benzylamine and tyramine. Subsequent work with this and other selective inhibitors confirmed this work and indicated that the two forms of monoamine oxidase could be detected in mitochondria from many, but by no means all, sources and that the relative proportions of these two forms varied widely between different organs and species (see e.g. Squires 1972; Knoll & Magyar 1972; Neff & Goridis 1972).

The use of polyacrylamide gel electrophoresis (Collins *et al.* 1968) and cellulose acetate electrophoresis (Kim & D'Iorio 1968) considerably complicated this situation in that up to five bands of monoamine oxidase activity could be separated when partly purified soluble preparations of the enzyme were treated in this way. In addition the properties of the separated forms did not correspond to those detected in intact mitochondria from the same sources. Thus differences in the sensitivities of the forms separated from rat brain mitochondria to a number of selective inhibitors were not nearly great enough to account for the differences observed with mitochondrial preparations from this source (Youdim *et al.* 1969; Squires 1972; Collins *et al.* 1972). Such anomalies led us to suggest that the procedures used in making soluble preparations of the enzyme might affect the properties and the apparent multiplicity of the enzyme (see e.g. Tipton 1972).

THE NATURE OF THE MULTIPLE FORMS

In the cell monoamine oxidase is tightly bound to the mitochondrial outer membrane and the techniques necessary to solubilize it for electrophoretic examination have usually involved prolonged sonication and treatment with a detergent. The possibility that those rather vigorous processes could lead to the formation of artifactually modified forms of the enzyme, by causing a single enzyme species to be released with varying amounts or types of membrane material bound to it, received support from our observation that the forms that could be separated by electrophoresis of soluble preparations of rat liver monoamine oxidase had widely different phospholipid contents (Tipton 1972; Houslay & Tipton 1973*a*). Further evidence in favour of this idea came from the results of experiments in which solubilized preparations of the enzyme were treated with chaotropic agents such as sodium perchlorate or sodium thiocyanate. When monoamine oxidase obtained from rat liver (Houslay & Tipton 1973*a*) or human brain (Tipton *et al.* 1973) was treated with a chaotropic agent under controlled conditions a considerable amount of lipid material, which could be separated by gel filtration, was liberated from the preparation and the resultant material migrated as a single band of activity on polyacrylamide gel electrophoresis. In addition the material treated in this way, which had lost no activity, did not respond in a biphasic manner to selective inhibitors

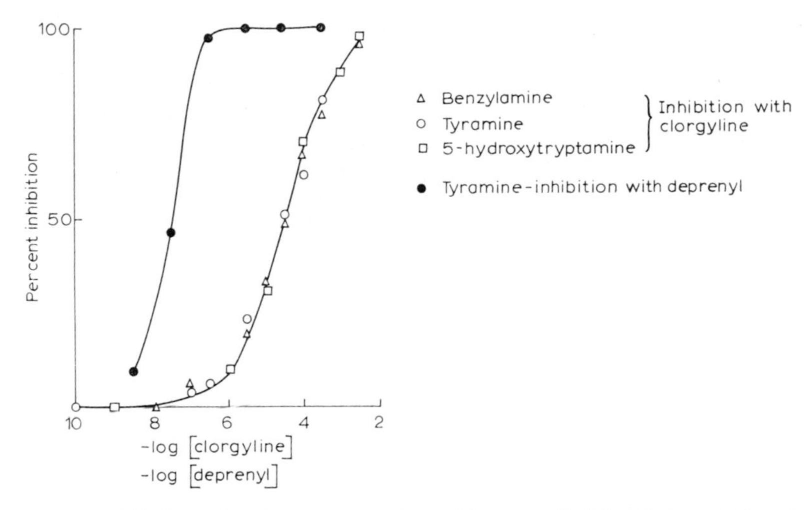

FIG. 1. Inhibition of rat liver monoamine oxidase, purified by Triton elution from DEAE-cellulose, by clorgyline and deprenyl. The enzyme was purified by a procedure involving elution from DEAE-cellulose by 0.075% Triton (Houslay & Tipton 1975*a*). Samples were incubated with the indicated concentrations of clorgyline or deprenyl for 45 minutes before the activity was assayed with the substrate indicated at a concentration of 1.0 mM at 30°C.

and showed no differential loss of activity towards different substrates when subjected to thermal denaturation.

The apparent multiplicity of rat liver mitochondrial monoamine oxidase can also be abolished by eluting the enzyme from DEAE-cellulose with the non-ionic detergent Triton X-100 (Houslay & Tipton 1975*a*). A partly purified preparation of rat liver mitochondrial monoamine oxidase may be absorbed on to DEAE-cellulose in 20 mM-phosphate buffer at pH 7.4 and inclusion of 0.075% Triton in the buffer causes the enzyme to be eluted as a sharp peak with the detergent front. The enzyme treated in this way migrates as a single band of activity on polyacrylamide gel electrophoresis and, whereas the enzyme applied to the DEAE-cellulose showed a selective response to the inhibitors clorgyline and deprenyl, the eluted material showed no such effect, as shown in Fig. 1. This effect is particularly interesting since it had been previously shown that electrophoresis of a partly purified preparation of the enzyme in the presence of Triton resulted in the abolition of the multiple forms of the enzyme (Tipton 1972). Further purification of the enzyme that had been eluted with Triton resulted in an enzyme preparation which appeared to be a homogeneous protein by the criterion of polyacrylamide gel electrophoresis.

These results strongly imply that the electrophoretically separable multiple forms of the enzyme result from the binding of lipid material and that treat-

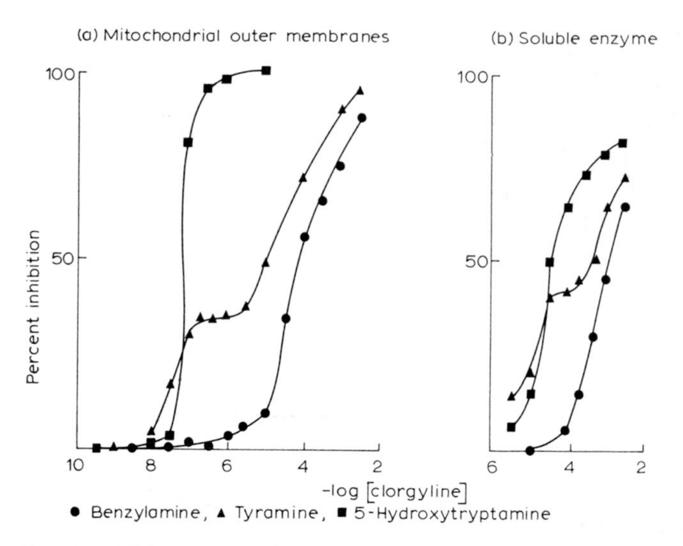

FIG. 2. Inhibition of rat liver monoamine oxidase by clorgyline. Two enzyme preparations were used: (*a*) a preparation of mitochondrial outer membranes (Houslay & Tipton 1973*b*) and (*b*) a preparation of enzyme that had been rendered soluble by prolonged sonication and treatment with Triton X-100 (see e.g. Houslay & Tipton 1973*a*). Other details were as described in the legend to Fig. 1.

ment with chaotropic reagents or elution from DEAE-cellulose with Triton causes release of bound lipid material and abolition of the multiple forms. Indeed the phospholipid associated with the single band of activity in monoamine oxidase preparations that had been treated in this way was considerably reduced. It is tempting to conclude that the vigorous procedures used in 'solubilizing' the enzyme do indeed result in the generation of these forms, although the fact that an enzyme preparation that is sufficiently negatively charged to bind to DEAE-cellulose at pH 7.2 nevertheless contains a component that is positively charged at pH 9.2 would argue that the electrophoretic procedure itself may generate some of the bands of activity. Indeed one of the forms has been shown to be an artifact of the loading procedure used in the electrophoresis experiments (Houslay & Tipton 1973*a*).

The relevance of these results to the situation within the cell is not so clear, since the soluble preparation constitutes only about half the activity that was originally present in the mitochondria and, although the response of such preparations to selective inhibitors is broadly similar to the response of membrane-bound preparations of the enzyme, the soluble enzyme is considerably less sensitive to such inhibitors, as shown in Fig. 2. This shows a comparison of the effects of clorgyline on the activity of rat liver monoamine oxidase in a

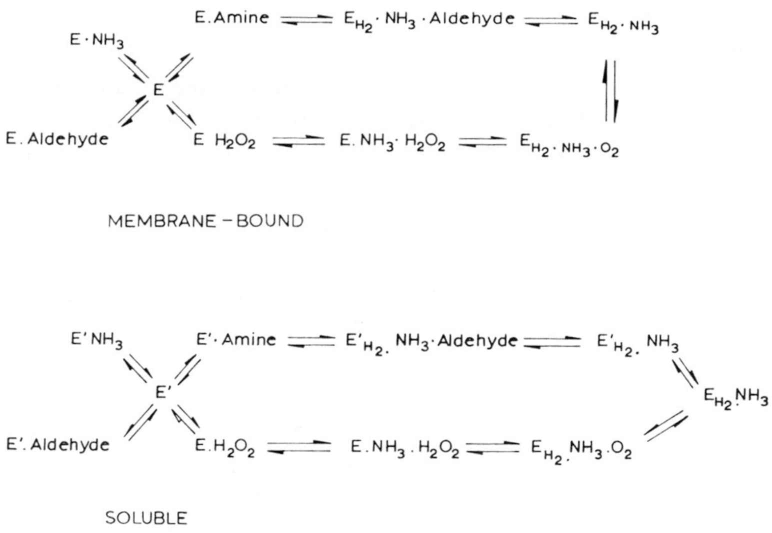

E.Amine ⇌ E_{H_2}·NH_3·Aldehyde ⇌ E_{H_2}·NH_3

E·NH_3

E

E.Aldehyde

E H_2O_2 ⇌ E.NH_3·H_2O_2 ⇌ E_{H_2}·NH_3·O_2

MEMBRANE – BOUND

E′NH_3

E′·Amine ⇌ E'_{H_2}.NH_3·Aldehyde ⇌ E'_{H_2}.NH_3

E′

E_{H_2}.NH_3

E′.Aldehyde

E.H_2O_2 ⇌ E.NH_3.H_2O_2 ⇌ E_{H_2}.NH_3.O_2

SOLUBLE

FIG. 3. The kinetic mechanisms of rat liver monoamine oxidase. The kinetic mechanisms of the enzyme bound to mitochondrial outer membranes are compared with those of a soluble preparation purified as described in the legend to Fig. 1. The species designated E and E′ present two different conformations of the enzyme and the subscript H_2 denotes the reduced (or partially reduced) form of the enzyme.

preparation of mitochondrial outer membranes and in a preparation that had been rendered soluble by sonication and treatment with Triton X-100. It is clear that the solubilization procedure has rendered both components less sensitive to the inhibitor and has also decreased the difference between the two activities in terms of inhibitor sensitivity.

This effect may be due to the vigorous procedures used in rendering the enzyme soluble causing a conformational change, to a conformational change occurring when the enzyme is released from a non-polar environment to one that is more polar, or simply to partitioning of the relatively lipophilic inhibitor between the aqueous phase and the less polar environment of the outer membrane.

Removal of the enzyme from mitochondrial outer membranes also causes a small but significant alteration in the kinetic mechanism obeyed by the enzyme (Houslay & Tipton 1973*b*, 1975*a*). The soluble and membrane-bound preparations obey broadly similar kinetic mechanisms but the kinetic data obtained with the former preparation are best interpreted in terms of a compulsory isomerization step in the reduced enzyme–ammonia complex before it can bind oxygen. A comparison between the kinetic mechanisms obeyed by the two preparations is shown in Fig. 3. It is suggested that this conformational change, which does not appear to occur with the membrane-bound preparation, allows the soluble enzyme to adopt a conformation that is similar to that existing in the membrane-bound form, which is constrained in the 'E' conformation throughout the catalytic cycle. A related difference between the soluble and membrane-bound preparations is that the former is less sensitive to inhibition by the products ammonia and benzaldehyde when K_i values are determined with respect to the amine substrate (Houslay & Tipton 1975*a*). This decreased sensitivity is also shown by the perchlorate-treated soluble enzyme. This effect may result from the different conformations of the two enzyme species (see Fig. 3), since in the soluble preparation these products will inhibit by binding to the free enzyme species 'E′' whereas in the membrane-bound preparation they will interact with 'E'.

Whatever their causes, these changes that accompany solubilization of the enzyme have led us to conclude that work with preparations of the enzyme that had been rendered soluble by the vigorous procedures frequently used (and with the multiple forms that may be separated from them) cannot be expected to provide useful information on the nature and control of the monoamine oxidase activities within the cell.

In an investigation of milder procedures for obtaining soluble preparations of the enzyme we have found that treatment of preparations of rat liver mitochondrial outer membranes with a chaotropic agent under mild conditions followed by gel filtration in the presence of Triton X-100 results in the release of enzyme from membranes. This soluble form of the enzyme can be obtained in very high yields ($\sim$ 90% of that originally present in the outer membranes), and it behaves in a very similar way to the membrane-bound enzyme in its response to selective inhibitors such as deprenyl (Fig. 4). More vigorous treatment with a chaotropic agent in the manner previously described (Houslay & Tipton 1973*a*) could be used to convert this material into a preparation that showed no indication of multiplicity in its response to selective inhibitors without loss of activity. These enzyme preparations are compared in Fig. 4. Clearly it is possible to convert the membrane-bound enzyme into a soluble form that still retains the inhibitor sensitivity associated with the starting material. This result may suggest that the decreased inhibitor sensitivity in preparations that have been rendered soluble by sonication and detergent treatment is due to these procedures damaging the enzyme, although the possibility that the difference may result from different amounts or types of lipid material being bound to the two soluble preparations cannot be excluded.

The observation that the membrane-bound enzyme could be converted into a species that appeared to be homogeneous, in terms of monoamine oxidase activity, without loss of enzyme activity would suggest that, like the soluble

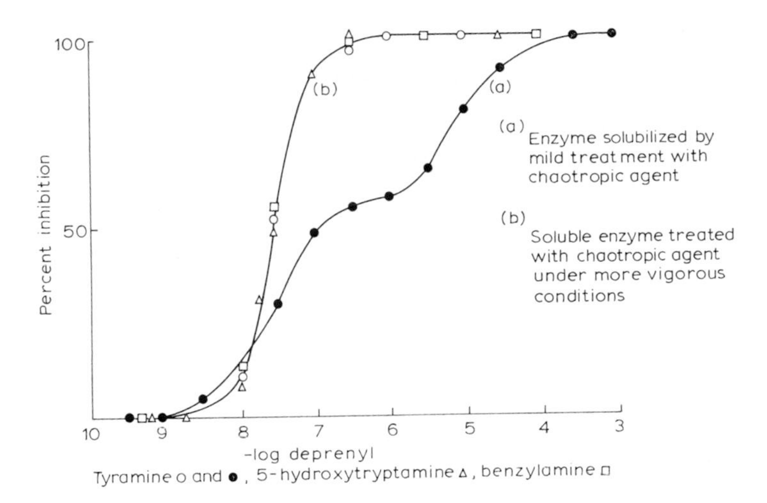

FIG. 4. Inhibition of soluble preparations of monoamine oxidase obtained from rat liver mitochondrial outer membranes by deprenyl. The preparations used were (*a*) a soluble enzyme prepared by treatment of the outer membranes with a chaotropic agent under mild conditions followed by gel filtration in the presence of Triton (assayed with tyramine as the substrate) and (*b*) the material resulting from treatment of this preparation with a chaotropic agent as previously described (Houslay & Tipton 1973*a*), assayed with (○) tyramine, (△) 5 hydroxytryptamine or (□) benzylamine. Other details are as in the legend to Fig. 1.

enzyme, the multiple forms of the membrane-bound enzyme result from the action of environmental factors on a single enzyme species. The conclusion that there is only a single species of monoamine oxidase in a given organ and that its activity may be modified by binding to membrane material indicates that any changes in the specificities of the enzyme that accompany changes in mental state should best be interpreted in terms of changes in the environment of the enzyme rather than in the enzyme protein itself.

MULTIPLICITY IN MEMBRANE-BOUND PREPARATIONS

The use of selective inhibitors has provided useful information on the situation that is likely to exist within the cell, and there is a considerable body of evidence to indicate that there are two major species of enzyme activity in mitochondria from a number of sources (see e.g. Squires 1972; Neff *et al.* 1973; Houslay & Tipton 1974; Mantle *et al.* 1975). The proportions of these two forms have been found to vary widely between different organs and in some organs it has been found that there is essentially only one species present. In order to build up a picture of the situation in a single animal species we have

TABLE 1

The distribution of the two major species of monoamine oxidase activity in the rat

Organ	*Species A* %	*Species B* %	*References*
Kidney	~70	~30	Squires (1972)
Intestine	~70	~30	Squires (1972)
Intestinal mucosa	60–70	30–40	This work
Brain	~55	~45	Squires (1972)
Spleen	>95	< 5	This work
Lung	~50	~50	This work
Testis	<90	>10	This work
Liver	~40	~60	Squires (1972) Hall *et al.* (1969)
Liver parenchymal cells	~50	~50	This work
Superior cervical ganglia	~90	~10	Neff & Goridis (1972)
Pineal gland	~15	~85	Neff & Goridis (1972)
Denervated pineal gland	< 5	>95	Neff & Goridis (1972)
Vas deferens	~50	~50	Jarrott (1971)
Denervated vas deferens	~35	~65	Jarrott (1971)

investigated the occurrence of the major forms in a number of organs from the rat. These results extend those of previous workers, whose data are included with our own in Table 1.

The proportions of the two forms do indeed appear to vary widely and some organs, including the spleen, seem to be largely composed of the A species. Since the spleen shares an embryological origin with a component of the non-parenchymal cells of liver it was of interest to see if the parenchymal cells contributed one form to the duality of monoamine oxidase in liver, whereas the other form was contributed by the non-parenchymal cells. Parenchymal cells from rat liver were prepared by the method of Howard & Pesch (1968). The table shows that the parenchymal cells closely resemble whole liver in their content of the two forms.

Since there is an adrenergic innervation in the non-parenchymal cells of rat liver (Ungváry & Donáth 1969), it was of interest to investigate the effect of adrenergic denervation on the activities of the two species. This was attempted by treating male rats with 6-hydroxydopamine (injections of 30 mg/kg on day 1, day 2 and day 8 before killing on day 9). Rats treated in this way showed typical features of chemical sympathectomy in that electron microscopy revealed an absence of nerve terminals in the heart and noradrenaline in the heart was depleted. This treatment, however, resulted in no significant change in the specific activity of the liver monoamine oxidase towards tyramine as compared with controls (rats injected with saline instead of 6-hydroxydopamine) and the ratio of the enzyme species determined using clorgyline was unchanged (M. D. Houslay & G. Lyles, unpublished observations). This evidence, together with the results obtained with separated parenchymal cells, suggests that these two enzyme species may exist in similar proportions in all major cell types in rat liver. Such a distribution would be in agreement with the results of Jarrott (1971) who demonstrated in denervation experiments that both species existed in neuronal and extraneuronal cells of rat vas deferens, and with those of Goridis & Neff (1973) who reached similar conclusions in the case of rat mesenteric arterial monoamine oxidase, although in this case the A species appeared to predominate in the neuronal cells. However, the amount of activity associated with the adrenergic innervation may be too small to be significant.

It is interesting that testis contains predominantly the A species of monoamine oxidase activity, since it has been reported that spermatogenesis and testicular weight are adversely affected by administered 5-hydroxytryptamine (see e.g. Salgado & Green 1955; Boccabella *et al.* 1962; O'Steen 1963), which is a substrate for this form of the enzyme.

In an attempt to see if enzyme activities of predominantly the A type responded to chaotropic agents in the same way as the A/B mixtures previously used, a soluble preparation of monoamine oxidase was made from rat spleen mitochondria in the way previously described for rat liver monoamine oxidase (Houslay & Tipton 1973*a*). This soluble preparation resembled the membrane-bound material in spleen in being predominantly (greater than 95%) of the A type. Preliminary experiments indicated that treatment of this material with perchlorate under the conditions previously described (Houslay & Tipton 1973*a*) resulted in its conversion to a form that was similar to the perchlorate-treated enzymes from liver and brain in bearing a closer resemblance to the B species in terms of its inhibitor sensitivity.

THE SPECIFICITIES OF THE MULTIPLE ACTIVITIES IN RAT LIVER

We have made a detailed study of the specificities of the two major forms of activity in rat liver mitochondrial outer membranes using the inhibition of the enzyme by clorgyline to assess whether a given amine was a substrate for both activities or for only one. The results obtained in this way were validated by use of the reversible inhibitors benzyl cyanide and 4-cyanophenol. These compounds act as competitive inhibitors of the B enzyme and mixed inhibitors of the A enzyme, whereas with respect to amines that are substrates for both enzyme activities, inhibition plots are non-linear (Houslay & Tipton 1974). The specificities worked out in these ways, which are shown in Table 2, confirm

TABLE 2

The specificities of monoamine oxidase species in rat liver mitochondrial outer membranes

Substrate	*Oxidized by species A*	*Oxidized by species B*
5-Hydroxytryptamine	+	—
Adrenaline	+	—
3-*O*-Methyladrenaline	+	—
Noradrenaline	+	—
3-*O*-Methylnoradrenaline	+	—
Octopamine	+	—
Tyramine	+	+
Dopamine	+	+
3-Methoxytyramine	+	+
Benzylamine	—	+
4-Hydroxybenzylamine	—	+
Vanillylamine	—	+
2-Phenylethylamine	—	+
3,4-Dimethoxy-2-phenylethylamine	—	+
4-Methoxy-2-phenylethylamine	—	+
Tryptamine	—	+
5-Methoxytryptamine	—	+

+, activity; —, no activity.

and extend those of previous workers (Johnston 1968; Hall *et al.* 1969; Neff *et al.* 1973; Neff & Goridis 1972). Since the multiple activities of monoamine oxidase preparations result from a single enzyme species being bound in different lipid environments, it follows that different lipid compositions of mitochondria from different species and organs could result in the specificities of the major forms differing from source to source and thus the results obtained with other systems may differ from those reported here for rat liver.

An interesting observation that has been made by a number of workers is that the method of mixed substrates (Dixon & Webb 1964), which should provide a sensitive indication of the presence of more than one enzyme species, fails to provide any indication of multiplicity in the case of monoamine oxidase. We have shown that this paradox may be resolved by the assumption that any substrate will bind to both forms of the enzyme with similar affinity constants but that in most cases it will act as a substrate for one species and an inhibitor of the other (Houslay *et al.* 1974). This conclusion, that substrates for one species will act as inhibitors of the other, imposes certain restrictions on the likely effects of selective inhibition of one species (Houslay & Tipton 1975*b*). Thus one would expect a 'damping out' of the specific effects of selective inhibitors *in vivo*, since inhibition of one species of the enzyme would be expected to cause an elevation of the levels of the substrates for that species and thus result in reversible inhibition of the other species.

ACKNOWLEDGEMENTS

K.F.T. is grateful to the Medical Research Council for a project grant. We are grateful to Mr N. J. Garrett for skilled technical assistance.

References

BOCCABELLA, A. V., SALGADO, E. D. & ALGER, E. A. (1962) Testicular function and histology following serotonin administration. *Endocrinology 71*, 827-837

COLLINS, G. G. S., YOUDIM, M. B. H. & SANDLER, M. (1968) Isoenzymes of human and rat liver monoamine oxidase. *FEBS Lett. 1*, 215-218

COLLINS, G. G. S., YOUDIM, M. B. H. & SANDLER, M. (1972) Multiple forms of monoamine oxidase. Comparison of *in vitro* and *in vivo* inhibition patterns. *Biochem. Pharmacol. 21*, 1995-1998

DIXON, M. & WEBB, E. C. (1964) *Enzymes*, pp. 84-87, Longman, London

GORIDIS, C. & NEFF, N. H. (1973) Selective localisation of monoamine oxidase forms in rat mesenteric artery, in *Frontiers in Catecholamine Research* (Snyder, S. H. & Usdin, E., eds.), pp. 157-170, Pergamon Press, London & New York

HALL, D. W. R., LOGAN, B. W. & PARSONS, G. H. (1969) Further studies on the inhibition of monoamine oxidase by M&B 9302 (clorgyline). I. Substrate specificity in various mammalian species. *Biochem. Pharmacol. 18*, 1447-1454

HOUSLAY, M. D. & TIPTON, K. F. (1973*a*) The nature of the electrophoretically separable multiple forms of rat liver monoamine oxidase. *Biochem. J. 135*, 173-186

HOUSLAY, M. D. & TIPTON, K. F. (1973*b*) The reaction pathway of membrane-bound rat liver mitochondrial monoamine oxidase. *Biochem. J. 135*, 735-750

HOUSLAY, M. D. & TIPTON, K. F. (1974) A kinetic evaluation of monoamine oxidase activity in rat liver mitochondrial outer membranes. *Biochem. J. 139*, 645-652

HOUSLAY, M. D. & TIPTON, K. F. (1975*a*) Rat liver mitochondrial monoamine oxidase: a change in the reaction mechanism on solubilization. *Biochem. J. 145*, 311-321

HOUSLAY, M. D. & TIPTON, K. F. (1975*b*) Amine competition for oxidation by rat liver mitochondrial monoamine oxidase. *Biochem. Pharmacol. 24*, 627-631

HOUSLAY, M. D., GARRETT, N. J. & TIPTON, K. F. (1974) Mixed substrate experiments with human brain monoamine oxidase. *Biochem. Pharmacol. 23*, 1937-1944

HOWARD, R. B. & PESCH, L. A. (1968) Respiratory activity of intact, isolated parenchymal cells from rat. *J. Biol. Chem. 243*, 3105-3109

JARROTT, B. (1971) Occurrence and properties of monoamine oxidase in adrenergic neurones. *J. Neurochem. 18*, 7-16

JOHNSTON, J. P. (1968) Some observations upon a new inhibitor of monoamine oxidase in brain tissue. *Biochem. Pharmacol. 17*, 1285-1297

KIM, H. C. & D'IORIO, A. (1968) Possible isoenzymes of monoamine oxidase in rat tissue. *Can. J. Biochem. 46*, 295-297

KNOLL, J. & MAGYAR, K. (1972) Some puzzling pharmacological effects of monoamine oxidase inhibitors, in *Monoamine Oxidases–New Vistas* (Costa, E. & Sandler, M., eds.) *(Adv. Biochem. Psychopharmacol. 5)*, pp. 393-408, Raven Press, New York and North-Holland, Amsterdam

MANTLE, T. J., WILSON, K. & LONG, R. F. (1975) Studies on the selective inhibition of membrane-bound monoamine oxidase. *Biochem. Pharmacol. 24*, 2031–2038

NEFF, N. H. & GORIDIS, C. (1972) Neuronal monoamine oxidase: specific enzyme types and their rates of formation, in *Monoamine Oxidases–New Vistas* (Costa, E. & Sandler, M., eds.) *(Adv. Biochem. Psychopharmacol. 5)*, pp. 307-323, Raven Press, New York and North-Holland, Amsterdam

NEFF, N. H., YANG, H-Y. T. & GORIDIS, C. (1973) Degradation of the transmitter amines by specific types of monoamine oxidases, in *Frontiers in Catecholamine Research* (Snyder, S. H. & Usdin, E., eds.), pp. 133-137, Pergamon Press, London & New York

O'Steen, W. K. (1963) Serotonin and histamine: Effects of a single injection on the mouse testis and prostate gland. *Proc. Soc. Exp. Biol. Med. 113*, 161-163

Salgado, E. & Green, D. M. (1955) Renal necrosis induced in rats by serotonin. *Am. J. Physiol. 183*, 657

Squires, R. F. (1968) Additional evidence for the existence of several forms of mitochondrial monoamine oxidase in the mouse. *Biochem. Pharmacol. 17*, 1401-1409

Squires, R. F. (1972) Multiple forms of monoamine oxidase in intact mitochondria as characterized by selective inhibitors and thermal stability: a comparison of eight mammalian species, in *Monoamine Oxidases–New Vistas* (Costa, E. & Sandler, M., eds.) *(Adv. Biochem. Psychopharmacol. 5)*, pp. 355-370, Raven Press, New York and North-Holland, Amsterdam

Tipton, K. F. (1972) Some properties of monoamine oxidase, in *Monoamine Oxidases–New Vistas* (Costa, E. & Sandler, M., eds.) *(Adv. Biochem. Psychopharmacol. 5)*, pp. 11-24, Raven Press, New York and North-Holland, Amsterdam

Tipton, K. F. (1975) Monoamine oxidase, in *Handbook of Physiology*, Section 7: *Endocrinology*, vol. 6: *Adrenal Gland* (Blaschko, H. K. F. & Smith, A. D., eds.), pp. 667-691, American Physiological Society, Washington, D.C.

Tipton, K. F., Houslay, M. D. & Garrett, N. J. (1973) Allotopic properties of human brain monoamine oxidase. *Nature (Lond.) 246*, 213-214

Ungváry, G. Y. & Donáth, T. (1969) On the monoaminergic innervation of the liver. *Acta Anat. (Basel) 72*, 446-459

Youdim, M. B. H. (1973) Multiple forms of mitochondrial monoamine oxidase. *Br. Med. Bull. 29*, 120-122

Youdim, M. B. H., Collins, G. G. S. & Sandler, M. (1969) Multiple forms of rat brain monoamine oxidase. *Nature (Lond.) 233*, 626-628

Discussion

Oreland: Dr Ekstedt and I have done experiments on rat liver monoamine oxidase rather similar to yours, Dr Tipton, but with somewhat different results (Ekstedt & Oreland 1975). We delipidated the mitochondrial membranes by extraction with aqueous methyl ethyl ketone. Originally this was a two-step procedure (Hollunger & Oreland 1970), rendering the enzyme soluble, but by using only the first step we obtained delipidated membranes with the enzyme still bound, which then served as the material for our studies. This bound, delipidated form of the enzyme, however, seems to be very similar to the soluble form (B. Ekstedt & L. Oreland, unpublished work). After delipidation with the ketone extraction procedure we recovered the B form activity of the enzyme completely, but almost nothing of the A form. We have shown that there is no transformation from the A form to the B form during the extraction procedure, but that the A form is probably inactivated.

Our results suggest that if there are two forms of the enzyme, and not just two areas of the active site, as proposed by Severina (1973), it seems likely that there is a difference in the enzyme protein rather than just differences in the phospholipids close to the enzyme.

Tipton: It is interesting that it is the B species that you extract, since treatment with chaotropic agents converts the enzyme into a form that resembles the B species in its sensitivity to selective inhibitors. The possibility that you mentioned of two areas of the active site of a single enzyme would be consistent with a single enzyme having its specificity modified by lipid binding. Have you tried the effect of selective inhibitors on your preparations?

Oreland: Yes. Clorgyline and deprenyl were used, with the clear-cut result I mentioned.

Youdim: In our studies on the effect of clorgyline on the activity of the electrophoretically separated forms of monoamine oxidase we have observed that the bands containing the least lipid appear to be more resistant to clorgyline, which agrees with your results.

Tipton: This is what might be expected if the overall sensitivity to clorgyline and other lipophilic inhibitors is affected by a partitioning effect.

Neff: We have also used perchlorate, in an attempt to solubilize MAO of rat brain. We have some indication that both A and B enzyme activity is lost as a result of treatment with perchlorate, but the A activity is lost much more quickly than the B form. It is not a conversion, but just that one form of the enzyme is destroyed much faster than the other.

Tipton: Treatment with perchlorate does result in loss of activity unless the enzyme is protected by the presence of a substrate and mercaptoethanol, and even under these conditions prolonged incubation will result in loss of activity (Houslay & Tipton 1973). The conditions that one has to use appear to depend on the amount of lipid material (or perhaps the lipid:protein ratio) present in the sample. Most of our work has been on soluble preparations of the enzyme where the lipid:protein ratio is relatively small, and samples with higher lipid concentrations require more vigorous treatment. With any new preparation of the enzyme it will be necessary to determine the optimum perchlorate concentration and the necessary incubation time empirically. We have recently been trying to use the procedure with brain mitochondrial outer membrane fractions, which are extremely rich in lipid, but so far we have been unable to obtain yields of more than 40%.

Neff: I wonder if the A form is destroyed, rather than is converted to B?

Tipton: It is interesting that treatment with chaotropic agents does not result in a gain in activity. If we start with a mixture of the two species and convert it all to the B type species one might expect a decrease in the activity towards substrates for the A species and an increase in that towards B species substrates. Since neither happens, it would appear that the species generated is not identical to the B species in its specificity, although it resembles it in its inhibitor sensitivity.

Oreland: We tried to tackle this problem by first inhibiting the B form of the enzyme in the mitochondria with deprenyl and then extracting the membranes with ketone in the way I described (p. 16), to see if any B form activity was produced from the A form activity by the delipidation, but there was none. Thus we do not believe in a transformation from the A to the B form (Ekstedt & Oreland 1975).

Tipton: You were left with no activity towards any substrate?

Oreland: That's right. On the other hand, if we inhibited the A form activity with clorgyline and then delipidated the membranes, we got complete recovery of the B form.

Sourkes: You have clearly specified two major species of the enzyme, Dr Tipton. How does this correlate with the electrophoretic studies? Were the electrophoretically separated species mainly two, plus minor amounts of others, that could have been missed in Dr Oreland's study?

Tipton: The distinctions that one can make using clorgyline or deprenyl must be relatively gross, in that the errors in plots of this type would preclude the detection of any minor species that represented less than 5% of the total activity. However, if you solubilize using the normal vigorous procedures you will probably liberate the enzyme with varying amounts or types of lipid material associated with it (see e.g. Veryovkina *et al.* 1964). These artifactually modified enzyme species may give rise to multiple bands of activity on electrophoresis but these bands will have no direct relevance to the situation that exists within the cell or in the intact mitochondrion.

Youdim: I agree with you. I think the inhibitor studies are probably not as sensitive as electrophoretic studies and will not be able to distinguish between the various forms. We have to consider that the populations of mitochondria within cells of liver and brain are heterogeneous with regard to a number of enzymes, including monoamine oxidase. I wonder whether the various forms that we detect could be derived from different populations of mitochondria having different lipid compositions? I don't think anyone has studied the lipid composition of monoamine oxidase thoroughly except Dr Oreland, who showed that cardiolipin is associated with monoamine oxidase.

Tipton: It would be particularly useful to have data on the lipid compositions of the mitochondrial outer membranes from different organs and animal species. I think it would be a mistake to assume that the specificities of the two enzyme species from rat liver (see Table 2, p. 14) will necessarily be applicable to the enzyme from other sources, since differences in the lipid compositions of the mitochondrial outer membranes may cause changes in the specificities of the enzyme species present in other sources without affecting the gross differences in inhibitor sensitivities that allow the crude A and B distinction to be made.

Blaschko: I want to take up Dr Youdim's point. If one wants to bring this type of work to a level which is interesting to the psychiatrist, one question that needs answering is whether in a preparation of liver or any other organ you deal with a mixed population of mitochondria or with a more or less homogeneous group of mitochondria which contain the enzymes A and B. I haven't seen many studies relevant to this point, which may be technically difficult to tackle, but they would bring the problem closer to those who are interested in what happens in neurons.

I was interested in Dr Houslay's experiment with 6-hydroxydopamine. This differs from Dr Jarrott's results on the vas deferens (1971), where there was a difference in the substrate specificities before and after denervation.

Iversen: The difference is that the liver has only a very sparse sympathetic innervation, whereas the vas deferens has a dense one. The liver also has an extraordinarily high monoamine oxidase activity in the parenchyma and other cells, so I think Dr Tipton's explanation is the most probable one, namely that the proportion of enzyme activity associated with the sympathetic nerves in the liver is so small that losing it makes no detectable difference, whereas in the vas deferens approximately 50% of total tyramine activity is lost with denervation (Jarrott 1971).

Gorkin: Could you give more details of the treatment with the chaotropic agent, Dr Tipton? What do you mean by mild or severe treatment?

Tipton: I was defining severe treatment as that which we have previously reported (Houslay & Tipton 1973). The mild treatment that I referred to used a lower concentration of the chaotropic agent (0.3M) and incubations were done at between 0 °C and 4 °C for one minute in phosphate buffer pH 7.2. These conditions are milder than those we previously reported and phosphate ions have been shown to retard the action of chaotropic agents.

Gorkin: Is this treatment enough to bring the enzyme into the soluble form?

Tipton: After treatment of the mitochondrial outer membranes in this way the enzyme is gel-filtered on a column of Sepharose 4B equilibrated with Triton X-100. The enzyme is retarded by the column and is eluted after the bulk of the membrane material. In addition the enzyme does not form a pellet when centrifuged under conditions that would sediment mitochondrial outer membrane fragments.

Sandler: What happened to McCauley & Racker's (1973) approach as initiated by Hartman (1972)—the immunochemical separation of A and B?

Tipton: I don't think those immunochemical studies are of much help in clarifying the situation, since the results would be equally consistent with the presence of a single enzyme that was modified in different ways by binding to membrane material as with the presence of more than one enzyme. The possibility that membrane binding may mask the antigenic response of proteins is well known (see e.g. Tooze 1973).

Sandler: There are other possible ways of tackling this problem. Baker & Hemsworth's (1975) approach by affinity chromatography seems promising and may also lead to some degree of *in vitro* separation. But one ought to state that the whole concept of A and B forms of MAO may be an over-simplification. Gascoigne *et al.* (1975), for instance, using a histochemical approach, have claimed to find a C form of the enzyme localized to circum-ventricular areas of the brain. This third form is relatively insensitive to clorgyline but readily utilizes 5-hydroxytryptamine (5HT) as a substrate.

To turn to the question of organs that contain only a single form of MAO, Bathina *et al.* (1975) claim that human placenta is a rich source of A. Platelet MAO, on the other hand, is predominantly of the B type (Collins & Sandler 1971). Again we find that the classification breaks down: Di Chiara *et al.* (1974) have noted that apomorphine selectively inhibits dopamine oxidation, and we (M. Sandler, B. L. Goodwin, R. D. Johnson & C. R. J. Ruthven, unpublished) have obtained evidence to suggest that viloxazine selectively inhibits dopamine oxidation, *in vivo*. Dopamine, however, is supposed to be a substrate of both A and B forms (Neff *et al.* 1973).

Tipton: I am suspicious of the idea of a dopamine monoamine oxidase. We too have detected a monoamine oxidase species in soluble preparations that appears to have a relatively high activity towards dopamine and to be less sensitive to inhibition by phenethylhydrazine (Tipton 1972; Tipton *et al.* 1972) but since there have been, to my knowledge, no studies on the activity of the species towards the other catecholamines it would seem premature to term it dopamine monoamine oxidase. We have shown that this species vanishes when the enzyme is treated with chaotropic agents (Houslay & Tipton 1973) and thus we conclude that it is another modification due to lipid binding. Since most of the work on this enzyme has used soluble preparations of the enzyme, I don't think one can come to any firm conclusions about its importance in the intact cell.

The interpretation of *in vivo* inhibition data is very difficult because of the possibility that the amines are located at different sites and the existence of permeability barriers to the inhibitor.

Youdim: On the question of whether one can distinguish A and B forms *in vivo*, and whether the substrate of one form inhibits the other enzyme, we have begun to study monoamine oxidase in intact organs, including the rat lung, which we believe is closer to the *in vivo* situation.

TABLE 1 (Youdim)

Metabolism of phenylethylamine and 5-hydroxytryptamine in the presence of each other during passage through the pulmonary circulation of isolated rat lung

Labelled amine	*Competing amine*	*% metabolite (mean ± S.E.M. (no. of expts))*
[^{14}C]Phenylethylamine (1.25×10^{-5}M)	None	55.8 ± 3.6 ($n = 8$)
	5-Hydroxytryptamine (1.25×10^{-5}M)	55; 55; 54
[^{3}H]5-Hydroxytryptamine (1.25×10^{-5}M)	None	11.6 ± 1.3 ($n = 4$)
	Phenylethylamine (1.25×10^{-5}M)	11.0; 12.3
[^{3}H]5-Hydroxytryptamine (7×10^{-8}M)	None	53.6 ± 2.6 ($n = 7$)
	Phenylethylamine (1.25×10^{-5}M)	53.8 ± 4.1 ($n = 4$)

Isolated rat lungs were perfused with Krebs solution at a flow of 8 ml/min. The radioactive amine, [1-^{14}C]phenylethylamine or [G-^{3}H]5-hydroxytryptamine, was infused for 4 min. The competing non-radioactive amine was infused for 5 min before, during and for 25 min after the infusion of radioactive amine. Lung effluent was collected for 30 min after the start of the infusion of radioactive amine and analysed by ion exchange chromatography.

The work on the isolated perfused rat lung was done in collaboration with Dr Bahkle (Bahkle & Youdim 1975). We looked for competition between 5-hydroxytryptamine (5HT) and phenylethylamine (PHE) uptake by measuring the metabolism of perfused [^{3}H]5HT or [^{14}C]PHE in the presence of unlabelled 5HT or PHE. Competition for uptake should show itself as decreased metabolism. Our experiments are summarized in Table 1 and show that there was no competition between the amines at equimolar concentrations and that even when PHE was present at almost 200 times molar excess, the metabolism of 5HT was not inhibited. Although we conceived these experiments to test for competition for uptake, the results may be interpreted as showing further that there is no competition even at the enzyme site under these conditions. An interesting point to emerge from Table 1 is that at equimolar concentrations, PHE suffered about 50% metabolism, whereas 5HT was less affected.

Our experiments on PHE metabolism in isolated perfused rat lung have led us to the following conclusions: (1) PHE is inactivated in a single passage through this tissue by a process of uptake and catabolism by MAO type B. (2) This MAO, like others metabolizing PHE in the brain, is particularly susceptible to inhibition by deprenyl, whereas the MAO metabolizing 5HT is particularly sensitive to clorgyline. (3) The removal processes for PHE and 5HT are independent of each other and no competition could be observed either for uptake or for the active site of the enzyme. We feel that this study demonstrates the independent operation of MAO type A and B in an organized tissue (Bahkle & Youdim 1975).

Tipton: What matters, of course, is the concentrations of the amines inside the cell, and this will be dependent on uptake processes. The problem with perfusion experiments of the type you describe is that one has no idea of the concentrations of the amines in the vicinity of monoamine oxidase.

Our results (Houslay & Tipton 1975) indicate that when the two enzyme species meet substrates for both of them there will be a 'damping' effect due to cross-inhibition. The observation that the selective monoamine oxidase inhibitors do not have nearly as dramatic an effect *in vivo* as they do *in vitro* (see e.g. Collins *et al.* 1972) may be due, at least in part, to such an effect.

Oreland: The presence of A and B forms of MAO has not been demonstrated in all organs. Thus Hall *et al.* (1969) found only one form in pig liver using clorgyline. However, by using clorgyline and deprenyl we have found that pig liver also contains at least two functional forms of the enzyme (B. Ekstedt & L. Oreland, in preparation). In Fig. 1 it is shown, by inhibition with deprenyl,

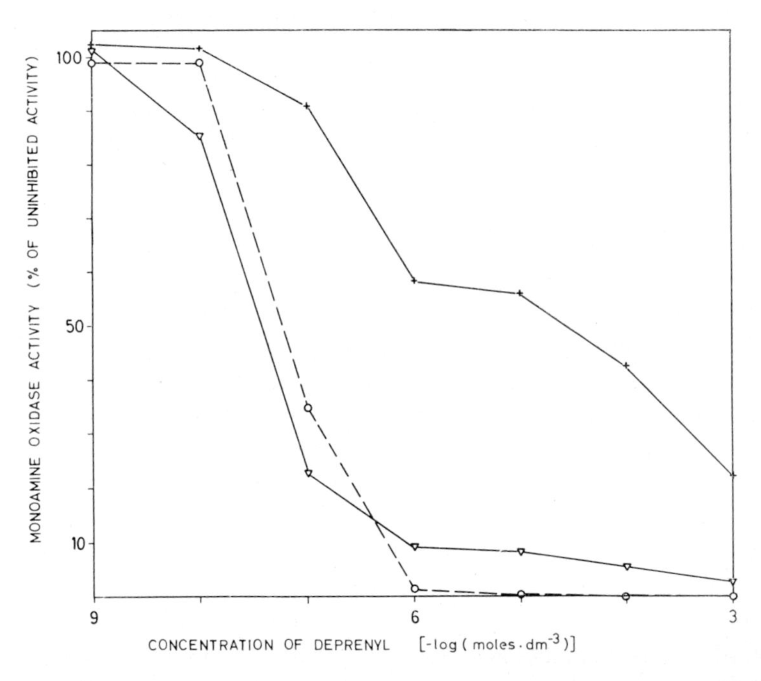

FIG. 1 (Oreland). Inhibition by deprenyl of monoamine oxidase in pig liver mitochondria. Samples of pig liver mitochondria were preincubated in the presence of the concentrations of deprenyl indicated, in a total volume of 275 μl of 10 mM-potassium phosphate (pH 7.4) for 20 min at 25 °C, prior to estimation of MAO activity. +—+, 5-hydroxytryptamine; ▽—▽, tyramine; ○ - - ○, β-phenylethylamine.

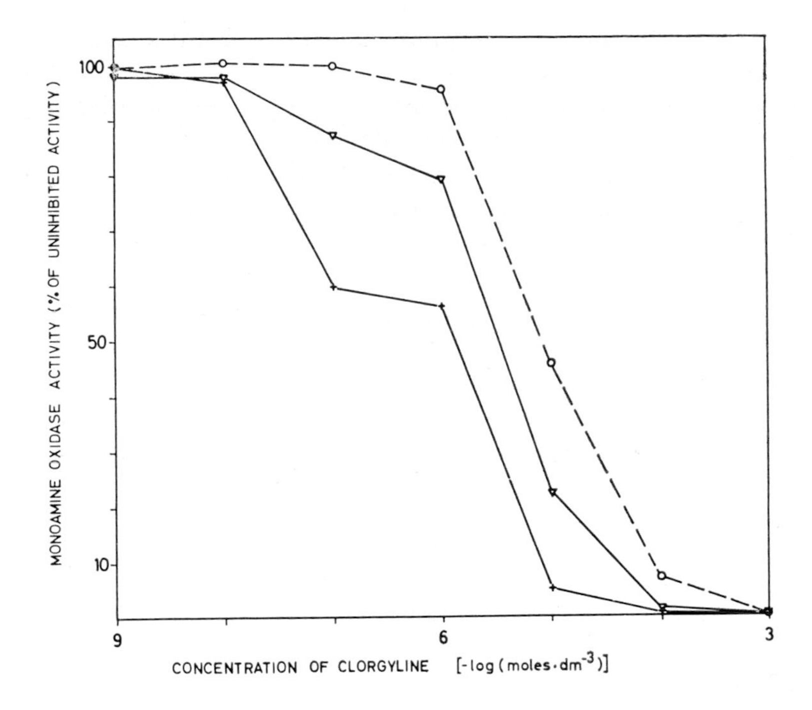

FIG. 2 (Oreland). Inhibition by clorgyline of monoamine oxidase in pig liver mitochondria. The experiments were done as described in the legend to Fig. 1 but with clorgyline instead of deprenyl as the inhibitor. +—+, 5-hydroxytryptamine; ▽—▽, tyramine; ○ - - ○, β-phenylethylamine.

that 5HT (serotonin) is oxidized by both the A and B forms, while tyramine, the classical A + B substrate, together with β-phenylethylamine, is mainly a substrate for the B form of the enzyme. Fig. 2 confirms these results by inhibition with clorgyline.

Tipton: I am surprised that you detected two species in pig liver using clorgyline and deprenyl, since Squires (1972) and Hall *et al.* (1969) reported that this organ contained only the B species, in terms of sensitivities to these two inhibitors. Perhaps differences in the preparative procedures used could have resulted in these differences in inhibitor sensitivity. I suggested earlier that the A and B species obtained from different organs might differ in their substrate specificities, and Squires (1972) has pointed out that some organs that only contain the B species, by the criterion of inhibitor sensitivity, are nevertheless able to oxidize substrates that are normally associated with the A species.

I think it would perhaps be safer to refer to the species as 'A type' and 'B type' on the basis of their inhibitor sensitivities, recognizing that this may not necessarily define their substrate specificities.

Oreland: In that case, perhaps we should come to an agreement on whether 'A type' means clorgyline-sensitive rather than 5HT-oxidizing.

Tipton: I would recommend the use of Johnston's nomenclature (Johnston 1968) and define the two species in terms of clorgyline sensitivity (although deprenyl sensitivity may be equally satisfactory). In testing for selective inhibition more than one substrate should be used and it would seem from our work and that of others that benzylamine, tyramine and 5HT would be a satisfactory trio of substrates to use.

Singer: Speaking as someone who has spent most of his life working on mitochondrial enzymes, it seems to me that the same series of fashions are developing in the monoamine oxidase field as regards what is 'harmful' and what is not 'harmful' as were in vogue in the inner membrane field a while back. Some people think that Triton is harmful, or that methyl ethyl ketone is, yet believe that chaotropic agents are mild.

We should remember that all these agents modify enzymes; it is simply a question of how carefully you look for signs of modification and how good your criteria are. All the so-called chaotropic agents are derangers of protein structure under the conditions you use—0.3 mol/l and slightly acid pH. If you look at more subtle criteria than retention of catalytic activity, criteria which are, in fact, important *in vivo*, such as regulatory properties, you are apt to detect modification by all these agents. In succinate dehydrogenase low concentrations of chaotropic agents alter the regulatory properties without affecting inhibitory or catalytic properties. So one should be cautious in taking any one method and claiming that it is the best. One has to anticipate that if one obtains the enzyme in purified form it will not be exactly the same as the membrane-bound form; it is just a question of *how* different it is.

Tipton: I think that the important conclusion from the experiments with chaotropic agents is that the 'solubilized' enzyme is an artifact in terms of its heterogeneity and its inhibitor sensitivity. Thus attempts to apply the results obtained with 'solubilized' preparations to the situation within the cell will probably not be fruitful. If one wants to do *in vitro* experiments that are relevant to the situation in the cell, the mitochondrial outer membrane is probably the smallest unit that one can trust. Would you agree, Professor Singer, that the mitochondrial outer membrane might be an adequate model?

Singer: Absolutely, but there is a problem there too, at least in the central nervous system, which I take it is really the point when one discusses monoamine oxidase. How are you going to make outer membrane from mitochondria from the CNS? Any technician in our laboratory can make it out of liver, kidney or placenta but no one has succeeded in making anything from brain that resembles outer membrane. The published methods don't work, in our hands. I am worried about what our reference material will be. Another problem is that even methods for making brain mitochondria from one species do not seem to work for another (e.g. beef as against pig).

Tipton: We have had such problems too, but recently we have had encouraging results with rat brain mitochondria using the method recommended by Bashford and his co-workers (Craven *et al.* 1969).

Singer: Is it an outer membrane in terms of the 'marker enzymes', so that the inner membrane enzymes are absent, and you have only the enzymes of the outer membrane? That is the problem. The separations don't work by Bashford's or any other published method because of the high lipid count.

Tipton: We have not yet characterized the outer membrane fractions in any detail since we are still at the stage of trying to obtain good yields of the enzyme that behave as outer membranes should on sucrose density gradient centrifugation.

Maître: To return to the A and B forms in terms of deprenyl and clorgyline sensitivity, Fig. 1 (*upper*, p. 26) shows dose–response curves (percentage MAO activity) 24 hours after giving deprenyl to rats. With phenylethylamine as substrate we find extensive inhibition with less than 1 mg/kg, injected subcutaneously, without inhibition of the deamination of 5HT or tyramine. But with doses of deprenyl higher than 1–3 mg/kg, 5HT and tyramine deamination are also inhibited. So the inhibition by deprenyl is preferential below approximately 1 mg/kg. Fig 1 (*lower*, p. 26) shows the inverse relationship with clorgyline, showing the protective effect on 5HT and the low protection of phenylethylamine deamination. Here again the preferential effect of the inhibitor is obvious but it is not complete, since phenylethylamine deamination is protected as well, at a much higher concentration of inhibitor.

Tipton: I don't think one would expect to see any close correlation between 'dose–response' curves *in vivo* and *in vitro* when a selective inhibitor is used. It should be remembered that clorgyline and deprenyl are irreversible inhibitors of monoamine oxidase. Thus if the enzyme is incubated with a molar excess of such an inhibitor one will eventually attain complete inhibition. The selectivity of these inhibitors arises because the two species are inhibited at markedly different rates (Tipton 1971) and one selects an incubation time such that one species is virtually completely inhibited whereas the other is scarcely affected at the 'plateau' inhibitor concentration. These restrictions indicate that the relative amounts of the two species determined in a given preparation may be, to some extent, dependent on the preincubation conditions used.

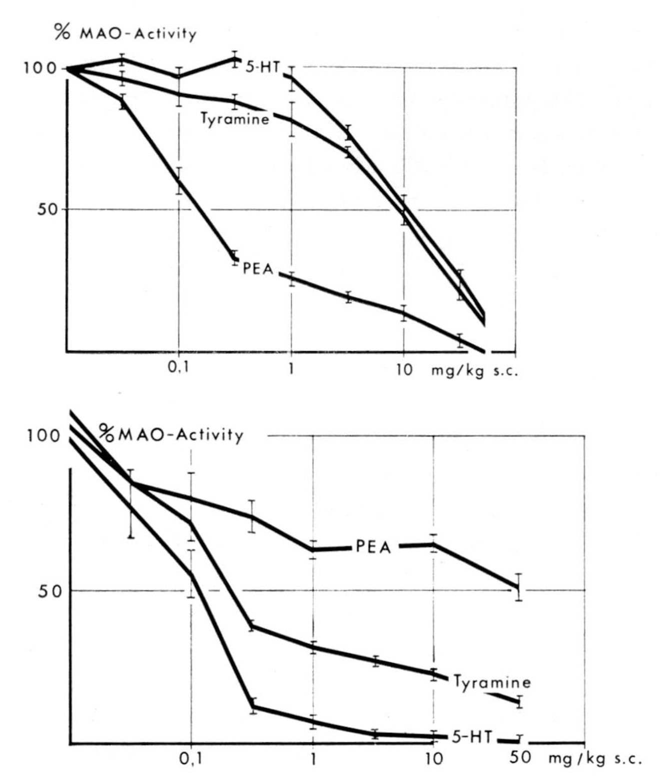

FIG. 1 (Maître). Inhibition of rat brain MAO by deprenyl (*upper*) and clorgyline (*lower*), showing dose–response relationships. The MAO inhibitors were injected subcutaneously 24 hours before removal of the brains. MAO activity in the whole brain was determined essentially as described by Wurtman & Axelrod (1963). ^{14}C-Labelled 5-hydroxytryptamine (5-HT), tyramine and phenylethylamine (PEA) were used as substrates, at concentrations of 6.25 nM (10 nCi) for a final incubation mixture of 0.3 ml. Values represent means ± S.E.M. of four or eight determinations.

In *in vivo* studies we have the added complications that arise from an uneven distribution of the enzyme and the existence of permeability barriers. Thus in the central nervous system both the presynaptic and postsynaptic cells contain monoamine oxidase but they may be able to take up the inhibitor and concentrate it to different extents, and thus the extents of inhibition of the two species may be different in pre- and postsynaptic systems.

Murphy: I would like to mention some recent work that complements your story about the parenchymal cell issue in studies of liver MAO. Donnelly, Richelson and I (Murphy *et al.* 1976) tried to approach the brain in a similar way, looking at rat C6 glial cells and mouse neuroblastoma cells in tissue culture preparations. Since the MAO A type has been suggested to be an intraneuronal MAO, we thought it would be interesting if the MAO subtypes were localized in brain with the MAO A in neurons, and the MAO B in glial cells. However, inhibition studies with clorgyline and deprenyl revealed sigmoid curves with plateaus, suggesting the presence of nearly exclusively MAO A type activity with tyramine in both the neuroblastoma and C6 glial cell preparations, in contrast to the mixed A plus B curves in brain and the essentially pure B type responses in human platelet preparations. Data on MAO specific activities with different substrates in the C6 glial cell preparation complemented the inhibitor data in showing relatively higher specific activities for the A type substrates, whereas with phenylethylamine and benzylamine (the preferential substrates for the B type enzyme form) there was a much lower proportion of activity in the glial cell preparations. These data are similar to those obtained with mouse neuroblastoma cells, which again seemed to contain primarily MAO A enzyme. One implication is that in terms of making a brain preparation or at least a glial cell preparation for further studies (as Dr Singer was suggesting), which would be relatively free of some of the lipid contaminants in whole-brain preparations, one might consider using glial cell preparations as a model, since they may resemble the majority of brain MAO activity.

Sourkes: Dr Tipton implied that it is difficult to get anything but a parenchymal cell preparation from liver; but there are two drugs that have been used to affect one cell type or the other. Carbon tetrachloride seems to affect the parenchymal tissue adversely; and the cells making up the bile canaliculi are affected by α-naphthylisothiocyanate, which stimulates their proliferation. Either of these drugs given chronically could seriously affect the functioning of one or the other cell, perhaps with the loss of mitochondrial enzymes. This is a possible technique for trying to separate these out (cf. Lal & Sourkes 1970).

Fuller: Dr Tipton, you have classified tryptamine as a type B substrate (Table 2, p. 14). What is the basis for that? From Dr Murphy's data and from our *in vivo* data (Fuller & Roush 1972), it should seem more like a type A substrate, and Neff & Yang (1974) have classified it as a mixed substrate. Have you information that points to it being a type B substrate?

Tipton: The evidence for our classification of tryptamine as being a substrate for the B enzyme has been discussed in some detail (Houslay & Tipton 1974). It is based on the use of the reversible inhibitors benzyl cyanide and 4-cyanophenol as well as on the response to clorgyline and deprenyl.

Youdim: I agree with Dr Tipton on the classification of tryptamine. Dr Green and I have been using a compound called quipazine, which is said to be an agonist of 5HT, and we have found that it is also an inhibitor of monoamine oxidase. It appears to be an A type enzyme inhibitor and, using it with a number of substrates, tryptamine falls into the B type group of substrates, with brain homogenate preparations.

Green: I notice that in your classification in Table 2 (p. 14), Dr Tipton, 5HT is A type and a related compound, 5-methoxytryptamine, is B type. Would you like to speculate on any structural requirements of the A and B forms for substrates?

Tipton: We have speculated on the structural requirements of species A and species B substrates on the basis of the results we obtained with the rat liver enzyme (Houslay & Tipton 1974). Substrates for the A species would appear to require a *p*-hydroxyl group and presumably the distance of this group from the amino group is important, since *p*-substituted benzylamines are not substrates for this species. In addition it would appear that the presence of a β-hydroxyl group may prevent the B species from acting. The specificities of the species for tryptamine (B) and 5HT (A) are somewhat more difficult to explain, unless a site of steric hindrance is postulated for the indole nucleus which is only encountered when the hydroxyl group also interacts with its binding site.

Blaschko: Dr Vane (1959) described the effects of amine oxidase inhibitors (non-selective ones) on the response of the rat stomach to tryptamine derivatives. He found an enormous potentiation of the response to tryptamine, 5-methoxytryptamine and methyltryptamine, and an entire absence of potentiation of the response to 5HT. I have speculated on this in a recent review (Blaschko 1974). But from your results, Dr Tipton, it occurs to me that perhaps in the rat's stomach there is a discriminatory effect on these two types of amine oxidase by the MAO inhibitors that Vane used (iproniazid and phenylisopropylhydrazine).

van Praag: What is the influence of tricyclic antidepressants on MAO (if any), and could this action be of therapeutic importance? Secondly, is there information on a possible correlation between the effect of the classical MAO inhibitors that are therapeutically used on the A and B forms, and their clinical efficacy? There have been some rather good MAO inhibitors from the clinical point of view, and some rather bad ones. Is there a relationship with their effects on different MAO types? And finally, is there an essential difference between MAO inhibitors and tricyclics with regard to their action on MAO?

Murphy: We looked through the clinical data and have reviewed and done some studies with the different drugs. All the clinically used MAO-inhibiting antidepressants are mixed A and B inhibitors, with nearly equal effects on the two enzyme forms. Pargyline is somewhat more specific for MAO B, but has been used in only a few studies of depressed patients. The more specific inhibitors, deprenyl and clorgyline, have not been compared in the same study; both have been reported to have clinical antidepressant potency, but only a very small number of patients have been studied. In animal investigations, it has been suggested that inhibition of both A and B forms is needed for some of the behavioural effects produced by MAO inhibition (Squires & Lassen 1975). Tricyclic drugs have some MAO-inhibiting potency, although relatively high concentrations were required in our studies (see p. 345).

Youdim: The tricyclic drug imipramine has been reported by Roth & Gillis (1974) to be an inhibitor of the B form *in vitro.*

Sandler: The *in vivo* effect of these drugs was demonstrated in platelets by Edwards & Burns (1974).

Fuller: There are no data to support an inhibition of MAO in the intact brain, in the animal, by tricyclic antidepressant drugs. I think it would be premature to conclude that MAO inhibition plays a part in the pharmacology of these drugs.

Maître: If we treat rats with imipramine or with desmethylimipramine up to doses of 50 mg/kg, orally, we find no inhibition of MAO, using phenylethylamine or 5HT as substrate.

Sandler: Dr Murphy mentioned pargyline in connection with the inhibition of both A and B forms being necessary for activity in depression. The drug may, in fact, be predominantly a B inhibitor (McCauley & Racker 1973). Even so, it seems to share the ability of other MAO inhibitors to bring about a lightening of affect (Bucci & Saunders 1961; Bucci *et al.* 1962; Stern 1963), although this action has been somewhat upstaged by its antihypertensive effect.

Tipton: The efficiency of the tricyclic antidepressants as monoamine oxidase inhibitors *in vivo* will depend on how effectively they are taken up into the cells and concentrated. Are there any data on this?

Pletscher: The tricyclic compounds do not just act on the outer cell membrane; some of the drug enters. This has been found in platelets and, I think, in other cells. The storage granules of platelets have a relatively high affinity for compounds like imipramine. But we agree with Dr Maître; we have never seen any inhibitory effect on monoamine oxidase, although these compounds do enter the cell.

Rafaelsen: Tricyclics are highly lipophilic substances; in the rat they pass the blood–brain barrier as fast as does water.

References

BAHKLE, Y. & YOUDIM, M. B. H. (1975) Inactivation of phenylethylamine and 5-hydroxytryptamine in rat isolated lungs: evidence for monoamine oxidase A and B in lung. *J. Physiol. (Lond.) 248*, 23-25*P*

BAKER, S. P. & HEMSWORTH, B. A. (1975) Some studies on the purification of monoamine oxidase by affinity chromatography. *Br. J. Pharmacol. 54*, 264P-275*P*

BATHINA, H. B., HUPRIKAR, S. V. & ZELLER, E. A. (1975) New approaches to the characterization of mitochondrial monoamine oxidase (MAO) types A and B. *Fed. Proc. 34*, 293

BLASCHKO, H. (1974) The natural history of amine oxidases. *Rev. Physiol. Biochem. Pharmacol. 70*, 83-148

BUCCI, L. & SAUNDERS, J. C. (1961) A unique monoamine oxidase inhibitor for depression. *Am. J. Psychiatry 118*, 255-256

BUCCI, L., HENDERSON, C. T. & SAUNDERS, J. C. (1962) Pargyline: a paragon of affective therapy. *Psychosomatics 3*, 308-311

COLLINS, G. G. S. & SANDLER, M. (1971) Human blood platelet monoamine oxidase. *Biochem. Pharmacol. 20*, 289-296

COLLINS, G. G. S., YOUDIM, M. B. H. & SANDLER, M. (1972) Multiple forms of monoamine oxidase. Comparison of *in vitro* and *in vivo* inhibition patterns. *Biochem. Pharmacol. 21*, 1995-1998

CRAVEN, P. A., GOLDBLATT, P. J. & BASHFORD, R. E. (1969) Brain hexokinase. The preparation of inner and outer mitochondrial membranes. *Biochemistry 8*, 3525-3532

DI CHIARA, G., BALAKLEEVSKY, A., PORCEDDU, M. L., TAGLIAMONTE, A. & GESSA, G. L. (1974) Inhibition by apomorphine of dopamine deamination in the rat brain. *J. Neurochem. 23*, 1105-1108

EDWARDS, D. J. & BURNS, M. O. (1974) Effects of tricyclic antidepressants upon human platelet monoamine oxidase. *Life Sci. 14*, 2045-2058

EKSTEDT, B. & ORELAND, L. (1975) Effect of lipid-depletion on the different forms of monoamine oxidase in rat liver mitochondria. *Biochem. Pharmacol. 25*, 119–124

FULLER, R. W. & ROUSH, B. W. (1972) Substrate-selective and tissue-selective inhibition of monoamine oxidase. *Arch. Int. Pharmacodyn. Ther. 198*, 270-276

GASCOIGNE, J. E., WILLIAMS, D. & WILLIAMS, E. D. (1975) Histochemical demonstration of an additional form of rat brain MAO. *Br. J. Pharmacol. 54*, 274*P*

HALL, D. W. R., LOGAN, B. W. & PARSONS, G. H. (1969) Further studies on the inhibition of monoamine oxidase by M & B 9302 (clorgyline). *Biochem. Pharmacol. 18*, 1447-1454

HARTMAN, B. K. (1972) The discovery and isolation of a new monoamine oxidase from brain. *Biol. Psychiatr. 4*, 147-155

HOLLUNGER, G. & ORELAND, L. (1970) Preparation of soluble monoamine oxidase from pig liver mitochondria. *Arch. Biochem. Biophys. 139*, 320-328

HOUSLAY, M. D. & TIPTON, K. F. (1973) The nature of the electrophoretically separable multiple forms of rat liver monoamine oxidase. *Biochem. J. 135*, 173-186

HOUSLAY, M. D. & TIPTON, K. F. (1974) A kinetic evaluation of monoamine oxidase activity in rat liver mitochondrial outer membranes. *Biochem. J. 139*, 645-652

HOUSLAY, M. D. & TIPTON, K. F. (1975) Amine competition for oxidation by rat liver mitochondrial monoamine oxidase. *Biochem. Pharmacol. 24*, 627-631

JARROTT, B. (1971) Occurrence and properties of monoamine oxidase in adrenergic neurones. *J. Neurochem. 18*, 7-16

JOHNSTON, J. P. (1968) Some observations on a new inhibitor of monoamine oxidase in brain tissue. *Biochem. Pharmacol. 17*, 1285-1297

LAL, S. & SOURKES, T. L. (1970) The effect of chronic administration of carbon tetrachloride and alpha-naphthylisothiocyanate on tissue copper levels in the rat. *Biochem. Med. 4*, 260-276

McCauley, R. & Racker, E. (1973) Separation of two monoamine oxidases from bovine brain. *Mol. Cell. Biochem. 1*, 73-81

Murphy, D. L., Donnelly, C. H. & Richelson, E. (1976) Substrate and inhibitor-related characteristics of monoamine oxidase in C6 rat glial cells. *J. Neurochem.*, in press

Neff, N. H. & Yang, H.-Y. T. (1974) Another look at the monoamine oxidases and the monoamine oxidase inhibitor drugs. *Life Sci. 14*, 2061-2074

Neff, N. H., Yang, H.-Y. T. & Goridis, C. (1973) Degradation of the transmitter amines by specific types of monoamine oxidases, in *Frontiers in Catecholamine Research* (Usdin, E. & Snyder, S. H., eds.), pp. 133-137, Pergamon, New York

Roth, J. A. & Gillis, C. N. (1974) Deamination of β-phenylethylamine by monoamine oxidase: inhibition by imipramine. *Biochem. Pharmacol. 23*, 2537-2545

Severina, I. S. (1973) On the substrate-binding sites of the active centre of mitochondrial monoamine oxidase. *Eur. J. Biochem. 38*, 239-246

Squires, R. F. (1972) Multiple forms of monoamine oxidase in intact mitochondria as characterised by selective inhibitors and thermal stability: a comparison of eight mammalian species. *Adv. Biochem. Psychopharmacol. 5*, 355-370

Squires, R. F. & Lassen, J. B. (1975) Inhibition of both A and B forms of MAO required for production of characteristic behavioural syndrome in rats after tryptophan loading. *Psychopharmacologia 41*, 145–151

Stern, F. H. (1963) Pargyline hydrochloride: a new agent for the control of hypertension. *J. Am. Geriat. Soc. 11*, 670-672

Tipton, K. F. (1971) Monoamine oxidases and their inhibitors, in *Mechanisms of Toxicity* (Aldridge, W. N., ed.), pp. 13-27, Macmillan, London

Tipton, K. F. (1972) Some properties of monoamine oxidase. *Adv. Biochem. Psychopharmacol. 5*, 11-24

Tipton, K. F., Youdim, M. B. H. & Spires, I. P. C. (1972) Beef adrenal medulla monoamine oxidase. *Biochem. Pharmacol. 21*, 2197-2204

Tooze, J. (1973) *The Molecular Biology of Tumor Viruses*, p. 217. Cold Spring Harbour Laboratory, USA

Vane, J. R. (1959) The relative activities of some tryptamine derivatives on the isolated rat stomach strip preparation. *Br. J. Pharmacol. 14*, 87-98

Veryovkina, I. V., Gorkin, V. Z., Mityushin, V. M. & Elpiner, I. E. (1964) On the effect of ultrasonic waves on monoamine oxidase connected with the submicroscopical structures of mitochondria. *Biochemistry (Moscow) (Engl. Transl. Biokhimiya) 9*, 503-506

Wurtman, R. J. & Axelrod, J. (1963) A sensitive and specific assay for the estimation of monoamine oxidase. *Biochem. Pharmacol. 12*, 1439-1441

The action of acetylenic inhibitors on mitochondrial monoamine oxidase: structure of the flavin site in the inhibited enzyme

A. L. MAYCOCK*†, ROBERT H. ABELES*, J. I. SALACH† and THOMAS P. SINGER†

*Graduate Department of Biochemistry, Brandeis University, Waltham, Massachusetts and †Molecular Biology Division, Veterans Administration Hospital and Department of Biochemistry and Biophysics, University of California, San Francisco

Abstract *N,N*-dimethylpropargylamine, a suicide inactivator of mitochondrial monoamine oxidase, acts both by immediate, competitive inhibition and as a time-dependent, irreversible inactivator. The latter effect depends on conversion of the inhibitor by the enzyme to a form capable of reacting irreversibly with the covalently bound flavin at the active site. This inhibitor has been prepared with ^{14}C label either in the methyl group or in the methylene group. Both types of labelled inhibitor were then reacted with the highly purified enzyme from liver and the resulting ^{14}C-labelled cysteinyl flavin–inhibitor adduct was isolated and its structure determined. Combination of the adduct is with the N-5 group of the flavin and the structure thus differs from adducts derived from the inactivation of other flavin enzymes (e.g., lactate oxidase) by acetylenic suicide inhibitors.

With the recognition of the potential importance of selective inhibitors of monoamine oxidase (amine: oxygen oxidoreductase [deaminating] [flavin-containing]; EC1.4.3.4) (MAO) to clinical medicine, the past two decades have witnessed a plethora of studies of MAO inhibitors (Fuller 1972). Of particular interest among widely investigated MAO inhibitors are acetylenic analogues of the substrate, as represented by pargyline. Pargyline and related acetylenic MAO inhibitors are sometimes called 'suicide inactivators' because an irreversible inhibitor is actually produced by the action of the target enzyme from a relatively innocuous compound which is a substrate. In order for an inhibitor of this type to be useful in elucidating the structure of the active site it must be produced within bonding distance of an essential group at the active site with which the inhibitor reacts readily, forming a covalent adduct

† *Present address:* Merck, Sharp and Dohme Research Laboratories, Rahway, New Jersey 07065, USA.

(Rando 1974). Determination of the structure of the adduct isolated from the inactivated enzyme may then yield a clue to the normal catalytic mechanism. Acetylenic suicide inactivators have been, in fact, widely used as probes of the mechanisms of action of various enzymes (Kass & Bloch 1967; Bloch 1969; Walsh *et al.* 1972*a*,*b*; Hevey *et al.* 1973).

Among the many investigations of the action of acetylenic suicide inactivators on mitochondrial MAO, those of McEwen *et al.* (1969), Rando (1973, 1974), and of Hellerman's group (Hellerman & Erwin 1968; Chuang *et al.* 1974) are particularly relevant in the context of the present study. McEwen *et al.* (1969) studied the kinetics of the inactivation process and emphasized that the inhibition by acetylenic compounds (2-propynylamines) is both instantaneous and competitive (as expected from the fact that the parent compound of the irreversible inhibitor must be a substrate) as well as time-dependent and irreversible. Rando (1973, 1974) explored the mechanism of the formation of the irreversible inhibitor on oxidation by the flavin moiety. Hellerman & Erwin (1968), using [^{14}C]pargyline (labelled at C-7), showed that on complete inhibition of the bovine kidney enzyme one mole of inhibitor becomes covalently bound per mole of enzyme. They also reported that pargyline initially reduces MAO, resulting in bleaching of its absorbance at 450 nm and the appearance of a strong absorption at 412 nm. The inactivation by pargyline was said to be prevented by substrates and by dithionite under anaerobic conditions. Inactivation was thought to be the result of combination of the inhibitor with an essential amino acid residue. In later work these authors (Chuang *et al.* 1974) recognized the likelihood that the flavin must be the combining site of the inhibitor, because radioactivity was recovered in the flavin peptide digest after proteolysis of MAO inactivated with [^{14}C]pargyline. Further, the flavin in this fraction had an altered absorption spectrum, similar to a synthetic model compound produced photochemically by reacting *N*,*N*-dimethyl-2-propynylamine with 3-methyllumiflavin (Zeller *et al.* 1972).

Isolation of the adduct of the radioactive inhibitor with the flavin peptide was not undertaken in the studies mentioned and, consequently, the structure of the flavin site of the inactivated enzyme remained to be established. With methods available for obtaining purified MAO from liver mitochondria in substantial quantity and for the isolation of the pure flavin pentapeptide from the enzyme (Kearney *et al.* 1971), the time seemed right for undertaking characterization of the adduct. The results, using labelled *N*,*N*-dimethyl-2-propargylamine as the inhibitor, are described here and are thought to be generally applicable to acetylenic MAO inhibitors.

MATERIALS AND METHODS

MAO activity was determined by the methods of Tabor *et al.* (1954) but at 30 °C. MAO was purified from beef liver mitochondria either by the procedure of Kearney *et al.* (1971) or by the procedure outlined in the accompanying paper (Salach *et al.*, this volume, pp. 49–56). Two types of labelled inhibitor were used: in one type the ^{14}C was located in the methyl group [$HC{\equiv}C–CH_2–N(^{14}CH_3)_2$] and will be referred to as $^{14}CH_3$-INACT and in the other it was located in the methylene carbon [$HC{\equiv}C–^{14}CH_2–N(CH_3)_2$], which will be referred to as $^{14}CH_2$-INACT. The radioactive adduct was isolated, after complete inactivation of MAO with the labelled inhibitor, by modifications of the procedure for the isolation of the flavin pentapeptide (Kearney *et al.* 1971). Since the flavin peptide derived from the inactivated enzyme did not fluoresce, it was followed during purification by radioactivity. Details of these methods and of the synthesis of the labelled inhibitors will be published elsewhere (Maycock *et al.* 1976).

RESULTS AND DISCUSSION

Characteristics of the inhibition

It is known that liver MAO is inhibited by acetylenic inhibitors, including 2-propynylamines, both by immediate, competitive inhibition and by a time-dependent, irreversible inhibition (McEwen *et al.* 1969; Hellerman & Erwin 1968). In accord with this, *N,N*-dimethylpropargylamine, the inhibitor used in this study, appears to act as a competitive inhibitor in initial rate measurements (30 to 180 second assay period, depending on the inhibitor concentration) (Fig. 1). The K_i values derived from Fig. 1 are not truly constant, however, but vary with the concentration of inhibitor. The K_i values calculated from the data of Fig. 1 varied from 1.01 mM to 1.42 mM, as the concentration of inhibitor was decreased from 6.7 to 1.76 mM. When the slopes obtained at varying concentrations of the inhibitor, Fig. 1, are replotted against the concentration of the inhibitor, a 'parabolic inhibition' is noted (Fig. 2). It appears that, despite the brief assay period and the fact that the inhibitor is oxidized by the enzyme only at 3% of the rate of benzylamine at V_{max}, sufficient oxidation of the inhibitor occurs to a form capable of combining with the enzyme irreversibly to produce dead-end inhibition (Cleland 1970).

Irreversible inhibition develops as a first-order process (Fig. 3) but much more slowly than in the case of pargyline (Hellerman & Erwin 1968). As expected, benzylamine prevents the inactivation (Table 1). Once irreversible

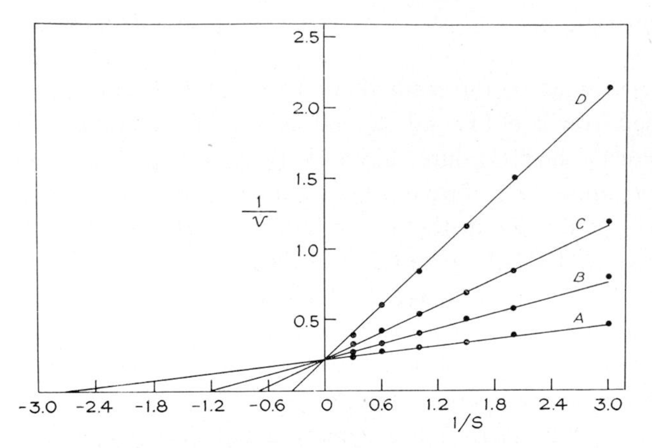

FIG. 1. Lineweaver–Burk plots indicating apparent competitive inhibition of MAO by *N*,*N*-dimethylpropargylamine. Substrate, in amounts indicated, together with inhibitor in 50 mM-NaP$_i$, pH 7.2, was brought to 30 °C and the reaction initiated by addition of 0.024 mg of enzyme (sp. act. = 2590 nmol/min/mg). Final pH was 7.1 in 3.0 ml total volume. A, in absence of inhibitor; B, in presence of 1.76 mM-inhibitor; C, 3.33 mM-inhibitor; and D, 6.70 mM-inhibitor. Abscissa, reciprocal concentration of benzylamine (mM); ordinate, reciprocal activity, expressed as μmol of substrate oxidized per min per ml.

TABLE 1

Protection of MAO from inactivation by *N*,*N*-dimethylpropargylamine

Addition	*Activity*	
	Substrate oxidized (μmol/min)	%
Enzyme	0.382	(100)
Same + 3.1 mM inhibitor	0.002	0.5
Same + 3.1 mM-benzylamine + 3.1 mM inhibitor	0.385	100

Two-side-arm Thunberg tubes contained 0.4 mg of MAO (sp. act. = 2790 nmol of benzylamine oxidized/min/mg) in one side-arm and sufficient 50 mM-sodium phosphate buffer, pH 7.2, in the main compartment to give a final volume of 1.29 ml. Benzylamine (4 μmol), where present, was added to the buffer. Inhibitor (4 μmol), where present, was in the second side-arm. The vessels were evacuated and filled with He at 0 °C. After brief equilibration at 30 °C, the contents of the tubes were mixed and incubation at 30 °C was continued for 90 min. After rapid chilling at 0 °C, the tubes were opened, an equal volume of oxygen-saturated buffer was added, and 0.2 ml aliquots were immediately assayed for remaining activity.

inactivation has occurred, neither extensive dialysis nor gel exclusion on Sephadex produces any reactivation. All this is as expected from previous studies with pargyline. In contrast to the report of Hellerman & Erwin (1968), however, that in anaerobiosis dithionite protects kidney MAO from inactiva-

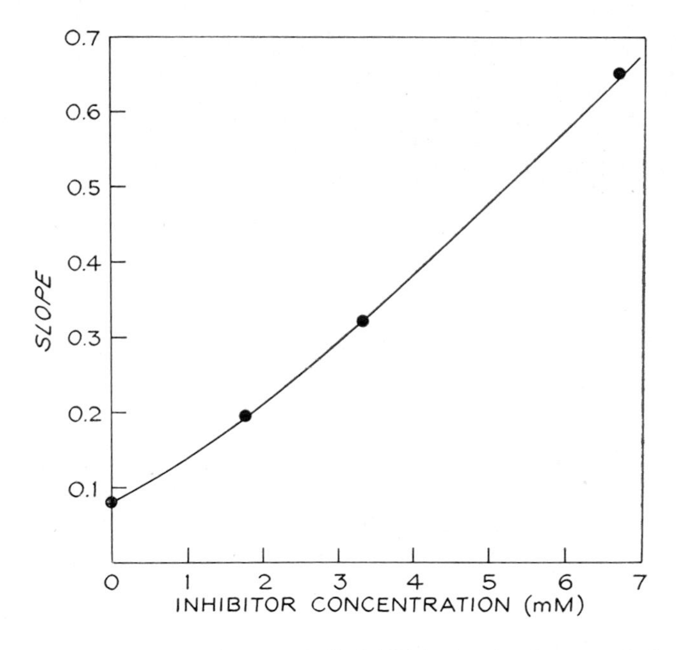

FIG. 2. Evidence for 'parabolic inhibition'. The data are derived from the primary plots given in Fig. 1. The numbers on the ordinate are the slopes of the lines in Fig. 1.

tion by pargyline, in our hands neither reduction with dithionite nor with light-EDTA under He protected the beef liver enzyme from inactivation by dimethylpropargylamine. In fact, the oxidation products of dithionite cause major loss of activity of purified liver MAO even in the absence of dimethylpropargylamine. It appears from these experiments that the substrate, benzylamine, protects from irreversible inactivation not because it reduces the enzyme but because it competes effectively with dimethylpropargylamine for the active site, preventing effectively its enzymic conversion to the form capable of producing irreversible inhibition. This is also seen in Fig. 4, which shows that, although < 0.003 mM-benzylamine should suffice to reduce the enzyme, for extensive protection against the inhibitor the presence of nearly 100 times more benzylamine was required.

Fig. 5 shows the spectral changes accompanying the inactivation of 2.3 μM-MAO by 5.3 mM-dimethylpropargylamine. Concurrently with the disappearance of the absorption of the enzyme in the 450 to 500 nm region, the absorbance at 410 nm rises as inactivation develops. On the basis of the cysteinyl flavin content of the enzyme, the molar absorbance of the inhibited species is $\varepsilon_{410\ nm} = 28\ 800$ (average value from three experiments), which may

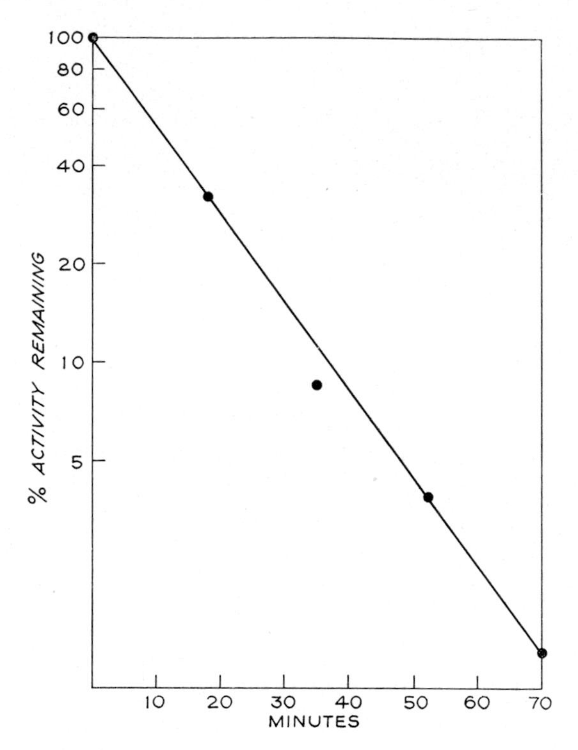

FIG. 3. Inactivation of MAO by *N,N*-dimethylaminopropargylamine. A partially purified preparation of liver MAO (1.7 ml enzyme, sp. act. = 260 nmol/min/mg, 4.2 mg of protein per ml) was mixed with 50 μl of a 17 mM solution of the inhibitor and 0.85 ml of 0.2 M-sodium phosphate buffer, pH 7.8, to give 3.3×10^{-4} M-inhibitor concentration. The solution was incubated at 25 °C and samples were periodically withdrawn for assay.

be compared with the values reported for the pargyline adduct of kidney MAO ($\varepsilon_{410\ nm}$ = 32 300, calculated from Fig. 1 of Chuang *et al.* [1974]) and of a proteolytic digest of that adduct ($\varepsilon_{398\ nm}$ = 29 900, pH not stated).

Purification and physical properties of the adduct

When MAO is inactivated with either methyl-labelled ($^{14}CH_3$-INACT) or methylene-labelled ($^{14}CH_2$-INACT) inhibitor, radioactivity becomes covalently bound to the protein, since neither dialysis, gel exclusion, nor precipitation with trichloroacetic acid, removes the radioactivity from the protein. When 1250 nmol of $^{14}CH_2$-INACT were reacted with 12.5 nmol of highly purified enzyme, after gel exclusion of the enzyme, precipitation with trichloroacetic acid, and proteolytic digestion, the preparation contained 20 to 35 nmol of

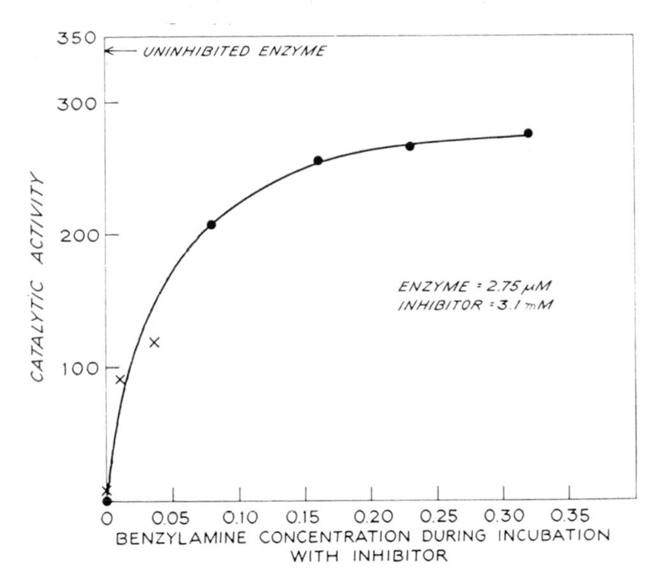

FIG. 4. Protection of MAO from inhibition by *N*, *N*-dimethylpropargylamine by substrate in anaerobiosis. Enzyme activity on ordinate is μmol of substrate oxidized per min per ml; substrate concentration on the abscissa is mM. Incubation with the inhibitor was for 60 min at 30 °C under He.

radioactivity; that is, 2.3 to 2.8 mol of inhibitor per mole of cysteinyl flavin. This contrasted to the mole-to-mole binding of pargyline to kidney MAO reported by Chuang *et al.* (1974). Although the determination of flavin and, hence, enzyme concentration in their study was an approximation based on bleaching of the enzyme by dithionite and an assumed absorption coefficient, rather than analysis for cysteinyl flavin, as in the present study, this would not account for the discrepancy in the amount of inhibitor bound. More likely, the slow oxidation of dimethylpropargylamine gives rise to a product, perhaps the aldehyde, capable of non-specific binding to proteins, while the oxidation of pargyline by MAO does not form such a product. However, also in the case of dimethylpropargylamine, only one mole of inhibitor is bound to flavin at the active site. This was shown by the fact that, after removal of unspecifically bound radioactivity in the first Florisil chromatography (Table 2), the molar extinction coefficient of the purified adduct (Table 3) agreed with the value reported in the literature (Chuang *et al.* 1974) for the adduct of pargyline, when the concentration of the flavin was calculated from the radioactivity in each fraction.

Table 2 also illustrates the fact that during isolation of the flavin peptide–inactivator adduct radioactivity was lost far more extensively when the label

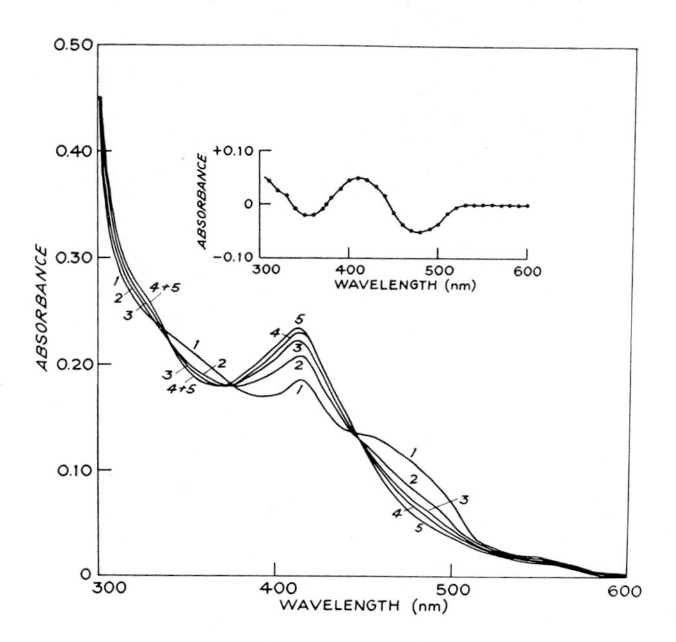

FIG. 5. Spectral changes accompanying inhibition of MAO by *N,N*-dimethylpropargylamine. MAO, 0.38 mg (sp. act. = 2810 nmol/min/mg) equivalent to 2.3 nmol cysteinyl flavin was incubated with 2.4 μmol of *N,N*-dimethylpropargylamine (5.23 mM final concentration) in 0.455 ml of 50 mM-NaP$_i$, pH 7.2, containing 1.1% (w/v) Triton X-100 at 22 °C. Spectra were taken at intervals of 10 min against a reference containing all additions except enzyme. 1: native enzyme; 2,3,4, and 5: 10, 20, 30, and 40 min after addition of inhibitor. After 40 min the enzyme was 88% inhibited. *Inset:* difference spectrum of native enzyme against inhibited enzyme after 50 min incubation.

was in the methyl group than when it was located in the methylene carbon. Suitable control experiments established that, after the removal of non-specifically bound radioactive inhibitor during the first Florisil step, further loss of radioactivity during isolation of the $^{14}CH_3$-INACT adduct occurred under alkaline conditions and was probably due to the liberation of dimethylamine.

The highly purified adduct is expected to be at the FMN level because the dinucleotide is hydrolysed in the initial steps (Kearney *et al.* 1971) but was found to migrate toward the anode in high voltage electrophoresis at pH 6.5 only half as quickly as FMN. It was virtually devoid of fluorescence but showed a characteristic absorption spectrum, with the maximum at slightly alkaline pH values shifted from 410 nm in the inactivated enzyme to 380 nm in the flavin peptide adduct (Fig. 6 and Table 3). The pH dependence of the maximum suggests an approximate pK_a value of 4 to 5.

TABLE 2

Isolation of cysteinyl flavin–inactivator adduct

Step	$^{14}CH_3$-*INACT*	$^{14}CH_2$-*INACT*	
	Exp. 1 (nCi)	*Exp. 2 (nCi)*	*Exp. 3 (nCi)*
Proteolytic digest[a]	26.8	100	294[c]
TCA extract of same	23.4	81	
Standard Florisil chromatography[b]	9.1		
Chromatography on acid-washed Florisil	8.4	57	30
First cellulose phosphate chromatography[d]	7.3	60	
Aminopeptidase digest	4.2		
Second cellulose phosphate chromatography			
20 mM-pyr-acetate eluate	0.45	37	
50 mM-pyr-acetate eluate	2.5		
High voltage electrophoresis[e]			
at pH 1.6	1.8		
at pH 6.5	0.41	35	
Second chromatography on acid-washed Florisil			
acetic acid eluate		24	
0.5–20% pyridine eluate		6.2	

[a] Highly purified MAO was incubated at pH 6.8, 0 °C, in the dark with an excess of the inhibitor until 90 to 99% inactivation occurred. The reaction mixture was then chilled, precipitated with 0.1 volume of 55% (w/v) trichloroacetic acid and the washed precipitate digested with trypsin–chymotrypsin as previously reported (Kearney *et al.* 1971). [b] The missing counts were in the 20% and 50% pyridine eluates and 0.05 M-NH_4OH eluate which contained little or no flavin. [c] Expected 73 to 97 nCi, based on flavin and assuming 1:1 stoichiometry. [d] 7.1 nCi applied in exp. 1, 57 nCi in exp. 2. [e] In exp. 1, 2.5 nCi 50 mM eluate from previous step were used at pH 1.6 and all 1.8 nCi recovered were applied to electrophoresis at pH 6.5.

TABLE 3

Spectral and acid–base properties of substituted dihydroflavins compared to those of the adduct[a]

Substitution	$\lambda_{max}^{H_2O}$ *(nm)*	ε*(M^{-1} cm^{-1})*	pK_a
N-5	296–355	5000–8000	6–7
C-4a	360–370	5000–9000	None 2–10
N-5, C-4a	325–360	5000–10 000	
Adduct	365 (pH < 3)	36 000	4–5
	380 (pH > 7)	34 000	

[a] Data from Ghisla *et al.* 1973; Porter *et al.* 1973; Hevesi & Bruice 1973; and Brustlein & Hemmerich 1968.

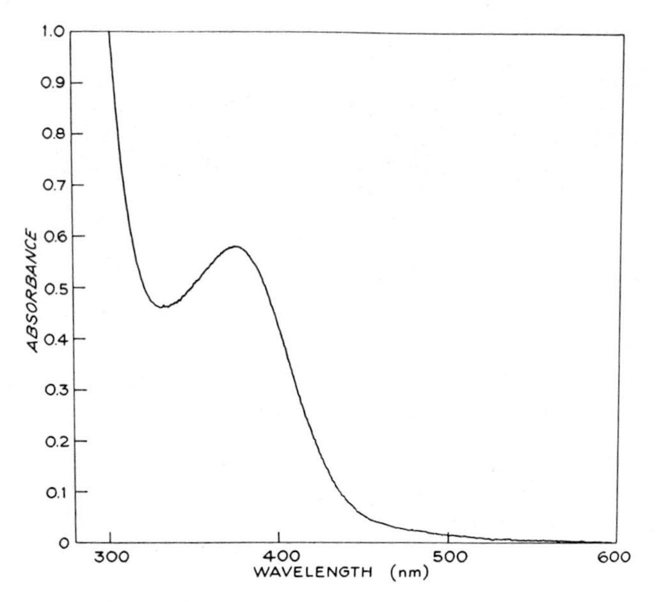

FIG. 6. Absorption spectrum of the cysteinyl flavin peptide adduct isolated from MAO after complete inactivation with *N,N*-dimethyl-^{14}C-propargylamine. The sample used is the acetic acid eluate at the terminal stage of purification (exp. 2, Table 2) after lyophilization and redissolving in water. The $\varepsilon_{375\ nm}$, based on radioactivity as a measure of the molar concentration of the adduct, was 36 100 in this experiment.

Chemical properties and determination of the structure of the adduct

A sample of the adduct prepared from treatment of the liver enzyme with $^{14}CH_2$-INACT, which had not been exposed to alkaline conditions and was judged to be free from non-specifically bound radioactivity, was used in chemical studies aimed at elucidating the structure of the adduct.

The purified adduct was stable in acid but labile in strong base (pH > 12), decomposing to yield oxidized flavin and an unidentified radioactive fragment. The fact that oxidized flavin can be produced from the adduct shows that the isoalloxazine ring system is retained in the adduct. Ozonolysis of the adduct produced [^{14}C]glyoxal in up to 65% yield. This suggests that the adduct contains one of the structures shown below:

$$(A)(B)C{=}CH{-}{}^{14}CH(E)(D) \quad \text{or} \quad (A)(B)C{=}CH{-}{}^{14}CH{=}G$$

where D and E would have to be heteroatoms. For reasons discussed elsewhere (Maycock *et al.* 1976), substituent A or B is hydrogen.

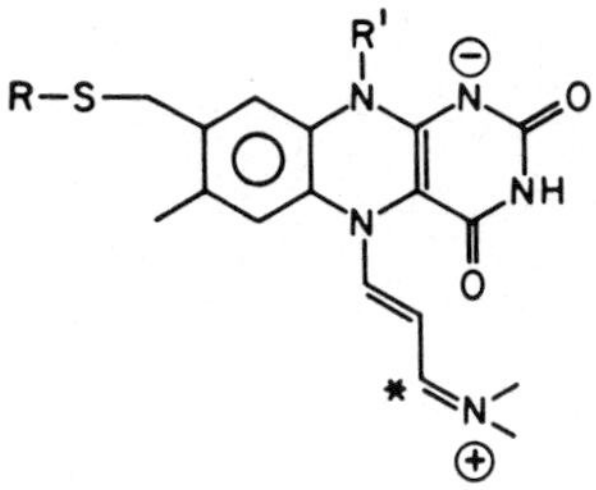

R–S R' N N O NH N O N ⊕ 1

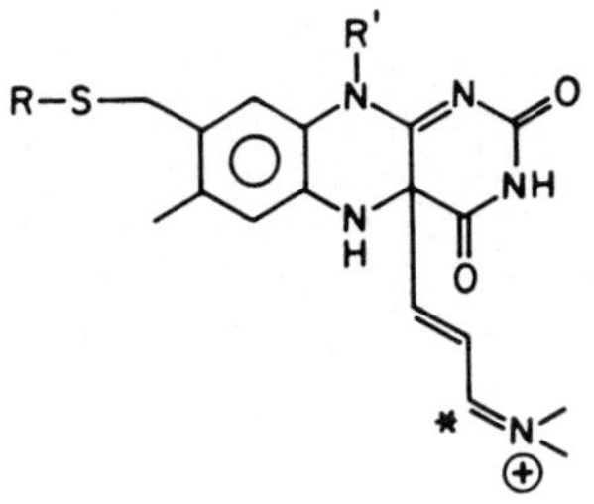

R–S R' N N O NH N H O N ⊕ 2

R–S R' N N O NH N O N 3

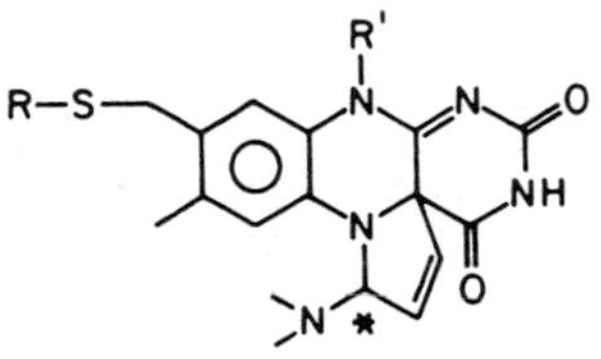

FIG. 7. Possible structures of the adduct.

It has been suggested (Bruice 1975) that in enzymic reactions involving covalent flavin–substrate intermediates the substrate combines at the N-5 or C-4a position. Since propargylamines are substrates of MAO, these positions also seem likely for attachment of the inactivator to the flavin nucleus. Among possible structures of the adduct involving N-5 and/or C-4a, only those shown in Fig. 7 are consistent with the chemical properties of the adduct discussed above.

We may narrow the choice further by considering the physical properties of the adduct. Thus, while N-5 and C-4a substituted dihydroflavins have molar absorbances not exceeding 10 000 M^{-1} cm^{-1}, that of the adduct is very much higher (Table 3). Of the structures shown in Fig. 7 only compound 1 has increased conjugation and thus might be expected to exhibit intense absorption. Moreover, in contrast to C-4a substituted dihydroflavins, N-5 substituted ones have pK values near that of the adduct.

Further evidence for the structure represented by compound 1 was obtained by comparing the adduct with several synthetic model compounds. Among these only compounds 4 and 5 (shown on p. 44) showed physical and chemical properties resembling those of the adduct (Table 4). Compound 5 was described by Zeller *et al.* (1972) and its structure was recently determined (Gärtner & Hemmerich 1975; Maycock 1975). In regard to absorption spectrum, pK, acid

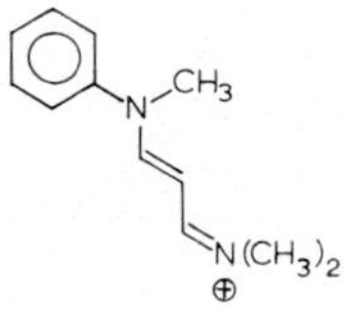

Compound 4 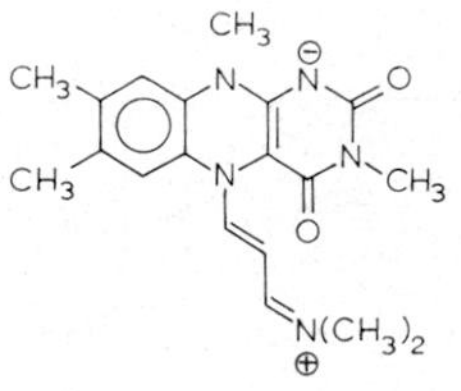 *Compound 5*

TABLE 4

Comparison of the properties of compounds 4, 5, and those of the adduct

Property	*Compound 4*	*Compound 5*	*Adduct*
Optical spectrum			
pH 3		375 nm (ε = 26 000)	365 (ε = 36 000)
pH 7	327 nm (ε = 45 000)	390 nm (ε = 24 000)	380 (ε = 34 000)
pK_a	–	5.3	4–5
H^+	Stable at pH 0.6	Stable at pH 1.3	Stable at pH 1.3
HO^-	Decomposes in 0.1 N-NaOH	Decomposes at pH 11.6	Decomposes at pH 12.6
BH_4^-	Slow reduction	Slow reduction	Slow reduction
O_3	Glyoxal + dimethyl-formamide + ?	Glyoxal + dimethyl-formamide + ?	Glyoxal + ?

and alkali stability and behaviour on ozonolysis and borohydride reduction, it parallels the properties of the adduct. We conclude, therefore, that the adduct formed on inactivation of mitochondrial MAO by *N,N*-dimethylpropargylamine is as shown in Fig. 8. It is interesting to note that in the covalent adduct formed on inactivation of lactate oxidase with an acetylenic analogue of the substrate, the inactivator is attached to *both* N-5 and C-4a of the flavin and that the carbon undergoing oxidation becomes attached to N-5 (Schonbrunn *et al.* 1976).

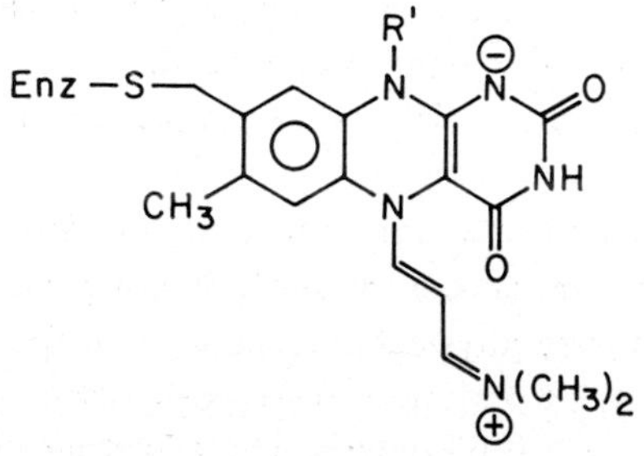

FIG. 8. Structure of the adduct formed when MAO is inactivated by $HC \equiv CCH_2N(CH_3)_2$.

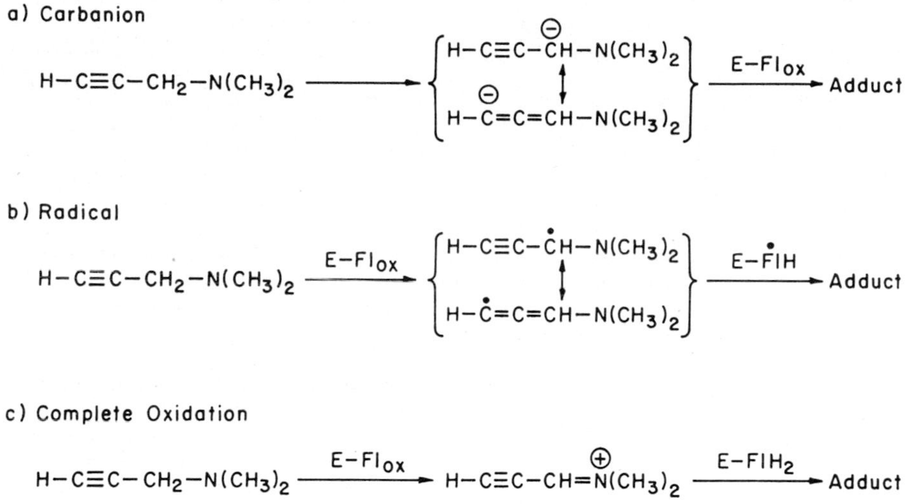

a) Carbanion

$$H-C\equiv C-CH_2-N(CH_3)_2 \longrightarrow \left\{ \begin{array}{c} H-C\equiv C-\overset{\ominus}{C}H-N(CH_3)_2 \\ \updownarrow \\ H-\overset{\ominus}{C}=C=CH-N(CH_3)_2 \end{array} \right\} \xrightarrow{E-Fl_{ox}} \text{Adduct}$$

b) Radical

$$H-C\equiv C-CH_2-N(CH_3)_2 \xrightarrow{E-Fl_{ox}} \left\{ \begin{array}{c} H-C\equiv C-\dot{C}H-N(CH_3)_2 \\ \updownarrow \\ H-\dot{C}=C=CH-N(CH_3)_2 \end{array} \right\} \xrightarrow{E-\dot{F}lH} \text{Adduct}$$

c) Complete Oxidation

$$H-C\equiv C-CH_2-N(CH_3)_2 \xrightarrow{E-Fl_{ox}} H-C\equiv C-CH=\overset{\oplus}{N}(CH_3)_2 \xrightarrow{E-FlH_2} \text{Adduct}$$

FIG. 9. Possible mechanisms of formation of the adduct.

Fig. 9 summarizes three possible mechanisms for the formation of the adduct from MAO and dimethylpropargylamine. In the first reaction the enzyme is shown to abstract the α-proton from the inhibitor, forming an allenyl carbanion, which then adds to N-5 of the flavoquinone. The second reaction shows the transfer of the α-proton and of an electron from the inhibitor to the flavoquinone, giving rise to the respective free radicals, which then collapse to the product. In the third reaction the enzyme is visualized as catalysing the oxidation of the inhibitor to the 'normal' product, $HC\equiv C-CH=\overset{+}{N}(CH_3)_2$, which then undergoes a Michael addition at N-5 of the reduced flavin. At present we cannot decide among these possibilities.

ACKNOWLEDGEMENTS

This investigation was supported by grants from the National Institutes of Health (HL 16251 and GM 12633) and the National Science Foundation (GB 36570X) and by a Special Fellowship from the National Institutes of Health (5 FOZ GM 40063) to A.L.M. The authors are grateful to Dr P. Hemmerich for a generous supply of compound 5.

References

BLOCH, K. (1969) Enzymatic synthesis of monounsaturated fatty acids. *Acc. Chem. Res. 2*, 193-202

BRUICE, T. C. (1975) in *Progress in Bio-organic Chemistry* (Kaiser, E. T. & Kezdy, F. J., eds.), vol. 3, Wiley, New York, in press

BRUSTLEIN, M. & HEMMERICH, P. (1968) Photoreduction of flavocoenzymes by pyruvic acid. *FEBS Lett. 1*, 335-338

CHUANG, H. Y. K., PATEK, D. R. & HELLERMAN, L. (1974) Mitochondrial monoamine oxidase. Inactivation by pargyline. Adduct formation. *J. Biol. Chem. 249*, 2381-2384

CLELAND, W. W. (1970) in *The Enzymes*, vol. II (Boyer, P. D., ed.), 3rd edn, pp. 1-65, Academic Press, New York

FULLER, R. W. (1972) Selective inhibition of monoamine oxidase, in *Monoamine Oxidase–New Vistas* (Costa, E. & Sandler, M., eds.) *(Adv. Biochem. Psychopharmacol. 5)*, pp. 339-354, Raven Press, New York and North-Holland, Amsterdam

GÄRTNER, B. & HEMMERICH, P. (1975) Inhibition of monoamine oxidase by propargylamine: structure of the inhibitor complex. *Angew. Chem. Int. Ed. Engl. 14*, 110-111

GHISLA, S., HARTMANN, U., HEMMERICH, P. & MÜLLER, F. (1973) Studien in der Flavin-Reihe, XVIII. Die reduktive Alkylierung des Flavinkerns; Struktur und Reaktivität von Dihydroflavinen. *Liebigs Ann. Chem. 8*, 1388-1415

HELLERMAN, L. & ERWIN, V. G. (1968) Mitochondrial monoamine oxidase. II. Action of various inhibitors for the bovine kidney enzyme. Catalytic mechanism. *J. Biol. Chem. 243*, 5234-5243

HEVESI, L. & BRUICE, T. C. (1973) Reaction of sulfite with isoalloxazines. *Biochemistry 12*, 290-297

HEVEY, R. C., BABSON, J., MAYCOCK, A. L. & ABELES, R. H. (1973) Highly specific enzyme inhibitors. Inhibition of plasma amine oxidase. *J. Am. Chem. Soc. 95*, 6125-6127

KASS, R. & BLOCH, K. (1967) On the enzymatic synthesis of unsaturated fatty acids in *Escherichia coli*. *Proc. Natl. Acad. Sci. U.S.A. 58*, 1168-1173

KEARNEY, E. B., SALACH, J. I., WALKER, W. H., SENG, R. L., KENNEY, W., ZESZOTEK, E. & SINGER, T. P. (1971) The covalently-bound flavin of hepatic monoamine oxidase. 1. Isolation and sequence of a flavin peptide and evidence for binding at the 8α position. *Eur. J. Biochem. 24*, 321-327

MAYCOCK, A. L. (1975) Structure of a flavoprotein-inactivator model compound. *J. Am. Chem. Soc. 97*, 2270-2272

MAYCOCK, A., ABELES, R. H., SALACH, J. I. & SINGER, T. P. (1976) Structure of the flavin-inhibitor adduct from monoamine oxidase. *Biochemistry 15*, 114-125

MCEWEN, C. M., JR, SASAKI, G. & JONES, D. C. (1969) Human liver mitochondrial monoamine oxidase. III. Kinetic studies concerning time-dependent inhibitors. *Biochemistry 8*, 3963-3972

PORTER, D. J., VOET, J. G. & BRIGHT, H. J. (1973) Direct evidence for carbanions and covalent N^5-flavin-carbanion adducts as catalytic intermediates in the oxidation of nitroethane by D-amino acid oxidase. *J. Biol. Chem. 248*, 4400-4416

RANDO, R. R. (1973) 3-Bromoallylamine induced irreversible inhibition of monoamine oxidase. *J. Am. Chem. Soc. 95*, 4438-4439

RANDO, R. R. (1974) Chemistry and enzymology of k_{cat} inhibitors. *Science (Wash. D.C.) 185*, 320-324

SALACH, J. I., SINGER, T. P., YASUNOBU, K. T., MINAMIURA, N. & YOUDIM, M. B. H. (1976) This volume, pp. 49-56

SCHONBRUNN, A., WALSH, C. T., GHISLA, S., OGATA, H., MASSEY, V. & ABELES, R. H. (1976) in *Flavins and Flavoproteins* (Singer, T. P., ed), Elsevier Scientific Publishing Company, Amsterdam, in press

TABOR, C. W., TABOR, H. & ROSENTHAL, S. M. (1954) Purification of amine oxidase from beef plasma. *J. Biol. Chem. 208*, 645-661

WALSH, C. T., SCHONBRUNN, A., LOCKRIDGE, O., MASSEY, V. & ABELES, R. H. (1972*a*) Inactivation of a flavoprotein, lactate oxidase, by an acetylenic substrate. *J. Biol. Chem. 247*, 6004-6006

WALSH, C. T., ABELES, R. H. & KABACK, H. R. (1972*b*) Inactivation of D-lactate dehydrogenase and D-lactate dehydrogenase-coupled transport in *Escherichia coli* membrane vesicles by an acetylenic substrate. *J. Biol. Chem. 247*, 7858-7863

ZELLER, E. A., GÄRTNER, B. & HEMMERICH, P. (1972) 4a,5-Cycloaddition reactions of acetylenic compounds at the flavoquinone nucleus as mechanisms of flavoprotein inhibition. *Z. Naturforsch. 27B*, 1050-1052

[For discussion of this paper, see pp. 57-59]

Cysteinyl flavin in monoamine oxidase from the central nervous system

J. I. SALACH*, T. P. SINGER*, K. T. YASUNOBU†, N. MINAMIURA† and M. B. H. YOUDIM‡

*Molecular Biology Division, Veterans Administration Hospital and Department of Biochemistry and Biophysics, University of California, San Francisco, †Department of Biochemistry and Biophysics, University of Hawaii, Honolulu, and ‡MRC Clinical Pharmacology Unit, Radcliffe Infirmary, Oxford

Abstract Mitochondrial monoamine oxidases from mammalian liver and kidney are known to contain 8α-cysteinyl-FAD at the active centre; that is, the flavin is covalently linked to the peptide chain by way of a thioether linkage from the 8α-methyl group to a cysteine residue. In contrast, the enzyme from pig brain has been reported to contain FAD in non-covalent linkage. We have purified the enzyme from pig brain mitochondria by two alternative methods, one of which appears to yield a homogeneous preparation as judged by acrylamide electrophoresis. Both preparations contain cysteinyl flavin in significant amounts. The turnover number of the enzyme from brain, per mole of cysteinyl flavin, in apparently homogeneous samples is nearly the same as in the enzyme from kidney (based on $[^{14}C]$pargyline binding). A higher turnover number per cysteinyl flavin is obtained in preparations from pig brain isolated by an alternative and more rapid procedure: this value agrees with the turnover number of monoamine oxidase in outer membranes of liver and in highly purified preparations from liver mitochondria. It is suggested that at least one form of the enzyme in brain contains flavin in covalent linkage.

It has been known for several years that monoamine oxidase from liver and kidney mitochondria contains covalently bound flavin (Nara *et al.* 1966; Igaue *et al.* 1967; Erwin & Hellerman 1967; Sourkes 1968) (amine: oxygen oxidoreductase [deaminating] [flavin-containing]; EC 1.4.3.4). The structure of this covalent adduct has been shown to be 8α-cysteinyl-FAD and the peptide sequence at the flavin site of the enzyme has been determined (Kearney *et al.* 1971; Walker *et al.* 1971).

Despite this progress in the elucidation of the active site of the enzyme from liver, some basic questions have remained unanswered about the flavin site of the enzyme from the central nervous system. It has been reported, for instance, that monoamine oxidase from pig brain mitochondria contains FAD in non-covalent linkage (Tipton 1968*a*). This report was of great interest to us not only because of the physiological and pharmacological importance of the brain enzyme but also because of the major difference in the active sites of mitochondrial monoamine oxidase in different tissues which it implied. The conclusion that the FAD moiety is non-covalently bound in the enzyme from brain was based on two observations. First, at the end of a long purification procedure, but not at earlier stages, trichloroacetic acid released free FAD from the enzyme. Second, at the terminal stage the acid-ammonium sulphate procedure produced a virtually inactive, non-fluorescent protein which appeared to be an apoenzyme, and the activity was partially restored on adding back FAD: this was viewed as the reconstitution of a holoenzyme (Tipton 1968*a*). Tipton recognized the puzzling aspects of his observations, namely, that acid denaturation liberates FAD only at the terminal stage, while if the linkage were non-covalent, it should be released at all stages, and that K_D, the dissociation constant for FAD, determined from reconstitution of the activity, was too high to account for the seeming stability of the bond during isolation. Clearly, the problem called for further probing of the nature of the flavin site, preferably by following the cysteinyl flavin content of the enzyme during its isolation from brain mitochondria, now that methods have been elaborated for the analysis of covalently bound flavin in monoamine oxidase.

Initial attempts to confirm and extend these observations were hampered by difficulties experienced in all three of our laboratories in reproducing the purification procedure of the enzyme from pig brain (Tipton 1968*b*). The main problem seemed to be our inability to extract monoamine oxidase in significant yield from brain mitochondria by the sonication method previously used (Tipton 1968*b*) with the aid of a variety of ultrasonic equipment available in Britain and in the United States. It became necessary, therefore, to elaborate alternative procedures for purifying the enzyme from mammalian brain. This turned out to be a major undertaking because of the very high lipid content of the starting material. Methods which are dependable in purifying the enzyme from liver mitochondria fail to function with brain mitochondria. Thus the large-scale separation of inner and outer membranes (Kearney *et al.* 1971), devised to separate succinate dehydrogenase, an inner membrane enzyme and another source of covalently bound flavin, from monoamine oxidase, an outer membrane enzyme, failed to achieve satisfactory separation in the case of brain, as did alternative procedures for the purification of outer membranes (S. Gabay & J. I. Salach, unpublished work). A relatively short procedure devised for purification of monoamine oxidase from beef liver mitochondria (Table 1), which yields preparations of about the same or higher specific activity than other procedures, similarly failed to yield satisfactory results in the case of brain mitochondria.

TABLE 1

Summary of rapid purification of monoamine oxidase from beef liver mitochondria

Step	*Total protein (g)*	*Total units*	*Specific activity (nmol benzylamine/min/mg)*	*Yield (%)*
Mitochondria	18.3	4.5×10^5	24	(100)
After extraction with phospholipases A and C	5.9	3.4×10^5	58	76
Triton extract	2.3	3.5×10^5	170	77
Three-phase polymer partition: precipitate at lower interface[a]		1.2×10^5		27
37 000 × *g* supernatant of same	0.20	1.2×10^5	596	27
252 000 × *g* pellet from previous step	0.039	1.2×10^5	2900	27

[a] Polyethylene glycol 6000: Ficoll: dextran (252 000 mol. wt.); protein not determined in this fraction. Proteins were determined by the biuret method. One unit corresponds to the oxidation of one nmol of benzylamine/min at 30 °C, pH 7.2 in 3 ml total volume. The specific activity of 2900 attained here may be compared with the values of 2580 (beef liver enzyme, Gomes *et al.* (1969)) and 3000 (beef kidney enzyme, Chuang *et al.* (1974)), recalculated from published values for the assay conditions used in this study. The experimentally determined conversion factors are: activity under the conditions of Gomes *et al.* (1969) (3.0 ml, pH 7.4, 25 °C) × 1.29 = activity under present conditions (3.0 ml, pH 7.2, 30 °C); activity under the conditions of Chuang *et al.* (1974) (2.2 ml, pH 7.6, 37 °C) × 0.44 = activity under present conditions.

In order to cope with this problem, two procedures were devised. Procedure A, elaborated in Oxford, involves extraction of pig brain mitochondria with Triton X-100, ammonium sulphate precipitation, and chromatography on Sephadex G-200 and on DEAE-cellulose. It yields some 50-fold purification from mitochondria, less than the alternative procedure (cf. below), but it has the advantages of shortness and simplicity. It may for these reasons minimize inactivation; this *might* be the reason for the higher turnover number (per mole of cysteinyl flavin) than in the longer procedure B (cf. below).

Procedure B, summarized in Table 2, yields an enzyme of high specific activity and one which gives a single band on electrophoresis in polyacrylamide gels, in the presence of sodium dodecyl sulphate (SDS) and mercaptoethanol and, in samples which enter the gel, without SDS or mercaptoethanol (Fig. 1).

The second problem which had to be overcome in analysing the relation of activity to cysteinyl flavin content during isolation of the enzyme from brain mitochondria centres around the two facts that the cysteinyl flavin content is calculated by subtracting the histidyl flavin content arising from succinate dehydrogenase from the total covalently bound flavin (Scheme 1) and that brain mitochondria contain far more succinate dehydrogenase than monoamine oxidase. Taken together, these facts mean that the cysteinyl flavin content can

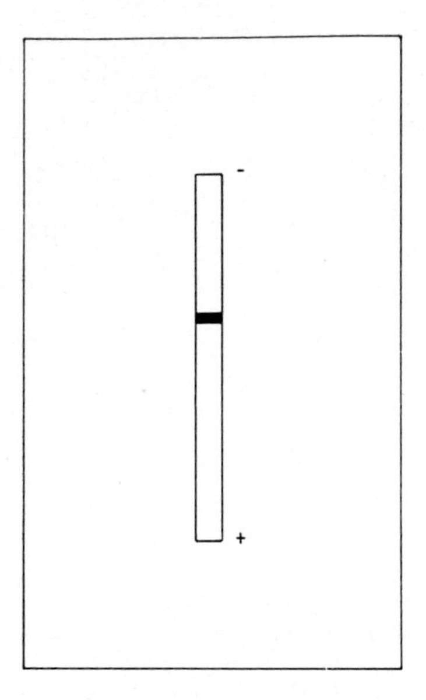

FIG. 1. Sodium dodecyl sulphate (SDS)-disc electrophoresis of purified pig brain mitochondrial amine oxidase (procedure B). The enzyme was first incubated in 0.1 M-sodium phosphate buffer, pH 7.2, containing 5% mercaptoethanol and 1% (w/v) SDS. About 12 μg of sample was then applied to a disc which had been equilibrated with 0.1% SDS–0.1 M-sodium phosphate buffer, pH 7.2. The current was 8 mA per tube and the time of the run was 4 hours.

TABLE 2

Summary of procedure B for purification of monoamine oxidase from pig brain mitochondria

Step	*Total protein*[a] *(mg)*	*Specific activity*[b]	*Yield (%)*
1. Mitochondria	12 500	1.0	100
2. Extract (Triton–cholate)	3900	3.5	108
3. Unabsorbed fraction from DEAE-cellulose treatment	2077	6.0	100
4. Calcium phosphate gel eluate	1000	8.4	68
5. DEAE-cellulose eluate	30	124	30
6. Hydroxylapatite eluate	17.5	356	22
7. Biogel A-1.5m eluate	2.8	803	19
8. Biogel A-0.5m eluate	0.81	1163	8

[a] Determined by the Lowry method.
[b] The enzyme was assayed in a volume of 3.0 ml, in the presence of benzylamine, potassium phosphate buffer, pH 7.4, and 0.5% Triton X-100 at 25 °C. Specific activity is nmol benzylamine oxidized/min/mg at 25 °C. Specific activity at 30 °C (as in Table 1) may be ~30% higher.

SCHEME 1

Determination of cysteinyl flavin in monoamine oxidase

Heat denaturation, trichloroacetic acid precipitation

↓

Wash with acetone-HCl, alkaline $CHCl_3$-CH_3OH, precipitate with acetone-trichloroacetic acid, wash with trichloroacetic acid

↓

Digestion with trypsin-chymotrypsin or pronase

↓

Lyophilization, sublimation of NH_4HCO_3, oxidation of thioether with performic acid to sulphone

↓

Removal of performic and formic acids by lyophilization; acid hydrolysis of remaining pyrophosphate linkage

↓

Fluorescence measurements at pH 3.4 and pH 7.0 before and after dithionite, with and without internal standards

↓

Subtraction of histidyl flavin from total flavin gives cysteinyl flavin content

be reliably measured only after the bulk of succinate dehydrogenase is removed —that is, after the cysteinyl flavin content equals or exceeds the histidyl flavin content.

Cysteinyl flavin was present in all brain preparations examined, but reliable values were obtained only in relatively highly purified samples. Thus, in the procedure of Table 2, the turnover number (moles of benzylamine oxidized/min/mol of cysteinyl flavin at 30 °C) remained reasonably constant (360 $\pm$ 50) over a 13-fold purification range (second DEAE-cellulose eluate to end of procedure). Moreover, in the last three steps the turnover number per cysteinyl flavin agreed well with the value (298) reported recently by Chuang *et al.* (1974) for the turnover number of the kidney enzyme, the latter being based on total flavin or pargyline binding. Considering the multiple sources of error in comparing different assay conditions and different bases for expressing cysteinyl flavin content, the agreement is gratifying and suggests that neither preparation contained much monoamine oxidase activity not associated with cysteinyl flavin (Table 3).

The picture becomes more complicated if one compares these values with the turnover number of the enzyme from liver mitochondria (Table 4). Although outer membrane preparations and the highly purified enzyme from liver (Table 1) show more of a variation in turnover number per cysteinyl flavin

TABLE 3

Turnover numbers of extensively purified preparations of monoamine oxidase

Source	*Preparation*	*Turnover number at 30 °C (moles of benzylamine/min/mol of cysteinyl flavin)*
Beef kidney	Chuang *et al.* (1974)	~298[a]
Pig brain	Table 2	~318[b]

[a] Value calculated for 30 °C and standard conditions from published assays by use of the factor 1.8 to 1.9. Turnover number is based on *total* flavin or on pargyline binding.
[b] Assays were at 25 °C. At 30 °C activity is somewhat higher.

TABLE 4

Turnover numbers of partially purified preparations of monoamine oxidase

Source	*Preparation*	*Turnover number at 30 °C (moles of benzylamine/min/mol of cysteinyl flavin)*
Beef liver	Outer membrane and soluble preparations[a]	715 ± 320
Beef liver	Highly purified preparations, as in Table I	1050
Pig brain	Procedure A, best preparations	860 ± 270

[a] J. I. Salach, unpublished data.

content (715 ± 320) than we would like to see, it is clear that the mean value is higher than that obtained at the end of the long isolation procedure from brain mitochondria or from kidney (Table 3). They do agree, however, reasonably well with the turnover number determined in monoamine oxidase samples of pig brain isolated by the short procedure (procedure A) (Table 4). Whether this suggests that monoamine oxidase is partly inactivated during the long procedures of Chuang *et al.* (1974) and that summarized in Table 2 for the brain enzyme remains to be proved but cannot be ruled out at this time.

What may we then conclude concerning the flavoprotein nature of monoamine oxidase from brain mitochondria? We believe that there is strong evidence suggesting that all purified preparations examined contain cysteinyl flavin. Hence, the least we can say is that one form—perhaps the predominant form—of monoamine oxidase from mammalian brain contains covalently bound cysteinyl flavin, just as do liver mitochondria. How are we to explain, on the other hand, the impressive data of Tipton (1968*a*)? One possibility is that, besides the form containing cysteinyl flavin, there is another form of monoamine oxidase in brain which contains flavin in non-covalent linkage.

Although we don't like this explanation, it cannot be ruled out, since several highly purified samples of the enzyme from brain mitochondria did contain significant amounts of acid-extractable flavin. Also, we cannot completely rule out the possibility that this had originated from impurities. It has also been suggested that the linkage of the flavin to the protein changes in the course of isolation (Tipton 1968*a*). The only observations we know of which may suggest this notion are those of Edmondson & Singer (1973) who have shown that the oxidation of cysteinyl flavin to the sulphone, followed by reduction with dithionite, liberates riboflavin. However, studies in this laboratory have shown that this reductive cleavage does not occur with the flavin pentapeptide or with the enzyme itself. Thus, at present we know of no chemical basis for the cleavage of cysteinyl FAD by acid denaturation at any stage of purification of the enzyme and, hence, cannot explain the observations of Tipton (1968*a*).

SUMMARY

Methods are described for the isolation of monoamine oxidase from brain and for the analysis of the cysteinyl flavin content of enzyme preparations. With the help of these procedures the question whether mitochondrial monoamine oxidase from mammalian brain mitochondria contains covalently bound flavin has been re-examined. One procedure for the isolation of the enzyme from brain, which appears to yield preparations monodisperse in polyacrylamide electrophoresis, gave a relatively constant ratio of activity to cysteinyl flavin content (i.e., turnover number) throughout the steps in which cysteinyl flavin could be accurately determined. The turnover number at the final stage, expressed on the basis of cysteinyl flavin content, was nearly the same as that of the enzyme from kidney mitochondria. An alternative procedure for purifying the enzyme from brain also yielded preparations containing cysteinyl flavin. The turnover number per cysteinyl flavin was higher in this case than in the kidney enzyme and in the range found for outer membrane and purified, soluble preparations from liver mitochondria. It appears that the form of monoamine oxidase isolated from brain by both of these procedures contains covalently bound flavin, as does the enzyme from kidney and liver mitochondria. The difference in turnover numbers seen between preparations isolated by long and relatively short procedures, respectively, might possibly reflect some inactivation in the former case.

ACKNOWLEDGEMENTS

This investigation was supported by the National Science Foundation (GB 36570X) and the National Institutes of Health (MH 21539).

References

CHUANG, H. K., PATEK, D. & HELLERMAN, L. (1974) Mitochondrial monoamine oxidase. Inactivation by pargyline. Adduct formation. *J. Biol. Chem. 249*, 2381-2384

EDMONDSON, D. E. & SINGER, T. P. (1973) Oxidation-reduction properties of 8α-substituted flavins. *J. Biol. Chem. 248*, 8144-8149

ERWIN, V. G. & HELLERMAN, L. (1967) Mitochondrial monoamine oxidase. Purification and characterization of the bovine kidney enzyme. *J. Biol. Chem. 242*, 4230-4238

GOMES, B., IGAUE, I., KLOEPPER, H. & YASUNOBU, K. T. (1969) Amine oxidase. XIV. Isolation and characterization of the multiple beef liver amine oxidase components. *Arch. Biochem. Biophys. 132*, 16-27

IGAUE, I., GOMES, B. & YASUNOBU, K. T. (1967) Beef mitochondrial monoamine oxidase, a flavin dinucleotide enzyme. *Biochem. Biophys. Res. Commun. 29*, 562-570

KEARNEY, E. B., SALACH, J. I., WALKER, W. H., SENG, R. L., KENNEY, W., ZESZOTEK, E. & SINGER, T. P. (1971) The covalently bound flavin of hepatic monoamine oxidase. I. Isolation and sequence of a flavin peptide and evidence for binding at the 8α position. *Eur. J. Biochem. 24*, 321-327

NARA, S., IGAUE, I., GOMES, B. & YASUNOBU, K. T. (1966) The prosthetic group of animal amine oxidases. *Biochem. Biophys. Res. Commun. 23*, 324-328

SOURKES, T. L. (1968) Properties of the monoamine oxidase of rat liver mitochondria. *Adv. Pharmacol. 6A*, 61-69

TIPTON, K. F. (1968*a*) The prosthetic group of pig brain mitochondrial monoamine oxidase. *Biochim. Biophys. Acta 159*, 451-459

TIPTON, K. F. (1968*b*) The purification of pig brain mitochondrial monoamine oxidase. *Eur. J. Biochem. 4*, 103-107

WALKER, W. H., KEARNEY, E. B., SENG, R. L. & SINGER, T. P. (1971) The covalently bound flavin of hepatic monoamine oxidase. 2. Identification and properties of cysteinyl flavin. *Eur. J. Biochem. 24*, 328-331

Discussion of the two preceding papers

Tipton: There is quite good evidence that the hydrazines function as inhibitors after they have been oxidized by the enzyme. Have you any data or surmise on the way in which methylenic inhibitors may act?

Singer: No. This is the next stage of our work. There is no reason *a priori* to believe that the irreversible inhibitors will all react with the flavin, much less that they will all react at the same site.

Youdim: I think the answer to Dr Tipton's question is that they also react with FAD, but not at the same site (Youdim 1976). Phenylethylhydrazine appears to react at a different site, because the ^{14}C adduct showed the same fluorescence and activation spectrum as that of pure flavin (FAD) but it was partially bleached.

Knoll: What about the *N*-cyclopropylamine group? Do you expect the same reaction?

Singer: We are working on this with Dr Abeles. I have no preconceived notions about its reaction. We may know soon what it is.

Tipton: Would a methylenic group be expected to act in a similar way?

Singer: I wouldn't expect it to, but it is difficult to make accurate predictions, and now that we have labelled inhibitors available we should be able to decide this experimentally.

Fuller: You say you are going to study the *N*-cyclopropylamines. I have been interested in this group of MAO inhibitors for several years. What particular *N*-cyclopropylamine do you plan to study? There is a wide difference in the selectivity of these inhibitors, depending on the nature of the other substituent on the nitrogen. For example, Lilly 51641 has a substituent similar to that of clorgyline and is a type A specific inhibitor, as is clorgyline. But we have studied other *N*-cyclopropylamines (for example, with the phenylisopropyl substituent like that of deprenyl) which seem to differ in their affinities towards types A and B MAO. We assume these inhibitors would react by the same mechanism but perhaps with different forms of MAO.

Singer: We are looking at a series of compounds, starting with labelled *trans*-phenylcyclopropylamine. Are you suggesting that both types of MAO should be tested? We would be happy to undertake that, if someone supplies the labelled inhibitor.

Gorkin: Did you study the substrate specificity of your purified enzyme? Or did you just measure the activity with benzylamine as a substrate?

Singer: Routinely we use benzylamine, but we have compared the activities on benzylamine and tyramine at various stages of purification, so as to permit comparison with Dr Tipton's experimental conditions.

Gorkin: I was also wondering whether the A form is killed during the extraction procedure, or whether you extract the B form selectively, because of course with benzylamine as a substrate you measure only the B form.

Singer: A systematic comparison of activity ratios from outer membrane to purified enzyme, using A- and B-specific substrates respectively, has not yet been made, although this is among our future plans.

Tipton: Preparations of the enzyme from rat liver that have been reported to be homogeneous have been shown to contain some 50% of the flavin in a non-covalently bound form (Sourkes 1968; Youdim & Sourkes 1972). Do you think there may be two forms of flavin binding?

Singer: There is nothing easier than to find free flavin in a so-called highly purified enzyme, no matter what the enzyme. Flavins are universal cell constituents and I don't believe Professor Sourkes claimed any kind of homogeneity for his preparations. The fact that free flavin was present is not surprising. It is present in everything, including Yasunobu's preparation, at the near-terminal stages, to the extent of about a third or half as much as the covalently bound flavin. This doesn't worry me because there is no strong reason to think that it comes from monoamine oxidase.

It is true that cysteinyl FAD is a labile compound and can undergo a variety of breakdowns which would yield free flavin. Thus in some unpublished work we have shown with Professor E. E. Snell that over a matter of minutes cysteinyl flavin is converted to pyruvate in fairly good yield, but this does not happen with the enzyme. It happens only with the cysteinyl flavin with no amino acids attached. Thus peptides of cysteinyl flavin are more stable. A second type of breakdown is reductive cleavage to yield riboflavin. This is a reaction involving oxidation of cysteinyl flavin to the sulphone form with performic acid, followed by dithionite reduction to give riboflavin. The enzyme doesn't undergo this type of reductive cleavage, nor does the pentapeptide; only free cysteinyl flavin does. It appears that in the enzyme the inherently labile cysteinyl flavin linkage is stabilized by secondary bonding, probably by tyrosine–flavin interactions. In fact there is a tyrosine vicinal to the flavin in the peptide. So I don't regard the cysteinyl flavin, as it occurs in monoamine oxidase, as a particularly labile linkage, much less one apt to undergo reductive cleavage. In our hands protein-bound cysteinyl flavin is completely stable to storage at —10 °C for long periods. Perhaps there is more than one type of enzyme in your preparation. The only way we shall know is to analyse for both free flavin and cysteinyl flavin at all stages of purification.

Youdim: Couldn't this be resolved by doing the experiments with ^{14}C-labelled inhibitors, where the enzyme is pretreated, fully inactivated, and then the ^{14}C adduct isolated? If there is any free FAD one should be able to see this.

Singer: That is an excellent idea and obviously the way to go on.

References

SOURKES, T. L. (1968) Properties of monoamine oxidase of rat liver mitochondria. *Adv. Pharmacol. 6A*, 61-69

YOUDIM, M. B. H. (1976) Rat liver mitochondrial monoamine oxidase: an iron requiring flavoprotein, in *Flavins and Flavoproteins* (Singer, T. P., ed.), Elsevier Scientific Publishing Company, Amsterdam, in press

YOUDIM, M. B. H. & SOURKES, T. L. (1972) The flavin prosthetic group of purified ratliver mitochondrial monoamine oxidase, in *Monoamine Oxidase—New Vistas* (Costa, E. & Sandler, M., eds.) (*Adv. Biochem. Psychopharmacol. 5)*, pp. 45-53, Raven Press, New York and North-Holland, Amsterdam

Monoamine oxidase inhibitors and the transformation of monoamine oxidases

V. Z. GORKIN

Institute of Biological and Medical Chemistry, Academy of Medical Sciences of the USSR, Moscow

Abstract The transformation of monoamine oxidases has been demonstrated in experiments with purified preparations of monoamine oxidases. Reversible and qualitative alterations in the catalytic properties of the enzymes were initiated by partial oxidation of their SH groups. The transformed monoamine oxidases acquired the property of deaminating diamines, ω-amino acids, nucleotides (AMP, for example) and other nitrogenous compounds which are not substrates of monoamine oxidases. Structural analogues of these compounds blocked the activity of the transformed enzymes. The transformation of monoamine oxidases was prevented *in vitro* by either irreversible or reversible monoamine oxidase inhibitors, sulphide-forming reagents or reagents alkylating SH groups.

In mitochondrial membranes the transformation of monoamine oxidase activity accompanied the stimulation of lipid peroxidation and was prevented either by monoamine oxidase inhibitors or by inhibitors of free-radical peroxidation processes.

The *in vivo* transformation of monoamine oxidase activity was observed in tissues of tumour-bearing animals, in experimental radiation injury and in hypervitaminosis D_2. These pathological conditions are characterized by the accumulation of lipid peroxides in tissues. The transformations of monoamine oxidases, resulting in marked increases in the deamination of AMP and other nucleotides, are probably of pathogenetic importance in these and related diseases. Structural analogues of AMP (adenosine 3′-monophosphate, for example) inhibited the deamination of nucleotides by the transformed enzymes (but did not inhibit deamination of biogenic monoamines) and had a definite beneficial effect in the tumour-bearing animals and in experimental radiation injury. Antioxidants prevented the transformation of monoamine oxidases *in vivo*. Some reducing agents caused a re-transformation of the enzyme. It is suggested that monoamine oxidase inhibitors may find new applications as drugs which prevent the transformation of monoamine oxidases *in vivo*.

The monoamine oxidases (amine: oxygen oxidoreductase [deaminating] [flavin-containing]; EC 1.4.3.4) (MAO) contain SH groups located mainly away from the catalytic sites (Hellerman & Erwin 1968). Partial oxidation of these SH groups has been shown to initiate reversible qualitative alterations (transformations) in the catalytic properties of monoamine oxidases, which acquire properties similar to those of diamine oxidases (pyridoxal-containing) (EC 1.4.3.6) (Gorkin & Tatyanenko 1957; Gorkin 1973).

TRANSFORMATION OF BACTERIAL MONOAMINE OXIDASE

Soluble monoamine oxidase (tyramine oxidase, or tyraminase) from *Sarcina lutea* (Yamada *et al.* 1967*a*) was isolated and purified without the use of detergents or organic solvents. The enzyme has a molecular weight of about 130 000, contains two moles of FAD per mole, and does not contain metals. Enzymic activity is inhibited by the MAO inhibitor iproniazid, but isoniazid, hydroxylamine, semicarbazide and hydrazine do not inhibit the activity. The enzyme is characterized by a sharp substrate specificity. Out of 27 nitrogenous compounds tested it catalysed the oxidative deamination of only tyramine and dopamine (Yamada *et al.* 1967*b*).

We have shown that highly purified preparations of this bacterial monoamine oxidase contained four SH groups per mole (Tatyanenko *et al.* 1971). The number of SH groups did not change in the presence of 6M-urea. Treatment of the enzyme with a reducing agent, $NaBH_4$ (at pH 8 for 10 minutes at 20 °C), also did not influence the content of SH groups, but treatment with an oxidizing agent, *o*-iodosobenzoate (0.1mM), decreased the number from four to one per mole of the enzyme. Incubation of the enzyme under aerobic conditions in the presence of a catalytic concentration (6 μM) of Cu^{2+} had a similar effect. The effect was partially reversible: treatment of the enzyme preparation with $NaBH_4$ partially restored the number of SH groups. These results suggest that atmospheric oxygen in the presence of a catalytic concentration of Cu^{2+} caused partial reversible oxidation of SH groups in the enzyme. Other oxidizing agents influenced the SH groups of the enzyme similarly.

What happens to the catalytic properties in these conditions? When the enzyme is incubated aerobically with catalytic concentrations of Cu^{2+}, the decreased content of SH groups is accompanied by a sharp decrease in tyramine-deaminating activity. At the same time, the enzyme acquired the qualitatively new property of deaminating lysine at a rapid rate—an activity that was absent in the native enzyme. Treatment with a reducing agent, $NaBH_4$, which partially restored the content of SH groups, also partially restored the tyramine-deaminating activity. At the same time, the activity towards lysine was inhibited. These data suggest that the modification in catalytic properties of bacterial monoamine oxidase (tyramine oxidase) which accompanied partial oxidation of the SH groups was reversible. The effect of $NaBH_4$ as a reducing reagent was completely unspecific; dithiothreitol, for example, had a similar effect. Is the effect of oxidizing agents specific?

We found (Tatyanenko *et al.* 1971) that oxygen (in the presence of catalytic concentrations of Cu^{2+}), *o*-iodosobenzoate, and ergosterol peroxide (0.1 mM) affected the properties of tyramine oxidase similarly. All these oxidizing agents decreased the rate of tyramine deamination and, at the same time, induced the property of deaminating lysine, which was absent in the native enzyme. Similar alterations in catalytic properties were observed when the enzyme was treated with H_2O_2 (Tatyanenko *et al.* 1974).

In the majority of our experiments reported here, partial oxidation of the enzyme was effected by incubation under aerobic conditions in the presence of catalytic concentrations of Cu^{2+} and unless otherwise stated the partially oxidized enzyme was prepared in this way.

Sulphide-forming reagents or compounds which alkylate SH groups decreased the content of SH groups in preparations of bacterial monoamine oxidase but did not affect the catalytic properties. Furthermore, they prevented transformation if the enzyme was treated with oxidizing reagents *after* the interaction with sulphide-forming or alkylating compounds.

Deamination of lysine by the partially oxidized tyramine oxidase preparations obeyed the conventional stoichiometry of oxidative deamination: one mole of NH_3 and one mole of H_2O_2 liberated per mole of O_2 consumed. Experiments with α- and ε-carbobenzoxylysines suggest that the ε-amino group of lysine was deaminated. When the pre-oxidized tyraminase was incubated with a preparation of lysine-rich histone, five moles of NH_3 were liberated per mole of the histone (Tatyanenko *et al.* 1971).

The lysine-deaminating activity of partially oxidized tyramine oxidase was only slightly inhibited by a monoamine oxidase inhibitor, tranylcypromine (*trans*-2-phenylcyclopropylamine) (1 mM); the same (relatively high) concentration of another monoamine oxidase inhibitor, pargyline, had no effect. The lysine-deaminating activity was inhibited by nucleophilic reagents (isoniazid, hydroxylamine, KCN) which did not influence the deamination of tyramine by native tyramine oxidase.

Since the activity of the modified enzyme was inhibited by carbonyl reagents we expected it to be active towards compounds that are substrates for diamine oxidase. This hypothesis was, however, only partly confirmed. After partial oxidation of tyramine oxidase the substrate specificity of the enzyme did indeed change: there was a dramatic decrease in tyramine-deaminating activity, paralleled by the appearance of the ability to deaminate lysine, putrescine, spermine and AMP at high rates (of the same order of magnitude as tyramine deamination by the native enzyme). The reaction with AMP obeyed the stoichiometry of hydrolytic deamination. Inosinic acid (one mole per mole of liberated NH_3) was identified among the reaction products but no formation of H_2O_2 could be detected. These data suggest that previously unrecognized relationships exist between the amine-oxidizing and adenylate-deaminating activities in nature. The transformed bacterial monoamine oxidase did not, however, deaminate histamine (Tatyanenko *et al.* 1971).

We found (Tatyanenko *et al.* 1974) that the transformations of amine oxidases were not accompanied by alterations in the quaternary structure (for example, dissociation of subunits) detectable by polyacrylamide gel electrophoresis; neither was there any separation of a flavin component from an apoenzyme.

In further studies on the molecular mechanisms of the transformation of bacterial monoamine oxidase we have used structural analogues of FAD (Khomutova *et al.* 1970; Shapiro *et al.* 1972). As shown in Fig. 1, these compounds differ in the length of the polyphosphate chain connecting the nucleotide residues which constitute the coenzyme molecule. Compound II in Fig. 1 is FAD, which has a pyrophosphate residue between the isoalloxazine and adenosine parts of the molecule. In compound I the pyrophosphate residue is substituted by a phosphate. In compound III the pyrophosphate residue is replaced by a chain of three phosphate residues. In compound IV the polyphosphate chain was elongated by the addition of a fourth phosphate residue. Compound IIa is derived from FAD by the attachment of an additional, third, phosphate to ribose in the adenosine part of the coenzyme molecule. This compound (0.1 mM) and FAD did not affect either the deamination of tyramine by the native bacterial monoamine oxidase or the deamination of lysine by the transformed enzyme. However, the structural analogues of FAD (compounds I and III) strongly inhibited the deamination of lysine by the transformed enzyme and, especially, deamination of tyramine by native bacterial monoamine oxidase. Compound IV, which resembles the natural coenzyme structurally much less than do compounds I and III, was also a less potent inhibitor of the activities of both native and transformed tyramine oxidase (Fig. 1).

Subsequent experiments were done with compounds I and III. These structural analogues of FAD caused completely or partially reversible inhibition of tyramine deamination (by native tyramine oxidase) or lysine deamination (by the modified form of this enzyme).

Studies of the inhibitory effects of compounds I and III as functions of substrate concentration revealed significant differences between the native and modified (partially oxidized) forms of tyramine oxidase. Compounds I and

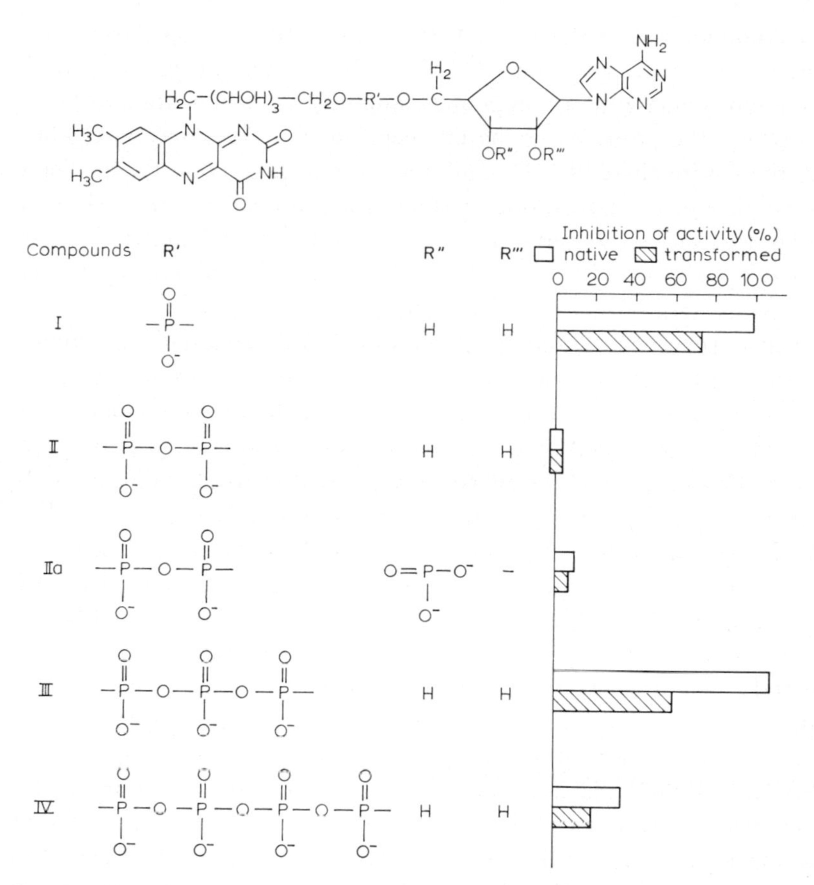

FIG. 1. Inhibition by structural analogues of FAD (final concentrations 0.1 mM) of the activity of the native (substrate tyramine) and transformed (substrate lysine) tyramine oxidase from *Sarcina lutea*. The samples (final volume 3 ml) contained 0.01M-phosphate buffer (pH 7.0), 1 mg of either native or transformed (by partial oxidation of the SH groups under aerobic conditions in presence of catalytic amounts of Cu^{2+} [Tatyanenko *et al.* 1971]) tyramine oxidase, and one of the FAD analogues (Khomutova *et al.* 1970; Shapiro *et al.* 1972). After preincubation for 30 min at 20 °C, tyramine (5 μmol) was added to the samples of the native enzyme and lysine (12 μmol) was added to the samples of the transformed enzyme. All samples were incubated for 30 min at 30 °C. Liberation of NH_3 in the control (without inhibitors) samples was 20 ± 0.5 nmol per mg protein per min. Mean values of 3–4 parallel experiments are presented (Data from Tatyanenko *et al.* 1974).

III (10^{-6}M and 10^{-7}M) inhibited deamination of tyramine by native tyramine oxidase competitively; only the apparent K_m but not the V_{max} values were altered. The structural analogues of FAD which were thus able to interact with the active site of tyramine oxidase, although it contained FAD firmly attached to the apoenzyme, were quite strong inhibitors of the enzymic activity. The K_i

values for compounds I and III were 4.10^{-7}M and 3.10^{-7}M, respectively (substrate, tyramine) (Tatyanenko *et al.* 1974). In similar experiments with transformed tyramine oxidase these compounds inhibited the deamination of lysine uncompetitively; the positions of the intercept but not the slopes in double-reciprocal plots were altered by the inhibitors (Tatyanenko *et al.* 1974). These data suggest that in the transformed enzyme (in contrast to native tyramine oxidase) the structural analogues of FAD interact not with the catalytic site but with an enzyme–substrate complex; this was demonstrated kinetically by the uncompetitive pattern of their inhibitory effects.

If the transformation in catalytic properties of the bacterial monoamine oxidase, initiated by partial oxidation of SH groups in the enzyme molecule, caused a particular alteration in the state of a flavin component of monoamine oxidase, it could be expected that although the structural analogues of FAD interact with the active site of the enzyme reversibly they would prevent modification of the catalytic properties of the monoamine oxidase.

Fig. 2 shows experimental data supporting this hypothesis. The first bar denotes the rate of deamination of lysine in samples containing partially oxidized tyraminase, but in the absence of structural analogues of FAD. In these control samples the average rate of lysine deamination constituted 20 nmol of NH_3 liberated per minute per mg of protein (the native enzyme deaminated tyramine at approximately the same rate). The addition of FAD (0.1 mM) to the samples just before their preincubation under aerobic conditions in the presence of Cu^{2+} did not influence the rate of the lysine deamination reaction induced in the monoamine oxidase. However, if FAD was replaced by its structural analogues I or III (i.e. the most potent inhibitors of the activities of both native and transformed monoamine oxidases), modification of the catalytic properties of the monoamine oxidase was completely prevented (Fig. 2). Compounds IIa and IV, when included in the experimental samples with tyramine oxidase before the enzyme was partially oxidized, only partly prevented the modification of the catalytic properties of monoamine oxidase.

These data also suggest that a certain functional relationship may exist between the SH groups, partial oxidation of which initiated the transformation of catalytic properties of the monoamine oxidase, and the flavin compounds, which prevented the transformation in catalytic properties.

In order to test this hypothesis we did the experiments illustrated in Fig. 3. Transformation of the catalytic properties of tyraminase by partial oxidation was accompanied by a decrease in the number of SH groups in the enzyme. Addition to the samples of compounds I or III (10^{-5}M) strongly inhibited the activity of both the transformed and, especially, the native tyraminase; we failed to measure the SH groups in these samples by titration with 5,5′-dithiobis-

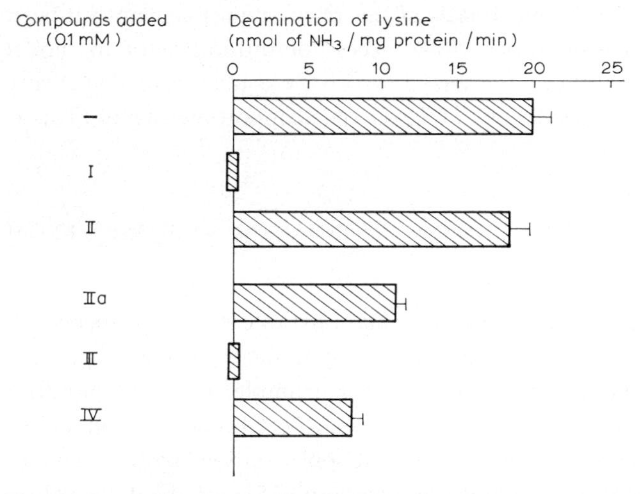

FIG. 2. Prevention of the transformation of tyramine oxidase by structural analogues of FAD. Composition of the samples and the experimental conditions as in Fig. 1. The FAD analogues were added to the samples containing the enzyme preparations, before the samples were incubated under aerobic conditions in presence of Cu^{2+}.

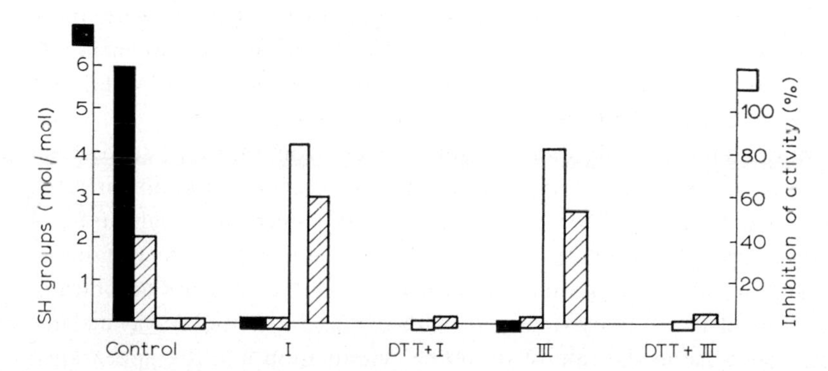

FIG. 3. The effect of FAD analogues (0.01 mM) on the activity and content of SH groups in native (open bars) and transformed (hatched bars) tyramine oxidase. Effect of dithiothreitol (DTT, 1.3 mM). For the composition of samples and experimental conditions, see Fig. 1. The content of SH groups was estimated by means of 5,5′-dithiobis(2-nitrobenzoate) (Tatyanenko *et al.* 1974).

(2-nitrobenzoate), or with *p*-chloromercuribenzoate. Addition of an excess of the reducing agent dithiothreitol, before the addition of the structural analogues of FAD to these systems, completely prevented the inhibition by the latter of lysine deamination by the transformed enzyme and of tyramine deamination by the native enzyme. The chemical basis of this phenomenon has not yet been studied. Even with a considerable molar excess of dithiothreitol in model systems (i.e. in samples without the enzymes) we were unable to detect any reduction of the structural analogues of FAD spectrophotometrically (Tatyanenko *et al.* 1974).

TRANSFORMATION OF MITOCHONDRIAL MONOAMINE OXIDASES FROM RAT LIVER

Reversible qualitative alterations (transformations) in catalytic properties in samples treated with oxidizing reagents were also found with highly purified preparations of other monoamine oxidases—for example, the mitochondrial monoamine oxidases from beef liver, rat liver and beef brain (Gorkin 1973). But in all these cases the V_{max} values of the reactions catalysed by the enzymes pretreated with oxidizing reagents did not exceed 10–15% of the V_{max} values found for the native amine oxidases. The transformed monoamine oxidases from animal tissues, unlike the transformed bacterial monoamine oxidase, readily deaminated histamine. Except for these two features, the characteristic patterns of altered properties of soluble bacterial monoamine oxidase and the structure-bound monoamine oxidases from mammalian tissues were quite similar. (These data also suggest that non-ionic detergents present in highly purified preparations of the structure-bound monoamine oxidases did not influence their ability to be transformed by oxidizing agents.)

Highly purified mitochondrial monoamine oxidase from rat liver (Veryovkina *et al.* 1972) or a similar enzyme preparation from beef liver (Akopyan *et al.* 1971) contained about eight SH groups per 100 000 molecular weight units of protein; they rapidly deaminated tyramine but did not deaminate histamine. Partial oxidation of the SH groups (by preincubating the enzymes aerobically in the presence of Cu^{2+} or by treatment with oxidized oleic acid) was accompanied by a decrease in the rate of tyramine deamination and the appearance of the property of deaminating histamine. Treatment of these modified enzymes with reducing agents (such as reduced glutathione, $NaBH_4$ or sodium arsenite) partially restored the content of SH groups and the tyramine deamination rates but inhibited the deamination of histamine. These data suggest that transformation of the catalytic properties of monoamine oxidases is reversible. Sodium arsenite is claimed to selectively reduce sulphenic acid residues (-SOH) to SH groups but not sulphinic or sulphonic acids or -S-S- bonds. If these assumptions (Parker & Allison 1969) are correct, it would seem that treatment with oxidizing agents decreased the content of SH groups by oxidizing them to sulphenic acid residues. The appearance on the surface of a protein molecule of sulphenic acid residues, which carry a considerable negative charge, would make the catalytic surface formally similar to that of diamine oxidase (Bardsley *et al.* 1970).

When highly purified mitochondrial monoamine oxidases from mammalian tissues have been treated with oxidized oleic acid or other oxidizing agents they exhibit qualitatively new properties of catalysing the deamination not only of histamine but also of cadaverine, putrescine, lysine and AMP.

Reagents which alkylate SH groups caused a decrease in the number of SH groups in the enzyme but did not initiate any transformation in catalytic activity; on the contrary, these reagents prevented transformation if the enzyme was subsequently treated with oxidizing agents. Certain specific SH groups appear to be important in these processes. Thus, pretreatment of highly purified beef liver mitochondrial monoamine oxidase with 0.01mM-*N*-ethylmaleimide decreased the content of SH groups in the enzyme from eight to five (per 100 000 molecular weight units of protein) but did not influence the rate of deamination of histamine after subsequent treatment with oxidized oleic acid. Pretreatment with 0.1 mM-*N*-ethylmaleimide decreased the number of SH groups to three per 100 000 molecular weight units of protein and completely prevented transformation (Akopyan *et al.* 1971).

The monoamine oxidase inhibitors pargyline and tranylcypromine did not inhibit the histamine-deaminating activity induced in transformed preparations of highly purified monoamine oxidases. This histamine-deaminating activity was inhibited (apparently as a result of their reducing properties) by the carbonyl reagents isoniazid and hydroxylamine, which had no effect on the activity of native monoamine oxidases. Thus transformation of monoamine oxidases alters not only their substrate specificity but also their sensitivity to inhibitors (Gorkin 1973).

Pretreatment of monoamine oxidases with pargyline or tranylcypromine, which block the active sites of these enzymes, prevented their transformation when they were subsequently treated with oxidizing reagents. Thus, we are dealing with a *qualitative* alteration in the catalytic properties of the enzymes. This peculiarity of the process enables one to detect the transformation of monoamine oxidases not only in purified enzyme preparations but also in much more complicated systems containing numerous proteins and other compounds besides monoamine oxidases.

TRANSFORMATION OF MONOAMINE OXIDASE ACTIVITY IN FRAGMENTS OF MITOCHONDRIAL MEMBRANES

Treatment of mitochondrial membranes from mammalian organs (rat liver,

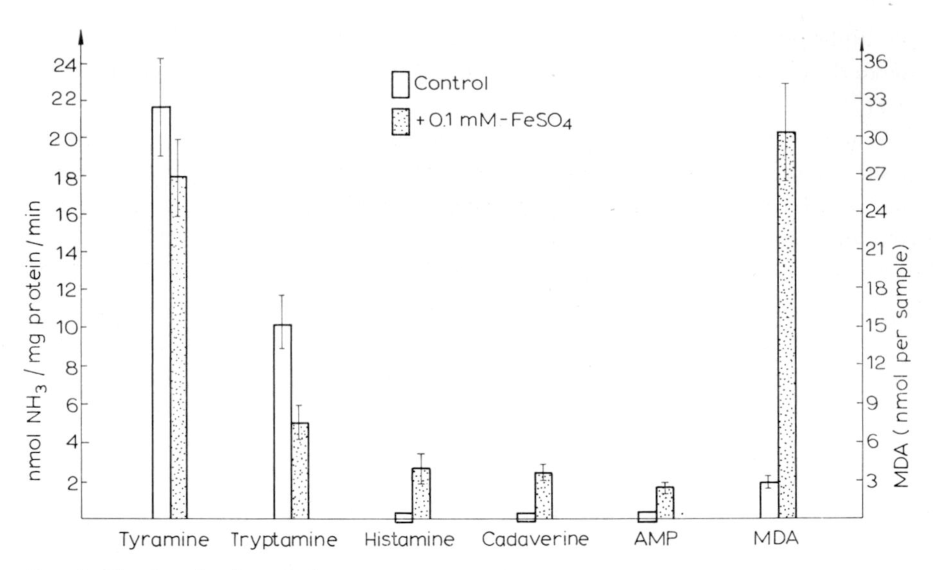

FIG. 4. The deamination of nitrogenous compounds and content of malondialdehyde (MDA) in fragments of rat liver mitochondrial membranes. The mitochondrial fractions (Hunter *et al.* 1963) were suspended in 0.025M–Tris HCl + 0.175M-KCl buffer (pH 7.4). Lipid peroxidation was stimulated by preincubation for 60 min at 37 °C with 0.1 mM-$FeSO_4$ (Vladimirov & Archakov 1972), after which deamination of the nitrogenous compounds was studied at the following optimal concentrations (mM): tyramine, 3.5; tryptamine, 3.4; histamine, cadaverine and AMP, 10. Incubations were for 60 min at 37 °C in oxygen. MDA was estimated by reaction with 2-thiobarbituric acid (Hunter *et al.* 1963). (Data from Garishvili *et al.* 1975.)

for example) with oxidized oleic acid or other oxidizing agents not only decreased the monoamine oxidase activity but also induced the appearance of the qualitatively new (for the given biological source) properties of deaminating diamines, histamine and other nitrogenous compounds (Gorkin 1973). However, comparatively high concentrations of oxidized oleic acids were required in these experiments (Gorkin & Tatyanenko 1967).

Stimulation of peroxidation of the unsaturated fatty acids present in mitochondrial membranes by preincubation with 0.1mM-$FeSO_4$ (Vladimirov & Archakov 1972) caused a significant increase in the amount of the lipid peroxidation product malondialdehyde (MDA) in fragments of mitochondrial membranes (Garishvili *et al.* 1975). This dramatic increase in MDA content was accompanied (as one would expect, on the assumption that lipid peroxidation in the membranes initiates transformation of mitochondrial monoamine oxidases) by a decrease in monoamine oxidase activity (substrates: tyramine and, particularly, tryptamine) and by the appearance of the ability to deaminate histamine, cadaverine and AMP (Fig. 4).

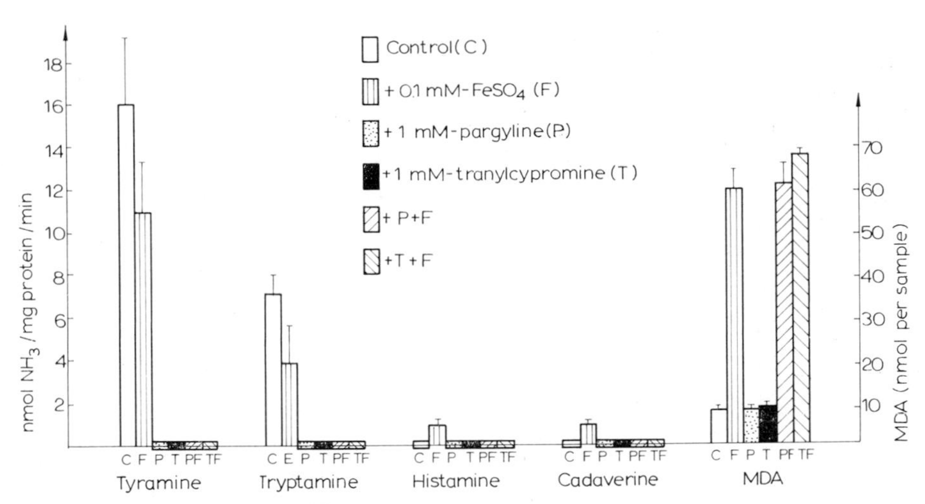

FIG. 5. The effect of monoamine oxidase inhibitors (pargyline and tranylcypromine) on the deamination of nitrogenous compounds and content of malondialdehyde (MDA) in fragments of rat liver mitochondrial membranes. The monoamine oxidase inhibitors were preincubated with the fragments of mitochondrial membranes for 30 min at 20 °C, after which $FeSO_4$ was added. Further treatment was as described in Fig. 4.

If these phenomena are due to alterations in the catalytic properties of monoamine oxidases, it would be expected that inhibitors which block the catalytic sites will prevent the transformation of monoamine oxidases. Fig. 5 shows that this hypothesis was confirmed. Preincubation with Fe^{2+} increased the content of MDA in mitochondrial membranes: pargyline or tranylcypromine did not alter the content of MDA either in samples without Fe^{2+} or in those preincubated with Fe^{2+}. Preincubation of mitochondrial membranes with Fe^{2+}, which decreased the deamination of tyramine or tryptamine, was accompanied by the appearance of the ability to deaminate histamine or cadaverine. But, if the mitochondrial membranes were first treated with the monoamine oxidase inhibitors pargyline or tranylcypromine and then preincubated with Fe^{2+}, these phenomena were not seen although lipid peroxidation was stimulated. These data suggest that the active sites of monoamine oxidases must be catalytically unaltered for transformation to occur.

Lipid peroxidation is a free-radical process which is inhibited by many inhibitors of radical processes (IRP). An IRP which neither altered monoamine oxidase activity nor bound Fe^{2+} ions could enable us to find out whether the changes in the deamination of nitrogenous compounds in mitochondrial membranes preincubated with Fe^{2+} were caused by stimulation of lipid peroxidation. Fig. 6 shows that both butylhydroxytoluene and propyl gallate

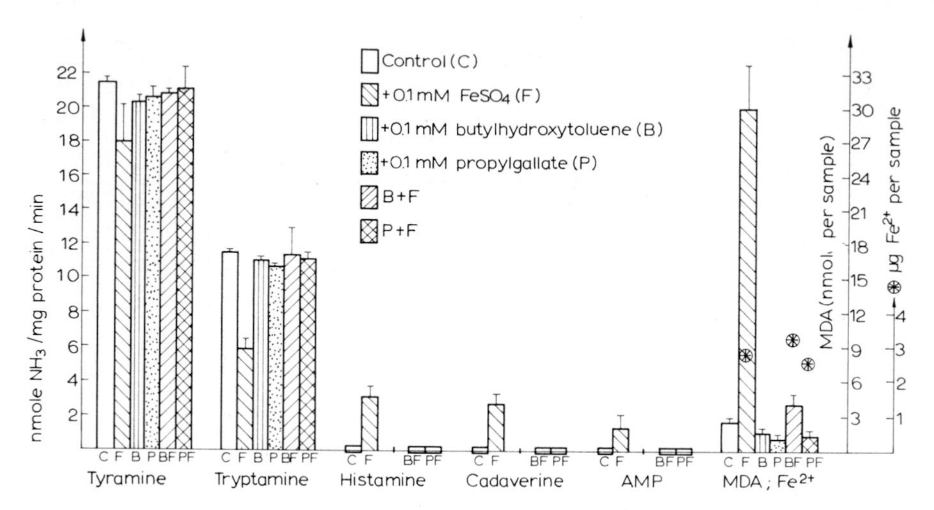

FIG. 6. Effect of inhibitors of free-radical processes (propyl gallate and butylhydroxytoluene) on the deamination of nitrogenous compounds, and the content of malondialdehyde (MDA) and Fe^{2+}, in samples of rat liver mitochondrial membrane fragments. Propyl gallate and butylhydroxytoluene were added to the samples before the addition of $FeSO_4$. Further treatment of the samples was as in Fig. 4. The content of Fe^{2+} was estimated colorimetrically using 4,7-biphenyl-1,10-phenanthroline (Ballentine & Burford 1957).

met these requirements completely. These IRP did not alter the monoamine oxidase activity (substrates tyramine or tryptamine) but they decreased the content of MDA in samples preincubated without Fe^{2+}, and prevented the dramatic increase in content of MDA when Fe^{2+} was added before the preincubation in order to stimulate lipid peroxidation. At the same time, neither IRP altered the content of Fe^{2+} in the samples. When mitochondrial membranes were preincubated with Fe^{2+} under conditions which favoured the stimulation of lipid peroxidation there was a decrease in monoamine oxidase activity (substrates tyramine or tryptamine) and the simultaneous appearance of the ability to deaminate histamine, cadaverine and AMP. But if fragments of mitochondrial membranes were incubated with Fe^{2+} in the presence of IRP, these alterations in the deamination of nitrogenous compounds did not take place although the catalytic sites of monoamine oxidases were not blocked (Garishvili *et al.* 1975).

TRANSFORMATION OF MONOAMINE OXIDASES IN VIVO

Stimulation of lipid peroxidation in mitochondrial membranes thus initiates the transformation in catalytic properties of mitochondrial monoamine

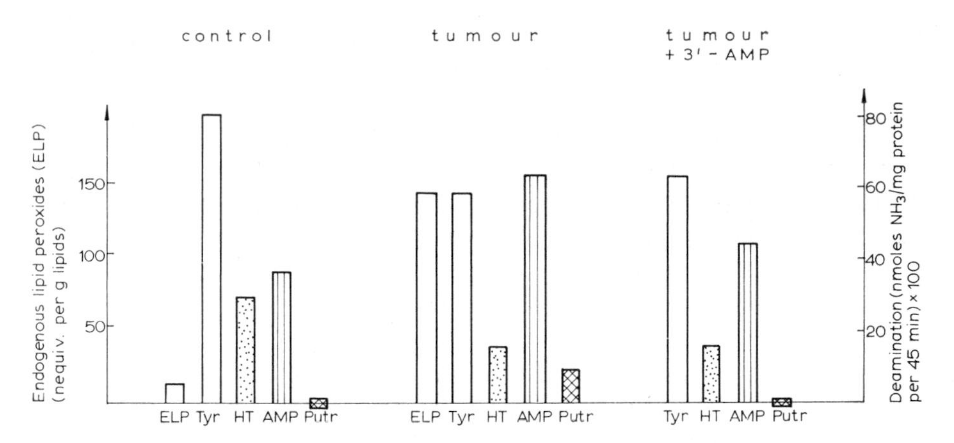

FIG. 7. The content of endogenous lipid peroxides (ELP) and deamination of nitrogenous compounds in mitochondrial fractions of mouse liver at the eighth day of development of Ehrlich ascites carcinoma, and the effect of adenosine 3′-monophosphate. The composition of samples and experimental conditions were as described by Khuzhamberdiev *et al.* (1973) and Gorkin *et al.* (1973). Adenosine 3′-monophosphate was injected (60 mg per 100 g body weight) intraperitoneally 2 hours before transplantation of the tumour cells. The injections of the nucleotide were repeated every 18 hours for 7–10 days. Tyr, tyramine; HT, 5-hydroxytryptamine; Putr, putrescine.

oxidases. Lipid peroxidation might be stimulated by various factors *in vivo* (Gorkin 1973). Considerable amounts of lipid peroxidation products accumulate in the liver in animals with tumours (Neifakh & Kagan 1969). If the transformation of structure-bound monoamine oxidases did take place, it could be expected that the monoamine oxidase activity in the membranes would be decreased with a simultaneous increase in or appearance of the ability to deaminate substrates of diamine oxidases or AMP. Fig. 7 shows the results of experiments in which cells of the Ehrlich ascites carcinoma were implanted into the peritoneal cavity of mice. Eight days after implantation there was considerable increase in the content of endogenous lipid peroxides (Neifakh & Kagan 1969) in the liver of the tumour-bearing animals. In the same organ the rates of tyramine or 5-hydroxytryptamine deamination were decreased (as compared with normal animals); at the same time, deamination of AMP was stimulated and the ability to deaminate putrescine appeared (Khuzhamberdiev *et al.* 1973). These data suggest that mitochondrial monoamine oxidases may undergo transformation *in vivo* under pathological conditions accompanied by the stimulation of lipid peroxidation.

Results of preliminary studies (by M. Khuzhamberdiev) of human liver tissue from patients who had died from the consequence of traumatic injuries, malignant tumours (the extrahepatic localization of which was confirmed by autopsy), or atherosclerosis showed that in cancer and atherosclerosis there was a tendency towards a decreased deamination of tyramine and 5-hydroxytryptamine; at the same time, in these cases the fragments of mitochondrial membranes acquired the ability to deaminate lysine or cadaverine. These observations support the hypothesis that transformation of monoamine oxidases in man can occur in pathological states that result in the stimulation of lipid peroxidation (Voskresenskii & Levitzkii 1970).

Studies of samples obtained by diagnostic biopsy of liver tissue in patients with hepatolenticular degeneration also suggested that the transformation of monoamine oxidases *in vivo* is possible in this disease (Veryovkina *et al.* 1972). Hepatolenticular degeneration is characterized by a significant increase in the tissue concentration of copper, which may catalyse the oxidation of SH groups in monoamine oxidases, thus initiating their transformation (Gorkin 1973).

The transformation of monoamine oxidases is, thus, at present considered mainly as a process which accompanies pathological states favouring oxidation of SH groups. Among these pathological states the most important, apparently, are the numerous diseases (radiation injury [Kudryashov & Goncharenko 1970] or hypervitaminosis D_2 [Spirichev & Blazheievich 1968], for example) accompanied by increased lipid peroxidation in tissues. Under these pathological conditions we observed transformations of monoamine oxidases *in vivo* (Gorkin 1973).

POSSIBLE PATHOGENETIC SIGNIFICANCE OF AND EFFECT OF DRUGS ON THE TRANSFORMATION OF MONOAMINE OXIDASES

The concept of the transformation of monoamine oxidases as a process accompanying pathological states poses questions about the participation of this transformation in the pathogenesis of the diseases as well as on the prevention of transformation or the elimination of its harmful consequences.

A major factor in the possible participation of the transformation of monoamine oxidases in the development of pathological conditions may be the induction of adenylate-deaminating activity (Gorkin 1973). The transformed monoamine oxidases might participate in deamination not only of adenosine phosphates but also of many vitally important coenzymes which contain an AMP residue (Gorkin *et al.* 1970). Accumulation of inosinic acid has been observed in tissues of irradiated animals (Tseveleva 1962).

Assuming that the transformation of monoamine oxidases not only accompanies the development of irradiation injury but plays a role in its pathogenesis, we would expect that transformation of monoamine oxidases: (1) will be manifested quantitatively differently in animals with different radiosensitivities, and (2) will be completely or partially prevented by radioprotective drugs (natural or synthetic). We observed that the transformation of mitochondrial monoamine oxidases with a significant increase in adenylate-deaminating activity took place in mitochondrial fractions of liver and intestine within two hours after X-irradiation (700 rad) of mice or rats. A similar transformation of monoamine oxidases in liver mitochondria of guinea pigs, which are much more radiosensitive than mice or rats, was noted after irradiation at a dose of 300 rad (Zeinalov *et al.* 1975). But in liver mitochondria of mongolian gerbils, which are characterized by extremely low radiosensitivity (Chang *et al.* 1964), irradiation even at a dose of 1500 rad did not cause either an increase in the deamination of AMP or other features of the transformation of monoamine oxidases. The remarkably low radiosensitivity of mongolian gerbils is ascribed to properties of the plants growing in the steppes of Central Asia that make up the food of these rodents under natural conditions (Ogryzov *et al.* 1974). Feeding mice with these plants or treating rats with a radioprotective drug, *S*-β-aminoethylisothiouronium, prevented the transformation of monoamine oxidases in mitochondrial fractions from liver or intestine after whole-body X-irradiation (Zeinalov *et al.* 1975). Treatment of rats with adenosine-2′(3′)-monophosphate, a competitive inhibitor of adenylate-deaminating activity, reversed qualitative alterations in the catalytic functions of amine oxidases in irradiated animals, and provided a certain radioprotective effect (Zeinalov *et al.* 1975). For example, when a group of rats were treated with adenosine-2′(3′)-monophosphate after irradiation with 600 rad, 10 animals (out of the total of 20) died within 30 days. In a control group of rats, injected with 0.9% NaCl solution after irradiation, 16 animals died within the 30 days (Zeinalov *et al.* 1975). In these experiments, if the animals were treated repeatedly (even at night!) with sufficient doses (60 mg per 100 g body weight) of the structural analogue of AMP to maintain a concentration of the analogue in the tissues which would inhibit the AMP-deaminating activity of transformed monoamine oxidase (Gorkin *et al.* 1971), the signs of transformation were suppressed and the death of a considerable number of irradiated animals was prevented (Gorkin 1973). Similar positive results were obtained in analogous experiments with mice with Ehrlich ascites carcinoma. Treatment of these animals with structural analogues of AMP (adenosine 3′-monophosphate, for example) suppressed the transformation of monoamine oxidases *in vivo*, prolonged survival and decreased the number of tumour cells in the peritoneal cavity (Gorkin *et al.* 1973). This beneficial effect of the administered nucleotides was not found in control experiments in which tumour-bearing mice were treated with AMP instead of its structural analogues (Khuzhamberdiev *et al.*

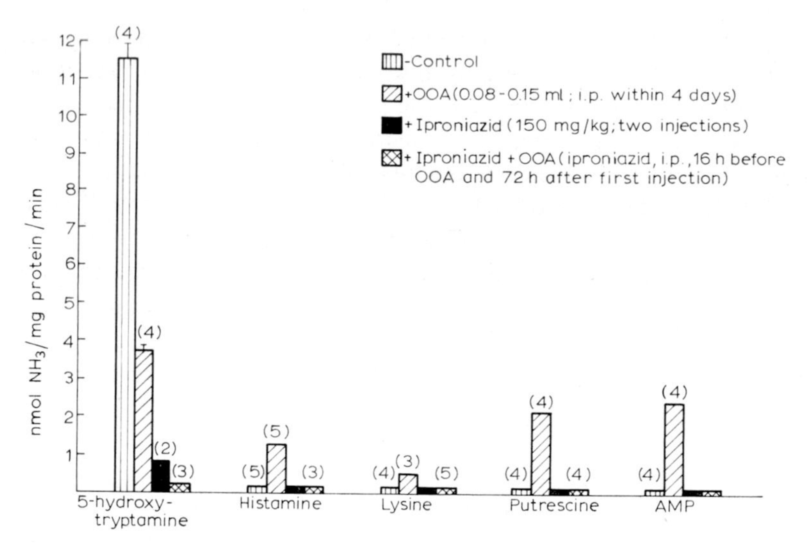

FIG. 8. The effect of iproniazid *in vivo* on the impairment of deamination of some nitrogenous compounds by mitochondria from liver tissue of rats treated with oxidized oleic acid (OOA). Numbers of samples in parentheses. (Data from Abdel Samed *et al.* 1971.)

1973; Gorkin 1973). These data suggest that the beneficial effect of the structural analogues of AMP was due to their property of inhibiting the adenylate-deaminating activity of transformed monoamine oxidases.

The development of more efficient inhibitors of adenylate-deaminating activity, which could cause irreversible inhibition, as has been demonstrated with acetylenic analogues of some substrates (Abeles & Walsh 1973), can be expected not only to foster this kind of experimental study but also to open up new approaches to clinical investigations. It seems reasonable, therefore, to undertake research programmes directed not simply at the design of new inhibitors that will selectively block the enzymic deamination of conventional substrates of monoamine oxidases; new efficient inhibitors of deamination of the numerous nitrogenous compounds that are attacked by transformed monoamine oxidases (Gorkin *et al.* 1970; Gorkin 1973) are also urgently required.

Using monoamine oxidase inhibitors (pargyline, indanamine, tranylcypromine and iproniazid) we could prevent the increase in AMP-deaminating activity and the other phenomena which accompany the transformation of monoamine oxidases, not only *in vitro*, but also *in vivo*, in animals with tumours (Khuzhamberdiev *et al.* 1973), in experimental radiation injuries, or in animals treated with dispersions of oxidized oleic acid (Fig. 8), which cause radiomimetic effects (Abdel Samed *et al.* 1971). Under these conditions there was no beneficial effect on the disease, even though an increase in AMP-deaminating activity was prevented (Khuzhamberdiev *et al.* 1973). Probably the beneficial effects in the experimental pathological states, described above, were possible only if the stimulation of AMP-deaminating activity was suppressed but monoamine oxidase activity, although decreased, was not less than 40–50% of the normal level. The monoamine oxidase inhibitors that we have tested prevented the stimulation of AMP-deaminating activity *in vivo* but at the same time completely inhibited the deamination of 5-hydroxytryptamine and tyramine (Abdel Samed *et al.* 1971; Gorkin 1973). It seems reasonable to continue research in this direction using selective inhibitors of monoamine oxidases A and B (Neff & Yang 1974).

Inhibitors of free-radical lipid peroxidation (Emanuel 1974), exemplified by propyl gallate or butylhydroxytoluene, prevented all manifestations of the transformation of monoamine oxidases (including the stimulation of AMP-deaminating activity) but did not inhibit the enzymic deamination of biogenic monoamines (Garishvili *et al.* 1975). However, the complete elimination of lipid peroxidation, the biological significance of which is still obscure (Vladimirov & Archakov 1972), might be a disadvantage *in vivo*. The selective normalization (re-transformation) of the functions of monoamine oxidases which undergo transformation of their catalytic properties when lipid peroxidation is stimulated may prove to be a more reasonable approach. The transformation of monoamine oxidases, initiated by 'mild' oxidation of some of the SH groups in these enzymes, is a reversible process (Gorkin 1973). It would be expected that nucleophilic compounds (for example, isoniazid or thiosulphate) that are of low toxicity and therefore could be used *in vivo*, will favour the re-transformation, or normalization, of the catalytic properties of monoamine oxidases, as a result of their reducing properties. This hypothesis has already been supported in part (Akopyan *et al.* 1972; Gorkin 1973). It thus seems probable that some known drugs might find new applications in the future. But special interest and hope are of course centred on research on new inhibitors of the reactions catalysed by the transformed monoamine oxidases and on compounds which could affect the ability of monoamine oxidases to undergo transformation. Further studies on selective monoamine oxidase inhibitors will be interesting in this respect also.

References

Abdel Samed, M. M., Akopyan, Z. I., Veryovkina, I. V., Kulygina, A. A. & Gorkin, V. Z. (1971) Effect of monoamine oxidase inhibitors on qualitative alterations in enzymatic properties of mitochondrial monoamine oxidases. *Biochem. Pharmacol. 20*, 2571-2577

ABELES, R. H. & WALSH, C. T. (1973) Acetylenic enzyme inactivators: inactivation of γ-cystathionase, *in vitro* and *in vivo*, by propargylglycine. *J. Am. Chem. Soc. 95*, 6124-6125

AKOPYAN, Z. I., STESINA, L. N. & GORKIN, V. Z. (1971) New properties of highly purified bovine liver mitochondrial monoamine oxidase. Reversible oxidation of sulfhydryl groups and reversible qualitative alteration (transformation) of the substrate and inhibitor specificity. *J. Biol. Chem. 246*, 4610-4618

AKOPYAN, Z. I., KULYGINA, A. A., TERZEMAN, I. I. & GORKIN, V. Z. (1972) Induced appearance of adenylate-deaminating activity in highly purified bovine liver mitochondrial monoamine oxidase. *Biochim. Biophys. Acta 289*, 44-56

BALLENTINE, R. & BURFORD, D. D. (1957) Determination of metals. VI. Iron. Spectrophotometric method. *Methods Enzymol. 3*, 1017-1018

BARDSLEY, W. G., HILL, C. M. & LOBLEY, R. W. (1970) A reinvestigation of the substrate specificity of pig kidney diamine oxidase. *Biochem. J. 117*, 169-176

CHANG, N. C., HUNT, D. M. & TURBYFIL, C. (1964) High resistance of mongolian gerbils to irradiation. *Nature (Lond.) 203*, 536-537

EMANUEL, N. M. (1974) General pattern of alteration in content of free radicals in malignant growth. *Dokl. Acad. Nauk SSSR 217*, 245-248

GARISHVILI, T. G., KRIVCHENKOVA, R. S. & GORKIN, V. Z. (1975) Deamination of nitrogenous compounds in mitochondrial membranes under conditions of stimulated lipid peroxidation. *Vopr. Med. Khim. 21*, 511-517

GORKIN, V. Z. (1973) Monoamine oxidases: versatility of catalytic properties and possible biological functions. *Adv. Pharmacol. Chemother. 11*, 1-50

GORKIN, V. Z. & TATYANENKO, L. V. (1967) 'Transformation' of mitochondrial monoamine oxidase into a diamine oxidase-like enzyme *in vitro*. *Biochem. Biophys. Res. Commun. 27*, 613-617

GORKIN, V. Z., AKOPYAN, Z. I., VERYOVKINA, I. V., GRIDNEVA, L. I. & STESINA, L. N. (1970) On deamination of some biogenic amines and other nitrogenous compounds in liver mitochondria. *Biokhimiya 35*, 140-151

GORKIN, V. Z., AKOPYAN, Z. I., KULYGINA, A. A. & ZEINALOV, T. A. (1971) On impairments in deamination of some nitrogenous compounds and an experimental approach to their normalization. *Biul. Eksp. Biol. Med. N11*, 42-44

GORKIN, V. Z., NEYFAKH, E. A., ROMANOVA, L. A. & KHUZHAMBERDIEV, M. (1973) Normalization by nucleotides of impairments in deamination of nitrogenous compounds in liver of tumour-bearing mice. *Experientia 29*, 22-23

HELLERMAN, L. & ERWIN, V. G. (1968) Mitochondrial monoamine oxidase. II. Action of various inhibitors for the bovine kidney enzyme. Catalytic mechanism. *J. Biol. Chem. 243*, 5234-5243

HUNTER, F. E., JR, GEBICKI, J. M., HOFFSTEN, P. E., WEINSTEIN, J. & SCOTT, A. (1963) Swelling and lysis of rat liver mitochondria induced by ferrous ions. *J. Biol. Chem. 238*, 828-835

KHOMUTOVA, E. D., SHAPIRO, T. A. & BEREZOVSKII, V. M. (1970) Nucleotides, coenzymes, phosphate esters. XXII. Synthesis of P′-(riboflavin-5′)-P³-(adenosine-5′)triphosphate and P′-(riboflavin-5′)-P⁴-(adenosine-5′)tetraphosphate. *Zh. Org. Khim. 40*, 470-474

KHUZHAMBERDIEV, M., ROMANOVA, L. A., NEIFAKH, E. A. & GORKIN, V. Z. (1973) Impairments and approaches to normalization of deamination of some nitrogenous compounds in liver of tumour-bearing animals. *Vopr. Med. Khim. 19*, 415-422

KUDRYASHOV, YU. B. & GONCHARENKO, E. N. (1970) Role of biologically active compounds (radiotoxins) in radiation injury. *Radiobiology (Moscow) 10*, 212-229

NEFF, N. H. & YANG, H.-Y. T. (1974) Another look at the monoamine oxidases and the monoamine oxidase inhibitor drugs. *Life Sci. 14*, 2061-2074

NEIFAKH, E. A. & KAGAN, V. E. (1969) Estimation of lipid peroxides in organs of normal animals *in vivo*. *Biokhimiya 34*, 511-517

OGRYZOV, N. K., GILEV, YU. V., KUCHINSKII, B. D. & PETROV, L. N. (1974) On the radioprotective effect of some components of the vegetable food of radioresistant rodents from Middle and Central Asia. *Radiobiology (Moscow) 14*, 437-440

PARKER, D. J. & ALLISON, W. S. (1969) The mechanism of inactivation of glyceraldehyde 3-phosphate dehydrogenase by tetrathionate, *o*-iodosobenzoate, and iodine monochloride. *J. Biol. Chem. 244*, 180-189

SHAPIRO, T. A., KHOMUTOVA, E. D. & BEREZOVSKII, V. M. (1972) Nucleotides, coenzymes, phosphate esters. XXVI. Synthesis of P′-(riboflavin-5′)-P²- adenosine-2′(3′)-phospho-5′-diphosphate and P′-(riboflavin-5′)-P′-(adenosine-5′)monophosphate. *Zh. Org. Khim. 42*, 1634-1638

SPIRICHEV, V. B. & BLAZHEIEVICH, N. V. (1968) On the mechanism of inhibitory effect of vitamin D_2 on ATPase activity. *Biokhimiya 33*, 1260-1263

TATYANENKO, L. V., GVOZDEV, R. I., LEBEDEVA, O. I., VOROBYOV, L. V., GORKIN, V. Z. & YAKOVLEV, V. A. (1971) Properties of tyramine oxidase from *Sarcina lutea:* oxidation of SH groups and qualitative alteration in substrate and inhibitor specificity. *Biochim. Biophys. Acta 242*, 23-35

TATYANENKO, L. V., SHAPIRO, T. A., YAKOVLEV, V. A., GORKIN, V. Z., KHOMUTOVA, E. D., PIVOVAROV, A. P., GVOZDEV, R. I. & LEBEDEVA, O. I. (1974) Studies on tyramine oxidase with the aid of structural analogues of flavin adenindinucleotide. *Mol. Biol. SSSR 8*, 871-878

TSEVELEVA, I. A. (1962) Effect of irradiation on metabolism of nucleotides in rabbit liver. *Radiobiology (Moscow) 2*, 674-680

VERYOVKINA, I. V., ABDEL SAMED, M. M. & GORKIN, V. Z. (1972) Mitochondrial monoamine oxidase of rat liver: reversible qualitative alterations in catalytic properties. *Biochim. Biophys. Acta 258*, 56-70

VLADIMIROV, YU. A. & ARCHAKOV, A. I. (1972) Lipid peroxidation in biological membranes, Nauka Publishers, Moscow

VOSKRESENSKII, O. N. & LEVITZKII, A. P. (1970) Lipid peroxides in living organism. *Vopr. Med. Khim. 16*, 563-583

YAMADA, H., UWAJIMA, T., KUMAGAI, H., WATANABE, M. & OGATA, K. (1967*a*) Bacterial monoamine oxidases. Part I. Purification and crystallization of tyramine oxidase of *Sarcina lutea. Agric. Biol. Chem. (Tokyo) 31*, 890-896

YAMADA, H., KUMAGAI, H., UWAJIMA, T. & OGATA, K. (1967*b*) Bacterial monoamine oxidases. Part II. Substrate and inhibitor specificities of tyramine oxidase of *Sarcina lutea. Agric. Biol. Chem. (Tokyo) 31*, 897-901

ZEINALOV, T. A., KULYGINA, A. A., RYASANOV, V. M., KUDRYASHOV, YU. B. & GORKIN, V. Z. (1975) Radioresistance of mammals and deamination reactions of some nitrogenous compounds. *Radiobiology (Moscow) 15*, 16-20

Discussion

Fuller: Have you tried the type A specific monoamine oxidase inhibitors, to see whether they prevent transformation?

Gorkin: Not yet, but these experiments are being planned. It follows from the experiment with iproniazid that selective inhibitors of A and B type would be suitable for these trials.

Pletscher: Would you expect a direct effect of MAO inhibitors on tumour growth, or as radio-protective agents?

Gorkin: It is difficult to see how MAO inhibitors could prevent tumour growth directly. I think these are secondary effects that are connected in some way with the growth of tumours or irradiation injury, as unspecific factors which lead to the stimulation of lipid peroxidation. It is known that tumours function as traps for natural antioxidants. The content of tocopherol-like compounds increases enormously in tumours, and falls in other tissues, and this is accompanied by an accumulation of lipid peroxides in these tissues. The same happens in radiation injury, but this is caused by another mechanism. Irrespective of the mechanism leading to the accumulation of lipid peroxides in the tissues, the lipid peroxides cause a decrease in SH groups. This initiates transformation. The transformed MAO destroys nucleotides, which are vitally important, and this determines the effect on the general state of the animals.

Tipton: Is the effect of 3′-AMP to prevent transformation or to cause re-transformation?

Gorkin: It causes re-transformation of the enzyme. *In vitro*, 3′-AMP just inhibits degradation by the transformed MAO of AMP, which is abnormally high *in vitro*. But if we use considerably higher concentrations of 3′-AMP, the enzyme is re-transformed. This effect could not be reproduced in control samples to which we added 5′-AMP instead of 3′-AMP. But any reducing agent will re-transform the enzyme *in vitro*.

Singer: It would be very valuable if your colleague who made these interesting FAD analogues could be persuaded to pull some more rabbits out of her hat, and make them available, so that one is able to define how the analogues inhibit and how they interfere with the transformation. I still don't understand what a flavin analogue can do to an enzyme that has a covalently bound flavin. One cannot even assume that iron is present and that somehow iron forms an inhibited chelate with the flavin, because if so, FAD should be just as inhibitory, or free riboflavin, and they are not. Have you any speculations as to what the analogues do?

Gorkin: Yes, we have speculations! We think the FAD analogues form charge-transfer complexes with the FAD in the enzyme. The idea is that in the enzyme there is firmly attached FAD sitting on the apoenzyme in the active site. When we add the structural analogue of FAD, it sticks to the FAD which is already in the active site, and so will inhibit it competitively. Some of us thought that this is impossible; others thought it was worth trying. The experiments show that these inhibitors inhibit very strongly. The K_i values, determined accurately, were 3×10^{-7} M for tyramine and with the native enzyme, and about the same order of magnitude for lysine with the oxidized enzyme. So the simple experiments show that these inhibitors work. But whether they form charge-transfer complexes or not, I do not know.

Singer: You should be able to see that very easily, because if you form a charge-transfer complex you should have an extremely intense absorption band in the long wavelength region, which is well described for flavin charge-transfer complexes. You need only a fraction of a milligram of one of these flavin analogues to decide the question.

Sourkes: Dr Gorkin, can you give us more details about the conditions in which you treat the whole animal and find the transformed enzyme?

Gorkin: We tried four approaches. The first was radiation damage. This is very easy and reproducible, much more so than transformation *in vitro*, because there one has to be careful that the oxidized oleic acid is prepared properly. If you add more, it doesn't work, and the idea is that transformation is initiated by oxidation of the SH groups with the formation of sulphenic acid (-SOH) residues, as is the case with glyceraldehyde-3-phosphate dehydrogenase. It can easily be reduced by $NaBH_4$ in mildly alkaline conditions. If one has more oxidized oleic acid than necessary, or if it is badly prepared, there is no transformation. But it works well *in vivo*; one irradiates the animal and within 6–8 days transformation is detectable. This was our first approach.

The second one was to treat rats with dispersions of oxidized oleic acid, which is a radiomimetic compound. This has the same effect. The third was to inject a suspension of Ehrlich ascites tumour cells into the peritoneal cavity of mice and wait for eight days, when the tumour develops; then we take out the liver and find transformation there.

The last approach involved studies of the effect of high doses of vitamin D, which also causes lipid peroxidation *in vivo*. In this case we also found transformation in the whole animal.

Sourkes: Did you also look at vitamin E deficiency? This might be interesting because one would be removing a natural antioxidant. In birds there are changes in the brain in vitamin E deficiency, as against muscular changes in mammals. One might find a transformed enzyme in the brain.

Gorkin: This is a logical extension of the project, but it hasn't yet been done.

Pletscher: Did I understand you to say that radiation-protecting agents inhibit lipid peroxidation due to X-rays?

Gorkin: No. Radio-protective drugs do not inhibit lipid peroxidation; but they prevent transformation. I don't know what the mechanism is.

Pletscher: Did you try 5-hydroxytryptamine? It is a potent protecting agent in these conditions.

Gorkin: We tried some indolylalkylamines which are good radio-protective compounds. They worked very well.

Nutritional requirements for amine metabolism *in vivo*

THEODORE L. SOURKES and KRYSTYNA MISSALA

Laboratory of Neurochemistry, Allan Memorial Institute of Psychiatry, Montreal

Abstract Nutritional requirements of the rat for the complete metabolism of the aliphatic amines *n*-pentylamine and putrescine (1,4-tetramethylenediamine) have been worked out by following the rate of conversion of the parenterally administered ^{14}C-labelled amine (labelled in the carbon atom vicinal to the amino group) to radioactive CO_2 in the expired gases. Oxidation of the monoamine *in vivo* is depressed in riboflavin and iron deficiencies, respectively, but not in pyridoxine or copper deficiency. These results correspond to the decreased monoamine oxidase activity in tissues of rats deficient in these nutrients, as well as to the presence of a flavin in the enzyme. Riboflavin and iron are required for a normal rate of catabolism of the diamine also; this may be related to the possibility that acetylputrescine, formed from putrescine, is metabolized by a pathway initiated through the action of monoamine oxidase. The important decrease in metabolism of putrescine in rats made deficient in copper or pyridoxine is regarded as evidence favouring the proposed roles of copper and pyridoxal phosphate in the functioning of diamine oxidase. The significance of nutritional studies for knowledge of the amine-oxidizing activities of mammalian species is summarized.

We have studied the nutritional requirements of the rat for the complete metabolism of *n*-pentylamine and 1,4-tetramethylenediamine (putrescine). Our aims have been to determine the physiological requirements for these representative amines; to compare these needs with what is known about the cofactors for enzymes involved in their catabolism, particularly in the initiation of this process through the action of monoamine oxidase (MAO; EC 1.4.3.4) and diamine oxidase (histaminase; DAO; EC 1.4.3.6) respectively; and, where the cofactors are not known, to try to shed some light on the possible candidate substances. Evidence accumulated over many years with an *in vivo* technique by Hawkins (1952) and in my laboratory (see summary in Sourkes 1972) in favour of a role of riboflavin in the action of mitochondrial MAO was ultimately

TABLE 1

Effect of iron deficiency on amine oxidase activity of human platelets

Measurement	*Iron status*	
	Normal	*Deficient*
Amine oxidase activity	100 ± 11.5	61 ± 9.1
Serum iron concentration	100 ± 9.9	17 ± 1.4
Haemoglobin concentration	100 ± 2.9	63 ± 1.7

Data shown have been calculated from values for means (± standard error) given by Callender *et al.* (1974) for 20 normal subjects and 16 iron-deficient patients, and are presented as percentages of the normal value.

confirmed through the demonstration of the presence of FAD in the enzyme of rat liver (Youdim & Sourkes 1972), beef liver (Erwin & Hellerman 1967; Walker *et al.* 1971) and pig brain (Tipton 1968). Interestingly, in a similar context, the fact that liver mitochondrial MAO contains iron (but not copper) (Youdim & Sourkes 1966; Oreland 1971) was matched by the findings that the MAO activity of the liver of iron-deficient rats is subnormal (Symes *et al.* 1969) and that the rate of metabolism of pentylamine in the rat depends upon adequate iron nutrition (Symes *et al.* 1971). Although we still do not know what role iron plays in the metabolism of this monoamine, its significance for the chemistry of MAO is sharpened by the establishment of a nutritional requirement for the metal in monoamine catabolism.

In a few experiments thus far we have assessed the effect of iron deficiency on putrescine breakdown *in vivo*. In this case the rationale for the work stemmed not from any proposal that DAO contains iron but rather from the demonstration that putrescine can be acetylated to a product that is a substrate for MAO. In the brain the formation of monoacetylputrescine is the first step in the production of *N*-acetyl-γ-aminobutyrate. This is a normal metabolic pathway, although of low flux (Seiler & Al-Therib 1974). In a comparison of a group of six rats made iron-deficient over a period of 20 days with an equal number of iron-supplemented animals, there was a 23% lowering of the rate of catabolism of putrescine at one hour ($P < 0.025$) and 19% after two hours ($P < 0.01$). For rats fed a diet of the same composition the deficiency in the rate of catabolism of pentylamine at one hour was 38–42% (Symes *et al.* 1971). Thus it is conceivable that the monoacetyl pathway may be of importance in the metabolism of at least putrescine among the symmetrical diamines.

Recently it has been shown (Callender *et al.* 1974; Youdim *et al.* 1975) that platelet amine oxidase is deficient in anaemic humans (Table 1) and is restored to normal levels by effective therapy so that, even with the warning that this

TABLE 2

Partial correlation coefficients for human platelet MAO activity and other iron-dependent constituents of the blood

Substrate	*No. of pairs*	*MAO activity versus:*	
		Serum iron	*Haemoglobin*
Tyramine			
Both sexes	14	0.217	0.264
Males	8	0.029	0.381
Dopamine			
Both sexes	12	0.099	0.170
Males	6	0.374	—0.187
5-Hydroxytryptamine			
Both sexes	12	0.749[b]	0.604[b]
Males	6	0.950[b]	0.733
Phenethylamine			
Both sexes	14	0.741[b]	0.426
Males	8	0.787[a]	0.196

Data given by Youdim *et al.* (1975) have been analysed to yield partial correlation coefficients (Snedecor 1956). Values shown represent the coefficients of correlation of platelet MAO activity with serum iron concentration (haemoglobin concentration) when haemoglobin concentration (serum iron concentration) is held constant. The coefficients have been calculated for each of four monoamines, for men and women taken together, and for men alone. The data for women alone were considered too few for statistical evaluation (see the original paper).

[a] $p < 0.05$.
[b] $p < 0.01$ (both for $n—2$ degrees of freedom).

enzyme and MAO have quite different properties (Collins & Sandler 1971), this is the second type of amine oxidase for whose activity iron is needed. In their work these investigators measured not only the enzymic activity but also the level of serum iron and the concentration of haemoglobin. This was done at the time of haematological investigation and again later on after treatment was well under way. Four different substrates were tested; all gave qualitatively similar results. We calculated the partial correlation coefficients among the three variables, using the individual values tabulated in the paper by Youdim *et al.* (1975): the results are shown in Table 2. It is evident that the platelet amine oxidase bears a close relationship to the serum iron concentration, and much less to the haemoglobin. Moreover, this analysis reveals the fact that for this series two of the four substrates are related to the serum iron more distinctly than are the others. A larger series of patients would be desirable to determine whether 5-hydroxytryptamine and phenethylamine have some special characteristics in relation to the structural requirements of the platelet enzyme in its substrates. The crucial point, however, would be to determine

TABLE 3

Inhibition of metabolism of pentylamine and putrescine by the rat *in vivo*

Inhibitor	*Dose (mg/kg)*	*Pretreatment time (hours)*	*Percentage inhibition during first hour*	
			Pentylamine	*Putrescine*
Aminoguanidine	0.01	0.33		81
Iproniazid	100	16	97	
Iproniazid	10	16		66
Iproniazid	5	16		42
AB-15[a]	10	0.5	0	
AB-15	10	4	17	
Pargyline	65	22	78	
Isoniazid	150	0.5		93
Isoniazid	150	0.5		
+ Pyridoxine	25	0.25		92
α,α′-Dipyridyl	100	1	15	14
Sodium diethyldithiocarbamate	600	2	11	43

[a] AB-15 is the abbreviation for 2-(*m*-aminophenyl)-2′-hydroxy-*N*-cyclopropylamine dihydrochloride.

whether the purified enzyme contains iron in a form essential to the catalytic activity.

There is actually a third amine oxidase requiring iron in the haem form. The source is *Serratia marcescens*; its oxidase also contains FAD (Campello *et al.* 1965).

We have continued our studies with pentylamine but have also investigated the metabolism of putrescine. The former is a classical substrate for MAO (Zeller 1951), the second for DAO (Zeller *et al.* 1956). The actions of some characteristic inhibitors are illustrated by the experimental results of Table 3. Aminoguanidine, the most powerful inhibitor of DAO (Schuler 1952), acted strongly on putrescine catabolism. The same was true for isoniazid, another carbonyl reagent. Simultaneous injection of pyridoxine did not provide protection against the effect of this compound. Iproniazid, an inhibitor of both MAO and DAO, affected significantly the rate of metabolism of both amines. AB-15, reputedly a strong inhibitor of MAO (Huszti *et al.* 1969), was not effective in our study of pentylamine oxidation. Pargyline, which inhibits MAO by adduct formation with the flavin moiety (Oreland *et al.* 1973; Chuang *et al.* 1974), was a potent inhibitor of pentylamine catabolism *in vivo*.

Two chelating agents were tested (Table 3). α,α′-Dipyridyl, considered to have a pronounced effect on iron, had little action on either amine under our

TABLE 4

Comparison of DAO and MAO inhibition

Preparation	*Measurement*	D-*Tranylcypromine*		*Aminoguanidine*	
		DAO	*MAO*	*DAO*	*MAO*
Guinea pig liver (Shore & Cohn 1960)	% inhibition at 1 hour after administration of drug *in vivo*	0 (3 mg/kg)	100	91	0 (20 mg/kg)
Cat kidney (Burkard *et al.* 1962)	ED_{50} in μmol/kg given *in vivo*; enzyme activity of tissue homogenates	> 168[a]	3.6	0.4	> 560
Intact rat (Sourkes & Missala, current research)	% inhibition of the conversion *in vivo* of labelled amine to respiratory $^{14}CO_2$ in first hour	6 (5 mg/kg given 20 h before the amine)	32	96 (10 mg/kg given 20 min before the amine)	0

[a] Based on racemic drug given to one cat.

experimental conditions. Sodium diethyldithiocarbamate significantly affected the oxidation of putrescine but had little action on pentylamine. The small degree of inhibition in that respect agrees with the weak inhibitory effect of this reagent on MAO *in vitro* (Sourkes 1968).

The lack of specificity of iproniazid has long been recognized (Pletscher 1966) and this has been emphasized recently in regard to some other compounds previously considered to be quite specific for MAO (Kusche *et al.* 1973; Crabbe & Bardsley 1974). We therefore undertook a brief investigation on this point using two inhibitors, D-tranylcypromine and aminoguanidine. The results are shown in Table 4. It is evident that the former compound has minimal action on the catabolism of putrescine *in vivo*, whereas it is a powerful inhibitor of the breakdown of pentylamine. Aminoguanidine, in the dose administered, blocked the catabolism of putrescine almost completely, but had no action on pentylamine. These results paralleled the much earlier data of Shore & Cohn (1960) and of Burkard *et al.* (1962) with the same inhibitors employed in other ways.

In regard to the actual rates of catabolism of the two amines, that for the monoamine is considerably greater (*vide infra*). We have also compared the rates for putrescine and cadaverine (1,5-pentamethylenediamine); the higher homologue is metabolized even more slowly than putrescine, but at a rate consistent with the findings of Schayer *et al.* (1954).

Our investigation of the metabolism of pentylamine led us to the conclusion that *in vivo* the initial step is rate-limiting (Sourkes 1972). We have not yet established whether there is a critical rate-limiting reaction in the catabolism of putrescine. This diamine follows several pathways in metabolism. Oxidation by DAO leads to the formation of an aminoaldehyde, which is dehydrogenated to γ-aminobutyric acid (GABA). Indeed, GABA, originating from administered putrescine, has been detected in mouse brain and liver (Seiler *et al.* 1971) and fish brain (Seiler *et al.* 1973*a*), as well as from endogenous sources, perhaps involving putrescine or spermine, in mouse liver nuclei (Seiler *et al.* 1973*b*). Putrescine also serves in the biosynthesis of spermidine and spermine (Seiler 1973). Finally, the monoacetylation pathway has already been described (p. 84).

The strong inhibition of putrescine metabolism by sodium diethyldithiocarbamate as seen in Table 3 agrees with the convincing evidence for the presence of copper in the molecule of pig kidney DAO (Mondovi *et al.* 1967*b*). We therefore attempted to make rats deficient in copper by dietary means alone and then determine their ability to handle putrescine. This technique leads to a reduction of dopamine β-hydroxylase activity *in vivo* (Missala *et al.* 1967) but not in the rate of pentylamine metabolism (Sourkes 1972). Despite the difficulties of achieving very low values of copper in the tissues of rats once they have been weaned, the deficient diet resulted in significant decreases of the rate of putrescine catabolism (Fig. 1). We tested this technique with rats that received aminoguanidine on the 18th day of being fed the copper-deficient food. On being challenged with putrescine two days later both deficient and supplemented rats metabolized the diamine at especially low rates, but the rate for the deficient animals was nevertheless significantly lower than for those receiving a supplement of copper ($P < 0.05$). The rates of metabolism for the supplemented rats returned to the control range about eight days after they had received the inhibitor, and were very much higher than the rates observed for the deficient animals. The curves for the two groups, extrapolated back to the abscissa, both converge on the point at which the aminoguanidine was administered; this can be interpreted to mean that this compound completely inhibited putrescine metabolism in the initial period, and then its effect began to wear off. As it is presumably an irreversible inhibitor, the recovery seen in Fig. 1 would represent *de novo* synthesis of DAO (and perhaps other enzymes involved in the metabolic processes for the conversion of putrescine to carbon dioxide).

The other cofactor that has been claimed for DAO is pyridoxal phosphate (Sinclair 1952; Davison 1956; Mondovi *et al.* 1967*a*; Costa *et al.* 1971). The evidence is not complete, as emphasized by Watanabe *et al.* (1972). In our current work rats were fed a diet deficient in pyridoxine; the controls received the same food with the addition of pyridoxine hydrochloride at the level of

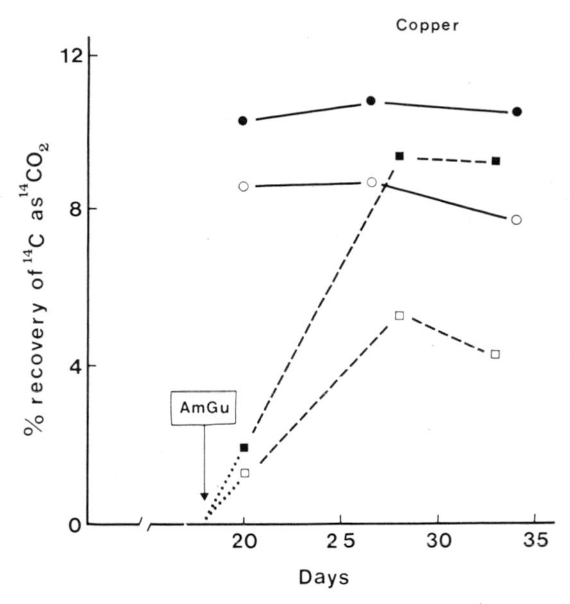

FIG. 1. Effect of copper deficiency on rate of catabolism of [^{14}C]putrescine to respiratory $^{14}CO_2$ in rats. Animals were fed a pelleted copper-deficient diet (ICN Nutritional Biochemicals, Cleveland, Ohio). Controls received a supplement of 1 mg of copper as the sulphate, dissolved in 30% sucrose solution, orally three times weekly (solid symbols); deficient animals received the sucrose solution only (hollow symbols). Each point represents the mean of six rats. Broken lines, animals received an intraperitoneal injection of aminoguanidine, 10 mg/kg, on day 18; solid lines, no aminoguanidine given. The ordinate represents the percentage of injected radioactivity as putrescine (5 or 10 μCi per kg body weight, suitably diluted with unlabelled putrescine) recovered in the first 2 hours after injection as $^{14}CO_2$. Abscissa indicates the number of days that the animals had received the experimental diets. For rats receiving aminoguanidine the differences between supplemented and deficient animals were statistically significant when tests were conducted at 20 days ($P < 0.05$), 28 days ($P < 0.005$), and 33 days ($P < 0.01$). For the series without aminoguanidine the differences were not significant ($P > 0.05$).

25 mg/kg. Within two weeks the deficient animals displayed very significantly lower ability to catabolize putrescine than the pyridoxine-supplemented rats, as shown in Fig. 2. Again, we tested the ability of the two groups to regenerate DAO by giving them all a large dose of aminoguanidine. In this case the respective groups of animals attained in about two weeks the same level of metabolic activity as their counterparts not treated with aminoguanidine (Fig. 2). The inhibitory effect of aminoguanidine is clear evidence for the functioning of DAO *in vivo* or of an enzyme like it in these circumstances, and the dietary requirement of pyridoxine for an optimal catabolism of putrescine favours the concept that pyridoxal phosphate is present in DAO. It is interesting to recall

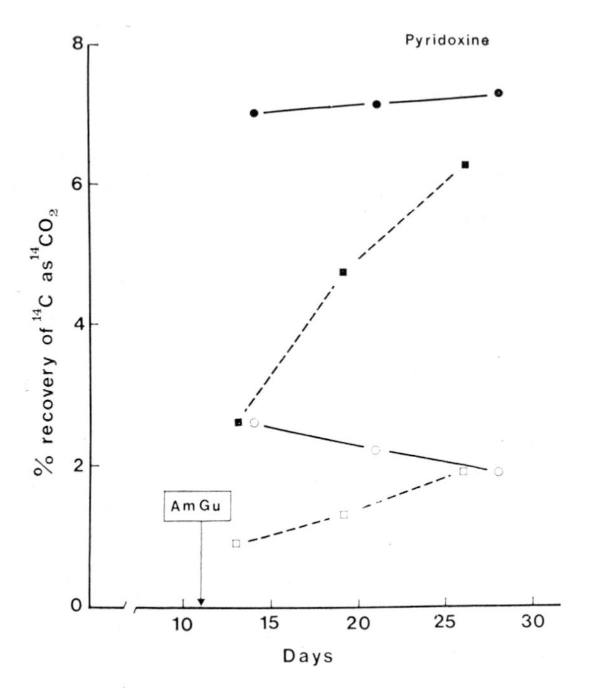

FIG. 2. Effect of pyridoxine deficiency on the rate of catabolism of [^{14}C]putrescine to respiratory $^{14}CO_2$ in rats. Animals were fed a pyridoxine-deficient diet (Sourkes *et al.* 1960) (hollow symbols). Controls received the same diet to which had been added 25 mg of pyridoxine hydrochloride per kg (solid symbols). The solid and broken lines have the same significance as in Fig. 1. Aminoguanidine (10 mg/kg) was injected intraperitoneally on day 11. The coordinates are as in Fig. 1. Each point represents the mean of six rats in the aminoguanidine series, nine rats in the other. Differences resulting from pyridoxine deficiency were statistically significant ($P < 0.01$) in each test period.

a suggestion made by Davison (1956) many years ago that pyridoxal phosphate may be tightly bound to the enzyme; in the light of later experience with this and other enzymes it may be that the coenzyme is, at least in part, covalently linked to the peptide chain. Of course, the carbonyl reagents that inhibit DAO may become affixed to some other essential group than that belonging to pyridoxal, but that seems unlikely, for Davison could reverse the inhibition of DAO *in vitro* to some extent with pyridoxal phosphate, but not with pyridoxal or with pyruvate.

Previously we reported on the effect of riboflavin deficiency in reducing the rate of metabolism of pentylamine *in vivo* (Sourkes 1972). We now have made a comparison of the two different states of riboflavin nutrition in rats that had received pargyline, an irreversible inhibitor of MAO. These results are shown in Table 5, along with data from our previous work (Symes *et al.* 1971) for rats

TABLE 5

Effect of riboflavin deficiency on pentylamine metabolism

Duration of deficiency (days)	*Riboflavin status*	*Pargyline*	*Mean body weight (g)*	*No. of rats*	*% injected ^{14}C recovered in first hour as $^{14}CO_2$*	*P*
7	+	—	70	3	36.1 ± 2.0	
	—	—	59	3	33.9 ± 0.7	> 0.05
7	+	—	90	3	36.6 ± 0.3	
	—	—	73	3	35.0 ± 2.5	> 0.05
10	+	+	100	5	20.6 ± 1.15	
	—	+	72	6	16.9 ± 0.6	< 0.025
15	+	—	107	6	34.7 ± 1.0	
	—	—	67	6	28.3 ± 0.9	< 0.001
17	+	+	117	6	21.9 ± 1.6	
	—	+	71	6	16.4 ± 0.5	< 0.01
21	+	—	146	6	36.1 ± 1.4	
	—	—	73	5	29.8 ± 1.7	< 0.025

Rats were fed a riboflavin-deficient diet beginning at day zero. Matched animals received a supplement of riboflavin, designated + in the table, consisting of 0.5 mg of the vitamin orally three times weekly. Pargyline was given to some rats on day 7, and these were tested on day 10 of the deficiency; other groups received the inhibitor on day 12 and were tested on day 17.

without an inhibitor. Although pargyline reduced the pentylamine-oxidizing capacity of both deficient and supplemented rats, the relative effect of the deficiency was approximately the same as in the non-pargyline-treated animals. A longer experimental period might have provided a greater differentiation in regard to the ability of the deficient animals to regenerate MAO.

Because of the analogous type of reaction catalysed by DAO and MAO the suggestion was made in the past that the former would probably require riboflavin in the same way as MAO. Some evidence for this was presented (Kapeller-Adler 1949; Zeller 1951; Goryachenkova 1956) but never suitably substantiated. Mondovi *et al.* (1967*b*) purified their pig kidney DAO to the point where the flavin content disappeared without loss of enzymic activity. Nevertheless, we decided to determine whether riboflavin was necessary in the catabolism of putrescine in the rat, and the results of these experiments with a brief deficiency period, lasting one and two weeks, are shown in Fig. 3. The rate of catabolism of putrescine was significantly depressed ($P < 0.001$) in each of the first three hours after injecting labelled putrescine. Thus, it seems clear that normal metabolism of this amine in the rat requires the participation of a flavin at some stage, whether that be in the DAO of this species or in some other

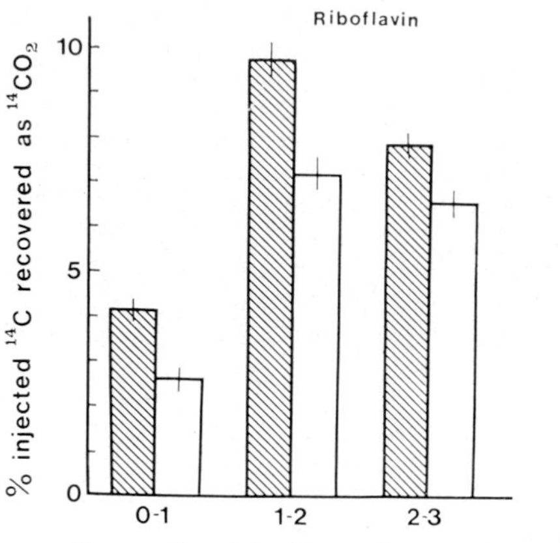

FIG. 3. Effect of riboflavin deficiency on the rate of catabolism of [^{14}C]putrescine to respiratory $^{14}CO_2$ in rats. Animals were fed a riboflavin-deficient (hollow bars) or riboflavin-supplemented (lined bars) diet (Symes *et al.* 1969) and were tested as described in the text and the legend to Fig. 1. Two series of 12 rats each were studied at different times; each rat was investigated at one and two weeks of deficiency, and the results were pooled so that each bar represents the mean of 24 determinations. Ordinate: percentage of injected ^{14}C recovered as expired $^{14}CO_2$ in each of the three indicated time periods following the injection of labelled putrescine. Differences resulting from the deficiency of riboflavin were statistically significant in each comparison ($P < 0.001$).

enzyme. Again, it is worth quoting Davison (1956), who suggested that a riboflavin coenzyme might play a role in an electron-transport chain that comes into play after the action of DAO. The absence of a flavin moiety from highly purified pig kidney DAO (Mondovi *et al.* 1967*b*) should not be taken as a general principle of this group of enzymes, for cephalopod histaminase is a yellow enzyme (Boadle 1969), and at least one bacterial DAO, that of *Micrococcus rubens*, contains FAD (Adachi *et al.* 1966; DeSa 1972).

It may be instructive to list the nutrients whose requirement for amine-oxidizing activities of mammalian species has been sought. These have been set out in Table 6, where it is clear that this technique has had a limited but important use. It produced the first evidence for a role of a flavin in MAO; it pointed to some function for iron in either the action or biosynthesis of that enzyme, and led directly to the testing of platelet amine oxidase in iron-deficient humans. It demonstrated that rats deficient in copper, as judged by tissue analysis and by severely depressed ceruloplasmin levels, were not incapacitated in their catabolism of a classical substrate for MAO (Sourkes 1968, 1972). It provides confirmatory evidence for the previously deduced roles of copper and pyridoxal phosphate in the action of DAO, as well as of copper in plasma amine oxidases of ruminants (Yamada & Yasunobu 1962) and pig

TABLE 6

Dietary deficiencies and mammalian amine oxidase activities

Nutrient	*Species*	*Measurements made*	
		In vitro	*Metabolism of substrates in vivo*
Riboflavin	Rat	MAO (liver)[a,b]	5-Hydroxytryptamine[c,d]; dopamine[a]; pentylamine[e]
Riboflavin	Rat	DAO (lung, intestinal mucosa)[f,g]	Putrescine[h]
Pyridoxine	Rat	MAO (liver)[i]	
Pyridoxine	Rat	DAO (lung, intestinal mucosa)[g]	Putrescine[h]
Copper	Rat	MAO (liver)[j]	Pentylamine[e,h]
Copper	Rat		Putrescine[h]
Copper	Pig	Benzylamine oxidase (plasma)[k]	
Copper	Sheep	Benzylamine oxidase (plasma)[l]	
Iron	Rat	MAO (liver)[j]	Pentylamine[e,h]
Iron	Rat		Putrescine[h]
Iron	Man	Platelet amine oxidase[m]	

[a] Hawkins (1952).
[b] Sourkes (1972).
[c] Wiseman & Sourkes (1961).
[d] Wiseman-Distler & Sourkes (1963).
[e] Symes *et al.* (1971). Copper deficiency did not significantly affect the initial rate of catabolism of pentylamine.
[f] Kapeller-Adler (1949).
[g] Goryachenkova (1956).
[h] Present results.
[i] Sourkes (1958). Pyridoxine deficiency appeared to attenuate the effect of lack of riboflavin on MAO activity of the liver.
[j] Symes *et al.* (1969). There was no effect of copper deficiency on hepatic MAO.
[k] Blaschko *et al.* (1965).
[l] Mills *et al.* (1966).
[m] Youdim *et al.* (1975).

(Buffoni & Blaschko 1964) plasma. Finally, it raises once again the question as to the mechanism of the diamine-oxidizing process and how a flavin coenzyme fits into this.

There is evidence, meagre as yet, for an interaction in pregnancy between iron and folic acid, on the one hand, and riboflavin, on the other, in relation to the riboflavin content of the erythrocytes (Clarke 1973), and also between iron and riboflavin in relation to the haemoglobin concentration (Weber *et al.* 1973). Even apart from the special conditions of the pregnant state, nutritional deficiency of iron is widespread in the world (WHO Scientific Group 1968). One is therefore eminently justified in proposing that problems of the metabolism of amines be given greater prominence in the study of the chemical pathology of blood disorders.

ACKNOWLEDGEMENT

Research in the authors' laboratory on amine biochemistry is supported by the Medical Research Council (Canada).

References

ADACHI, O., YAMADA, H. & OGATA, K. (1966) Purification and properties of putrescine oxidase of *Micrococcus rubens*. *Agric. Biol. Chem. (Japan) 30*, 1202-1210

BLASCHKO, H., BUFFONI, F., WEISSMAN, N., CARNES, W. H. & COULSON, W. F. (1965) The amine oxidase of pig plasma in copper deficiency. *Biochem. J. 96*, 4C-5C

BOADLE, M. C. (1969) Observations on a histaminase of invertebrate origin: a contribution to the study of cephalopod amine oxidases. *Comp. Biochem. Physiol. 30*, 611-620

BUFFONI, F. & BLASCHKO, H. (1964) Benzylamine oxidase and histaminase: purification and crystallization of an enzyme from pig plasma. *Proc. R. Soc. Lond. B Biol. Sci. 161*, 153-167

BURKARD, W. P., GEY, K. F. & PLETSCHER, A. (1962) Differentiation of monoamine oxidase and diamine oxidase. *Biochem. Pharmacol. 11*, 177-182

CALLENDER, S., GRAHAME-SMITH, D. G., WOODS, H. F. & YOUDIM, M. B. H. (1974) Reduction of platelet monoamine oxidase activity in iron deficiency anaemia. *Br. J. Pharmacol. 52*, 447P-448P

CAMPELLO, A. P., TABOR, C. W. & TABOR, H. (1965) Resolution of spermidine dehydrogenase from *Serratia marcescens:* requirements for flavin adenine dinucleotide and an additional electron carrier. *Biochem. Biophys. Res. Commun. 19*, 6-9

CHUANG, H. Y. K., PATEK, D. R. & HELLERMAN, L. (1974) Mitochondrial monoamine oxidase. Inactivation by pargyline. Adduct formation. *J. Biol. Chem. 249*, 2381-2384

CLARKE, H. C. (1973) In pregnancy: effect of iron and folic acid on riboflavin status. *Int. J. Vitam. Nutr. Res. 43*, 438-441

COLLINS, G. G. S. & SANDLER, M. (1971) Human blood platelet monoamine oxidase. *Biochem. Pharmacol. 20*, 289-296

COSTA, M. T., ROTILIO, G., AGRO, A. F., VALLOGINI, M. P. & MONDOVI, B. (1971) On the active site of diamine oxidase: kinetic studies. *Arch. Biochem. Biophys. 147*, 8-13

CRABBE, M. J. C. & BARDSLEY, W. G. (1974) The inhibition of human placental diamine oxidase by substrate analogues. *Biochem. J. 139*, 183-189

DAVISON, A. N. (1956) Pyridoxal phosphate as a coenzyme of diamine oxidase. *Biochem. J. 64*, 546-548

DESA, R. J. (1972) Putrescine oxidase from *Micrococcus rubens*. Purification and properties of the enzyme. *J. Biol. Chem. 247*, 5527-5534

ERWIN, V. G. & HELLERMAN, L. (1967) Mitochondrial monoamine oxidase. I. Purification and characterization of the bovine kidney enzyme. *J. Biol. Chem. 242*, 4230-4238

GORYACHENKOVA, E. V. (1956) On the nature of the active groups of diamine oxidase (histaminase). *Biokhimiya 21*, 247-257

HAWKINS, J. (1952) Amine oxidase activity of rat liver in riboflavin deficiency. *Biochem. J. 51*, 399-404

HUSZTI, Z., FEKETE, M. & HAJOS, A. (1969) Monoamine oxidase inhibiting properties of AB-15 – comparison with tranylcypromine and pargyline. *Biochem. Pharmacol. 18*, 2293-2301

KAPELLER-ADLER, R. (1949) Studies on histaminase. *Biochem. J. 44*, 70-77

KUSCHE, J., RICHTER, H., SCHMIDT, J., HESTERBERG, R., SPECHT, C. & LORENZ, W. (1973) Intestinal diamine oxidase: isolation, substrate specificity and pathophysiological significance. *Agents Actions 3*, 182-184

MILLS, C. F., DALGARNO, A. C. & WILLIAMS, R. B. (1966) Monoamine oxidase in ovine plasma of normal and low copper content. *Biochem. Biophys. Res. Commun. 24*, 537-540

MISSALA, K., LLOYD, K., GREGORIADIS, G. & SOURKES, T. L. (1967) Conversion of ^{14}C-dopamine to cardiac ^{14}C-noradrenaline in the copper-deficient rat. *Eur. J. Pharmacol. 1*, 6-10

MONDOVI, B., COSTA, M. T., FINAZZI-AGRÒ, A. & ROTILIO, G. (1967*a*) Pyridoxal phosphate as a prosthetic group of pig kidney diamine oxidase. *Arch. Biochem. Biophys. 119*, 373-381

MONDOVI, B., ROTILIO, G., COSTA, M. T., FINAZZI-AGRÒ, A., CHIANCONE, E., HANSEN, R. E. & BEINERT, H. (1967*b*) Diamine oxidase from pig kidney. Improved purification and properties. *J. Biol. Chem. 242*, 1160-1167

ORELAND, L. (1971) Purification and properties of pig liver mitochondrial monoamine oxidase. *Arch. Biochem. Biophys. 146*, 410-421

ORELAND, L., KINEMUCHI, H. & YOO, B. Y. (1973) The mechanism of action of the monoamine oxidase inhibitor pargyline. *Life Sci. 13*, 1533-1541

PLETSCHER, A. (1966) Monoamine oxidase inhibitors. *Pharmacol. Rev. 18*, 121-129

SCHAYER, R. W., SMILEY, R. L. & KENNEDY, J. (1954) Diamine oxidase and cadaverine metabolism. *J. Biol. Chem. 206*, 461-464

SCHULER, W. (1952) Zur Hemmung der Diaminooxydase (Histaminase). *Experientia 8*, 230-232

SEILER, N. (1973) Polyamine metabolism in the brain, in *Polyamines in Normal and Neoplastic Growth* (Russell, D. H., ed.), pp. 137-156, Raven Press, New York

SEILER, N. & AL-THERIB, M. J. (1974) Putrescine catabolism in mammalian brain. *Biochem. J. 144*, 29-35

SEILER, N., WIECHMANN, M., FISCHER, H. A. & WERNER, G. (1971) The incorporation of putrescine carbon into γ-aminobutyric acid in rat liver and brain *in vivo*. *Brain Res. 28*, 317-325

SEILER, N., AL-THERIB, M. J. & KATAOKA, K. (1973*a*) Formation of GABA from putrescine in the brain of fish (*Salmo irideus* Gibb). *J. Neurochem. 20*, 699-708

SEILER, N., KNÖGEN, B. & ASKAR, A. (1973*b*) The formation of γ-aminobutyrate in liver cell nuclei. *Hoppe-Seyler's Z. Physiol. Chem. 354*, 467-470

SHORE, P. A. & COHN, V. H., JR (1960) Comparative effects of monoamine oxidase inhibitors on monoamine oxidase and diamine oxidase. *Biochem. Pharmacol. 5*, 91-95

SINCLAIR, H. M. (1952) Pyridoxal phosphate as a coenzyme of histaminase. *Biochem. J. 51*, x-xi

SNEDECOR, G. W. (1956) *Statistical Methods*, 5th edn, Iowa State University Press, Ames

SOURKES, T. L. (1958) Oxidative pathways in the metabolism of biogenic amines. *Rev. Can. Biol. 17*, 326-366

SOURKES, T. L. (1968) Some properties of purified monoamine oxidase of rat liver mitochondria. *Adv. Pharmacol. 6A*, 61-69

SOURKES, T. L. (1972) Influence of specific nutrients on catecholamine synthesis and metabolism. *Pharmacol. Rev. 24*, 349-359

SOURKES, T. L., MURPHY, G. F. & WOODFORD, V. R., JR (1960) Effects of deficiencies of pyridoxine, riboflavin and thiamine upon the catecholamine content of rat tissues. *J. Nutr. 72*, 145-152

SYMES, A. L., SOURKES, T. L., YOUDIM, M. B. H., GREGORIADIS, G. & BIRNBAUM, H. (1969) Decreased monoamine oxidase activity in liver of iron-deficient rats, *Can. J. Biochem. 47*, 999-1002

SYMES, A. L., MISSALA, K. & SOURKES, T. L. (1971) Iron- and riboflavin-dependent metabolism of a monoamine in the rat *in vivo*. *Science (Wash. D.C.) 174*, 153-155

TIPTON, K. F. (1968) The prosthetic groups of brain mitochondrial monoamine oxidase. *Biochim. Biophys. Acta 159*, 451-459

WALKER, W. H., KEARNEY, E. B., SENG, R. & SINGER, T. P. (1971) Structure and sequence of a FAD-containing pentapeptide from monoamine oxidase. *Biochem. Biophys. Res. Commun. 44*, 287-292

WATANABE, K., SMITH, R. A., INAMASU, M. & YASUNOBU, K. T. (1972) Recent investigations on the prosthetic group of beef plasma amine oxidase, in *Monoamine Oxidases—New Vistas* (Costa, E. & Sandler, M., eds.) (*Adv. Biochem. Psychopharmacol. 5*), pp. 107-117, Raven Press, New York and North-Holland, Amsterdam

WEBER, F., GLATZLE, D. & WISS, O. (1973) The assessment of riboflavin status. *Proc. Nutr. Soc. 32*, 237-241

WHO Scientific Group (1968) Nutritional anemias. *WHO Tech. Rep. Ser.* no. 405

WISEMAN, M. H. & SOURKES, T. L. (1961) The effect of riboflavin and mepacrine on the metabolism of 5-hydroxytryptamine. *Biochem. J. 78*, 123-128

WISEMAN-DISTLER, M. H. & SOURKES, T. L. (1963) The role of riboflavin in monoamine oxidase activity. *Can. J. Biochem. Physiol. 41*, 57-64

YAMADA, H. & YASUNOBU, K. T. (1962) Monoamine oxidase. II. Copper, one of the prosthetic groups of plasma monoamine oxidase. *J. Biol. Chem. 237*, 3077-3082

YOUDIM, M. B. H. & SOURKES, T. L. (1966) Properties of purified soluble monoamine oxidase. *Can. J. Biochem. 44*, 1397-1400

YOUDIM, M. B. H. & SOURKES, T. L. (1972) The flavin prosthetic group of purified rat liver mitochondrial monoamine oxidase, in *Monoamine Oxidases—New Vistas* (Costa, E. & Sandler, M., eds.) (*Adv. Biochem. Psychopharmacol. 5*), pp. 43-53, Raven Press, New York and North-Holland, Amsterdam

YOUDIM, M. B. H., WOODS, H. F., MITCHELL, B., GRAHAME-SMITH, D. G. & CALLENDER, S. (1975) Human platelet monoamine oxidase activity in iron-deficiency anaemia. *Clin. Sci. Mol. Med. 48*, 289-295

ZELLER, E. A. (1951) Oxidation of amines, in *The Enzymes* (Sumner, J. B. & Myrbäck, K., eds.), vol. 2, part 1, pp. 536-558, Academic Press, New York

ZELLER, E. A., FOUTS, J. R., CARBON, J. A., LAZANAS, J. C. & VOEGTLI, W. (1956) Ueber die Substratspezifität der Diamin-oxydase. *Helv. Chim. Acta 39*, 1632-1644

Discussion

Pletscher: In your studies on copper deficiency and vitamin deficiency, did you compare deficient with pair-fed groups, or did you compare them only with groups fed *ad libitum*? We did similar experiments on ascorbic acid deficiency (Saner *et al.* 1975) and found that pair-feeding alone decreased the amine turnover of the brain, for example; so pair-feeding itself might have an effect.

Sourkes: It is possible. This is the eternal problem in animal nutrition studies, whether to pair-feed or free-feed. There are arguments on both sides. We did some pair-feeding experiments in the iron studies but we didn't pair-feed in the other experiments because we didn't see differences between the pair-fed groups and those fed *ad libitum*. But I agree that this is something that should be done in each case, because we are trying to account for specific effects, ultimately.

Blaschko: You discussed the pyridoxal phosphate content of diamine oxidase. Dr Yamada and his colleagues (Yamada *et al.* 1967) have probably gone beyond Mondovi, in having crystallized the pig kidney enzyme and having shown that it contains pyridoxal. They have also repeated the experiment that Professor Buffoni (1968) did, using the benzylamine oxidase, in showing, by the production of the reduced Schiff base between pyridoxal phosphate and histamine, that one can hydrolyse off the pyridoxal phosphate–histamine from the enzyme that has been incubated anaerobically with radioactive histamine (Kumagai *et al.* 1969). So there is no doubt that pyridoxal phosphate is present in these two enzymes. In addition, with benzylamine oxidase, you can split off a fraction that will reactivate an apo-enzyme that requires pyridoxal phosphate for activity (Blaschko & Buffoni 1965). And we showed that copper-deficient pigs do not have benzylamine oxidase activity (Blaschko *et al.* 1965).

Dr Sourkes also referred to experiments done by Miss Boadle (1969) on the octopod histaminase. This is a flavin-containing enzyme. She used what I think is still the best criterion for the flavin versus pyridoxal phosphate type of enzyme and found that ω-*N*-methylated histamine was oxidized almost as rapidly as histamine itself, with the *Eledone* renal appendage enzyme. You would not get this with a true histaminase preparation because the reaction with the carbonyl group of pyridoxal phosphate requires the presence in the substrate of a primary amino group; such a group is absent in the ω-*N*-methylated histamine.

Secondly, there are enzyme preparations from vertebrates in which what I call monoamine oxidase acts on putrescine. This was discovered by chance when I worked with Dr Sabina Strich and Dr Margaret Boadle (Blaschko *et al.* 1969). We studied a preparation from the liver of the sea lamprey, *Petromyzon marinus*. If you test this so-called monoamine oxidase, using aliphatic diamines, you find not only that the rate of oxidation of the diamines of long chain length has a very flat maximum between 10 and 15 CH_2 groups but that the enzyme acts on putrescine, cadaverine and hexamethylene diamine even more rapidly (Blaschko 1974). We haven't been able to convince ourselves that in this insoluble mitochondrial preparation different enzymes were at work. This study was done before the discovery of the MAO isoenzymes, and it would be interesting to see now whether one would be able to separate these two activities.

Sourkes: My scepticism about pyridoxal phosphate is not directed against the type of experiment done by Yamada's group and others, where labelled histamine is added to the purified enzyme, and presumably an adduct is formed from which pyridoxyl–histamine can then be extracted. My point is that from the chemical point of view one would want to isolate pyridoxal phosphate itself from that enzyme and to demonstrate that it is present. Of course, there is a possibility that just as the FAD in MAO is covalently linked, the pyridoxal phosphate may be covalently linked in the diamine oxidase also, and then it would be extremely difficult to get it free of other ingredients.

My scepticism is based also on the fact that for many years from 1965 it was stated, and is still being stated, that MAO is a copper-containing enzyme; since this was demonstrated to be wrong, and withdrawn a few years later, I think we should expect chemical proof of the coenzyme.

Blaschko: The idea that the diamine oxidase types of enzymes, including benzylamine oxidase, contain pyridoxal phosphate in some covalently linked form, is one which Professor Buffoni has always favoured.

Youdim: At the 1975 meeting on flavin and flavoproteins in San Francisco Dr Yasunobu mentioned that he does not now think that pyridoxal phosphate is a cofactor of diamine oxidase.

Blaschko: To my mind, Dr Yasunobu's very interesting paper on this (Inamasu *et al.* 1974) does not prove the case he wishes to make.

Singer: Dr Sourkes, could you summarize the chemical evidence that iron is present in stoichiometric quantities with cysteinyl flavin in highly purified monoamine oxidase, rather than being required for the incorporation of cysteinyl flavin into the enzyme?

Sourkes: I don't know where the iron is working. I have suggested that an iron enzyme may be required to convert the 8-methyl group of the iso-alloxazine side-chain of FAD into a hydroxymethyl compound so that it can form a covalent bond. For two or three years now I have been trying to work out how to detect and measure an enzymic activity of this kind.

Singer: To test this idea there is one very simple thing that you are in a position to do, namely to follow the decline of monoamine oxidase activity and of succinate dehydrogenase activity in iron deficiency.

Sourkes: The succinate oxidase approach for this iron-involved hypothetical enzyme may be easier than for monoamine oxidase; thank you for the suggestion.

Fuller: One could imagine that in a deficiency of a nutrient like iron there might be non-specific or secondary changes in a variety of enzymes, and I wonder if you have as a control compared MAO with other enzymes, such as another enzyme on the outer mitochondrial membrane or another enzyme in monoamine metabolism, and found a lack of change in iron deficiency?

Sourkes: No, we haven't done this.

Sandler: As we are considering a dietary cofactor, we might consider also the obverse of the coin—the possible presence of dietary MAO inhibitors. One thinks of nutmeg, not exactly an everyday item of diet, which contains such an inhibitory principle (Truitt *et al.* 1963), but I am sure there exist other foods containing MAO inhibitors if we look for them. The higher activity of hepatic MAO in germ-free than in normal chicks (Phillips *et al.* 1962) should be taken into account in this context. Presumably the gut flora, in this species at least, produce inhibitory factors that we don't know about.

Gorkin: Dr W. G. Bardsley has studied the inhibitory effect of MAO inhibitors on the activity of diamine oxidase, when the activity was measured with a synthetic substrate, *N*-(dimethylaminomethyl)benzylamine (Crabbe & Bardsley 1974). Do you think that your data are in agreement with this, Dr Sourkes, or do you still think that MAO inhibitors do not inhibit diamine oxidase?

Sourkes: As I said (Table 4 and p. 87), we have studied this in the intact rat and find a reasonably good separation of the pathways, as delineated using tranylcypromine and aminoguanidine. I think it is in line with Dr Pletscher's data (Burkard *et al.* 1962), and those of Shore & Cohn (1960). Schayer *et al.* (1954) have reported the same with histamine.

Tipton: We found clorgyline to be a reversible inhibitor of diamine oxidase (Houslay & Tipton 1975). In addition, the degree of inhibition does not respond linearly to the concentration of inhibitor and the effect could easily be confused with the biphasic irreversible effect seen with monoamine oxidase. Thus one should be careful in interpreting results obtained with impure preparations. Either the absence of diamine oxidase should be demonstrated or the inhibition by selective inhibitors should be shown to be irreversible.

van Praag: Professor Sourkes mentioned Dr Youdim's studies on diminished platelet MAO activity in iron-deficient people. Was the reduction sufficient to affect monoamine degradation to any considerable degree? It is interesting that Dr Murphy has evidence of reduced MAO activity in people with schizophrenia. One wonders what this reduction signifies.

Sourkes: A great deal remains to be done here, and whether the lower MAO level plays any role in the physiopathology of iron deficiency anaemias is a question for the future. It may be very important, however, because of the widespread occurrence of anaemia. One could approach this in a simple way by looking at the rate of excretion, or the amount of vanillylmandelic acid and 5-hydroxyindoleacetic acid excreted in the urine, in people with iron-deficiency anaemia.

Youdim: In the study of iron deficiency in man that we have just completed (Youdim *et al.* 1975) we have found that MAO activity is related to the level of serum iron rather than to haemoglobin levels (see Figs. 1 and 2). We have also looked at platelet function, namely platelet aggregation in response to 5HT and ADP. Platelet aggregation in response to 5HT in normal human subjects is transient. In iron-deficiency anaemia, platelet aggregation in response to 5HT is irreversible and similar to that produced by ADP. We think that iron deficiency is affecting the membrane structure of the platelet. Dallman & Goodman (1970, 1971) have provided evidence by electron microscopy that in iron deficiency the outer mitochondrial membrane is damaged (in the liver and heart).

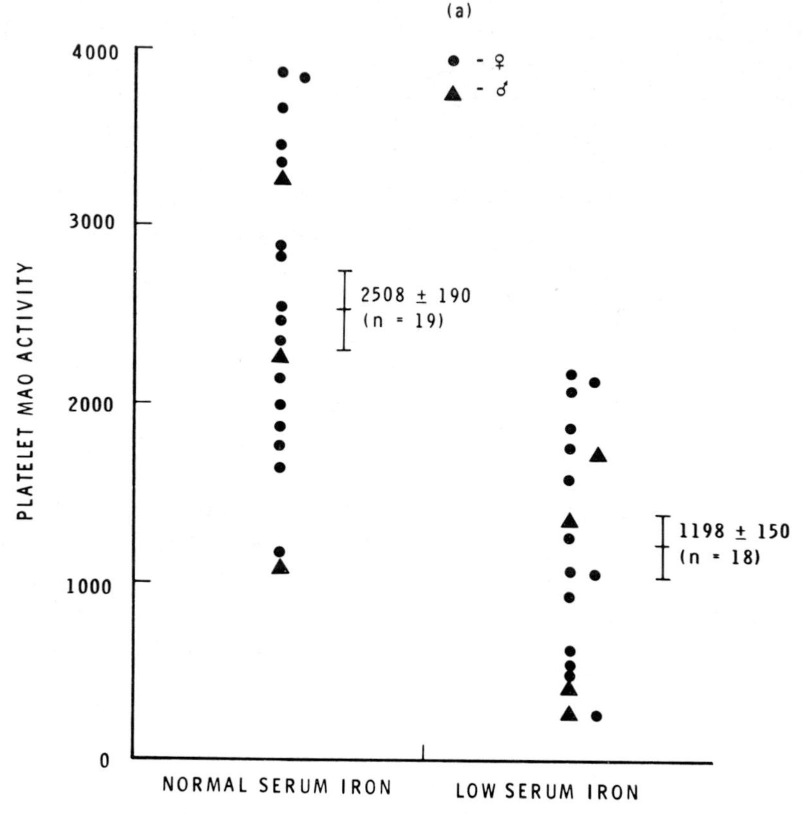

FIG. 1 (Youdim). The distribution of monoamine oxidase activity in platelets from subjects having a normal serum iron concentration and those with a low serum iron concentration. A low serum iron concentration is defined as being below 14.3 μmol/l for male patients (▲) and less than 10.7 μmol/l for female patients (●). Dopamine (1 mmol/l) was used as substrate and activity in the low serum iron group is significantly lower than that in the normal serum iron group ($P < 0.001$). (Youdim *et al.* 1975.)

As to the role of iron in MAO, we have now developed a method by which we can titrate the exact amount of monoamine oxidase in a tissue, using specific monoamine oxidase inhibitors labelled with ^{14}C. It has been reported that platelet MAO is a B type enzyme, using [^{14}C]deprenyl, kindly provided by Professor Knoll, and we have titrated monoamine oxidase activity in platelets from iron-deficient and normal subjects. We find less of the enzyme in iron deficiency (Youdim *et al.* 1976). This might be due to a lack of the co-factor (FAD), because it has been established that acetylenic inhibitors of monoamine oxidase bind covalently to the co-factor (see Maycock *et al.* this volume, pp. 33-47). It may also be possible that iron, as Professor Sourkes has said, is involved in the attachment of the co-factor to the apo-enzyme.

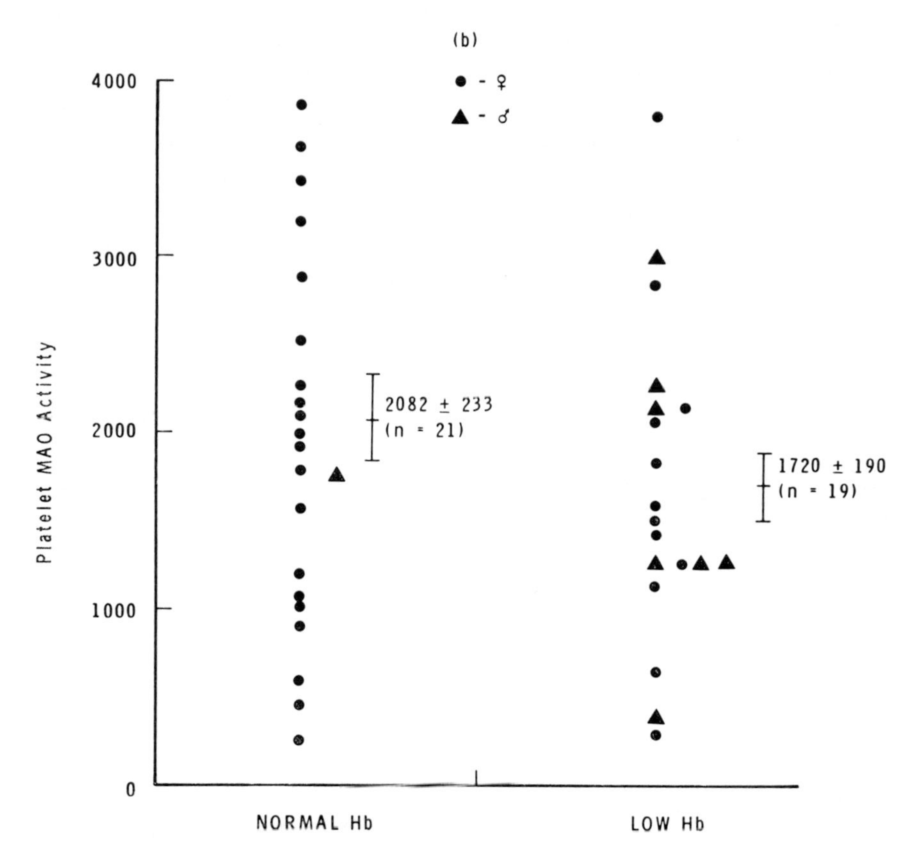

FIG. 2 (Youdim). The distribution of monoamine oxidase activity in platelets from subjects having a normal blood haemoglobin concentration and those with a low blood haemoglobin concentration. A low haemoglobin is defined as less than 11.5 g/dl for a female patient and less than 13.5 g/dl for a male patient.

Activity is expressed as d.p.m. radioactivity in the deaminated product per mg of platelet protein in 30 min incubation with dopamine (1 mmol/l). The results are mean values ± S.E.M. with the number of observations in parentheses. There is no significant difference between the activity in the low haemoglobin group and that in the normal haemoglobin group. ●, female subjects; ▲, male subjects. (Youdim *et al.* 1975.)

Rafaelsen: Källström (1954) studied changes in serum iron levels in cases of depression. I don't think it was established whether it was really iron deficiency, but there were changes in serum iron concentration accompanying the affective changes from depression back to a neutral state.

Coppen: I am slightly confused. You used the data of Youdim *et al.* (1975) to calculate partial coefficients, Professor Sourkes. 5HT and phenylethylamine, of the four substrates used, showed a significant correlation of platelet amine activity with serum iron level, when the haemoglobin level was kept constant. But I believe that platelets contain mainly type B MAO, and the characteristic of the platelet is that it is full of 5HT. The monoamine oxidase that seems to be most concerned with metabolizing 5HT isn't present in any amount.

Kety: That's why the platelet is full of 5HT!

Pletscher: I don't know whether platelet monoamine oxidase is of the A or the B type, but I do know that platelets actively metabolize 5HT, provided the granular storage of 5HT is impaired. Thus in reserpinized platelets there is a high rate of metabolism of 5HT.

Coppen: But why is that correlation found with 5HT as substrate?

Youdim: We didn't report this particular correlation. On plotting MAO activity from each group against serum iron concentration we obtained a correlation of about 0.66. But Dr Sourkes has used the individual values to calculate partial correlation coefficients. If the results are plotted on the basis of normal or low serum iron, there is a significant difference in MAO activity (Fig. 1, p. 100) between normal and iron-deficient subjects. However, if MAO activity is plotted on the basis of normal or low haemoglobin concentration, no difference in platelet MAO activity is observed between the two groups (Fig. 2). Our conclusion was strengthened by the platelet MAO activity in patients with carcinoid syndrome. We thought that perhaps in carcinoid one would find an induced monoamine oxidase activity, because there is so much 5HT, but in fact they had low MAO activity. In the three patients examined their haematological records showed normal haemoglobin values but low serum iron.

van Praag: Did you investigate the influence of administering iron on the lowered MAO activity in iron-deficient patients? The cause of the iron deficiency could have induced other kinds of deficiencies.

Youdim: Patients who were identified as being undoubtedly iron-deficient were treated with 200 mg of $FeSO_4$ three times daily. At intervals during therapy blood samples were obtained from them and platelet MAO activity was estimated. The interesting thing was that the activity of MAO increased as the serum iron level returned to normal. In two cases in which the serum iron did not return to normal, the MAO activity did not return to normal either.

References

BLASCHKO, H. (1974) The natural history of amine oxidases. *Rev. Physiol. Biochem. Pharmacol. 70*, 83-148

BLASCHKO, H. & BUFFONI, F. (1965) Pyridoxal phosphate as a constituent of the histaminase of pig plasma. *Proc. R. Soc. Lond. B Biol. Sci. 163*, 45-60

BLASCHKO, H., BUFFONI, F., WEISSMAN, N., CARNES, W. H. & COULSON, W. F. (1965) The amine oxidase of pig plasma in copper deficiency. *Biochem. J. 96*, 4*C*-5*C*

BLASCHKO, H., BOADLE, M. C. & STRICH, S. J. (1969) Enzymic oxidation of amines in Cyclostomes. *J. Physiol. (Lond.) 204*, 104-105*P*

BOADLE, M. C. (1969) Observations on a histaminase of invertebrate origin: a contribution to the study of cephalopod amine oxidases. *Comp. Biochem. Physiol. 30*, 611-620

BUFFONI, F. (1968) Pyridoxal catalysis in pig plasma benzylamine oxidase (histaminase), in *Pyridoxal Catalysis: Enzymes and Model Systems* (Snell, E. E., Braunstein, A. E., Severin, E. S. & Torchinsky, Y. M., eds.), pp. 363-374, Interscience, New York, London & Sydney

BURKARD, W. P., GEY, K. F. & PLETSCHER, A. (1962) Differentiation of monoamine oxidase and diamine oxidase. *Biochem. Pharmacol. 11*, 177-182

CRABBE, M. J. C. & BARDSLEY, W. G. (1974) Monoamine oxidase inhibitors and other drugs as inhibitors of diamine oxidase from human placenta and pig kidney. *Biochem. Pharmacol. 23*, 2983-2990

DALLMAN, P. R. & GOODMAN, J. R. (1970) Enlargement of mitochondrial compartments in iron and copper deficiency. *Blood 35*, 496-505

DALLMAN, P. R. & GOODMAN, J. R. (1971) The effect of iron deficiency on the hepatocyte; a biochemical and ultrastructural study. *J. Cell Biol. 48*, 79-90

HOUSLAY, M. D. & TIPTON, K. F. (1975) Inhibition of beef plasma diamine oxidase by clorgyline. *Biochem. Pharmacol. 24*, 429-431

INAMASU, M., YASUNOBU, K. T. & KONIG, W. A. (1974) Cofactor investigation of bovine plasma amine oxidase. *J. Biol. Chem. 249*, 5265-5268

KÄLLSTRÖM, B. R. (1954) *Serum Iron in Depressive States*, Almqvist & Wiksell, Uppsala

KUMAGAI, H., NAGATE, T., YAMADA, H. & FUKAMI, H. (1969) Characterization of sodium-borohydride-reduced histaminase–histamine intermediate. *Biochem. Biophys. Acta 185*, 242-244

MAYCOCK, A. L., ABELES, R. H., SALACH, J. I. & SINGER, T. P. (1976) This volume, pp. 33-47

PHILLIPS, A. W., NEWCOMBE, H. R., RUPP, F. A. & LACHAPELLE, R. (1962) Nutritional and microbial effects on liver monoamine oxidase and serotonin in the chick. *J. Nutr. 76*, 119-123

SANER, A., WEISER, H., HORNIG, D., DA PRADA, M. & PLETSCHER, A. (1975) Cerebral monoamine metabolism in guinea pigs with ascorbic acid deficiency. *J. Pharm. Pharmacol. 27*, 896-902

SCHAYER, R. W., SMILEY, R. L. & KENNEDY, J. (1954) Diamine oxidase and cadaverine metabolism. *J. Biol. Chem. 206*, 461-464

SHORE, P. A. & COHN, V. H., JR (1960) Comparative effects of monoamine oxidase inhibitors on monoamine oxidase and diamine oxidase. *Biochem. Pharmacol. 5*, 91-95

TRUITT, E. B., JR, DURITZ, G. & EBERSBERGER, M. (1963) Evidence for monoamine oxidase inhibition by myristicin and nutmeg. *Proc. Soc. Exp. Biol. Med. 112*, 647-650

YAMADA, H., KUMAGAI, H., KAWASAKI, H., MATSUI, H. & OGATA, K. (1967) Crystallization and properties of diamine oxidase from pig kidney. *Biochem. Biophys. Res. Commun. 29*, 723-727

YOUDIM, M. B. H., WOODS, H. F., MITCHELL, B., GRAHAME-SMITH, D. G. & CALLENDER, S. (1975) Human platelet monoamine oxidase activity in iron-deficiency anaemia. *Clin. Sci. Mol. Med. 48*, 289-295

YOUDIM, M. B. H., GRAHAME-SMITH, D. G. & WOODS, H. F. (1976) Some properties of human platelet monoamine oxidase in iron-deficiency anaemia. *Clin. Sci. Mol. Med.*, in press

Physiological aspects of the oxidative deamination of monoamines

M. B. H. YOUDIM* and MARGARETHE HOLZBAUER†

**Medical Research Council Unit and University Department of Clinical Pharmacology, Radcliffe Infirmary, Oxford, and †Agricultural Research Council Institute of Animal Physiology, Babraham, Cambridge*

Abstract Although much attention has been paid to the physicochemical nature of the enzyme monoamine oxidase, less is known about the physiological factors which affect and control the activity and synthesis of this enzyme. The paper discusses the influence of hormones, especially steroids, on MAO activity in various tissues and the possible physiological implications. In addition studies on the postnatal development of MAO in the rat are presented. The possibility of an independent development of a dopamine deaminating enzyme system in the striatum and hypothalamus is discussed. This might reflect the maturation of the neuronal system employing dopamine as a transmitter substance.

The deamination of most naturally occurring biologically active monoamines is achieved by monoamine oxidase (amine: oxygen oxido reductase [deaminating] [flavin-containing]; EC 1.4.3.4) (MAO) (Hare 1928). Although much attention has been paid to the physicochemical properties of this enzyme system (for review see Blaschko 1974) and its multiple forms (for review see Sandler & Youdim 1972) less is known about physiological factors which may affect its activity and consequently the disposition of monoamines in the organism. This is not least due to the difficulties involved in assessing accurately the enzyme activity *in vivo*. The technique most commonly used to measure MAO activity is the *in vitro* exposure of a monoamine to a homogenate of the tissue in question under conditions previously assessed to be optimal for the deamination to occur. The difficulties in the extrapolation from kinetic data derived from *in vitro* experiments to the conditions prevailing in the body have been discussed by Youdim & Woods (1975). Certain advantages are offered by techniques for perfusing isolated organs that have recently been used to study MAO activity (e.g. Youdim *et al.* 1974; Graham *et al.* 1975; Bakhle & Youdim 1975). Under these conditions the structural integrity of the tissues is preserved and the concentrations of coenzymes and substrates will be similar to those in the intact body. Yet another way to assess *in vivo* MAO activity is to study the tissue content of the metabolites formed from naturally occurring amines. All these techniques have obvious inherent weaknesses and any results obtained will only represent an approximation to the enzyme activity in the intact organism.

The results presented and discussed in this paper were derived from *in vitro* studies and the general physiological implications of the observed variations are thus subject to the limitations set by the technique.

Among physiological factors influencing tissue MAO activity the best-studied ones are hormones and the process of growth and development in the young animal.

HORMONES AND MAO ACTIVITY

A possible effect of thyroid hormones on MAO activity has been investigated for more than 15 years with varying results (e.g. Zile 1960; Novick 1961). Recently Lyles & Callingham (1974) reported that rat heart MAO activity can be stimulated by thyroxin and that this is due to an increase in the rate at which the enzyme is synthesized. The extent of the rise was dependent on the substrate used. In addition, thyroxin was found to be involved in the synthesis of flavin, a cofactor of MAO (Rivlin *et al.* 1976).

MAO is also affected by the steroid hormones produced in the ovaries and adrenal glands. Thus during the menstrual cycle large increases in human endometrial MAO coincided with the peak in the plasma progesterone concentrations (Southgate *et al.* 1968). In the rat, progesterone stimulated and oestradiol inhibited uterine MAO activity (Collins *et al.* 1970). Recently, Luine *et al.* (1975) have treated ovariectomized rats with oestradiol and found an inhibition of MAO by approximately 50% in the basomedial hypothalamus and the corticomedial amygdala.

In a study of the cyclic variations of MAO activity in the rat uterus, ovary and adrenal glands and also in four discrete brain regions (hypothalamus, striatum, septum and part of the amygdala complex), periods of high activity were found to coincide with periods in which progesterone secretion was high and *vice versa* (Holzbauer & Youdim 1972, 1973). In all tissues studied MAO activity was lowest in the evening of oestrus. The fall during oestrus was especially pronounced in the hypothalamus (about 70%). Maximum values were reached in late di-oestrus and pro-oestrus. In the uterus the rise (about five-fold) between late oestrus and early metoestrus was very steep. MAO activity fell again in di-oestrus when the values were about three times those in late oestrus. Our observations support the view that the catabolism of monoamines undergoes cyclic variations (for references see Holzbauer & Youdim 1973) which might be important for the possible interaction between monoamines and the release of pituitary hormones.

In studies on human blood platelets no positive correlation between MAO activity and serum progesterone concentrations during the oestrous cycle could be established (Belmaker *et al.* 1974). Confirmation of these observations has to be awaited since platelet MAO can be influenced by a number of factors (e.g. serum iron, see Youdim *et al.* 1975). This can lead to contradictory results, as exemplified by studies on platelet MAO in schizophrenia (Bailey *et al.* 1975).

In the pregnant rat, hepatic MAO activity was found to decrease to a minimum by day 18 and to rise again before parturition simultaneously with blood progesterone (Parvez *et al.* 1975).

In ovariectomized rats the enzyme activity in the brain, adrenal gland and uterus, ten days after the operation, was equal to the lowest values found in intact females during the oestrous cycle (Holzbauer & Youdim 1973). Ten days after castration of male rats MAO activity in the hypothalamus was 17% ($P < 0.02$) and that in the septum 30% ($P < 0.05$) higher than that in sham-operated controls. No changes occurred in the adrenal glands, heart and liver. Small increases in brain MAO activity after castration which could be antagonized by testosterone injections have also been described by Lemay & Berlinguet (1967).

After adrenalectomy Avakian & Callingham (1968) observed an increase in the MAO activity of the rat heart. This was confirmed by other workers who have also described a simultaneous rise in brain MAO activity (e.g. Ceasar *et al.* 1970; Holzbauer & Youdim 1972; Parvez & Parvez 1973). Fig. 1 shows results of experiments in which we measured MAO activity in the heart, liver and four brain regions eight days after bilateral adrenalectomy. The effect of adrenalectomy is probably caused by the absence of the adrenal cortex and not the adrenal medulla, since Avakian & Callingham (1968) were able to antagonize it with hydrocortisone and Parvez & Parvez (1973) found that metyrapone, a drug which inhibits glucocorticoid synthesis, can cause a rise in cardiac MAO. In addition we were not able to detect any increase in the MAO activity of the heart in rats bearing regenerated adrenal glands (i.e. glands which had been enucleated and then allowed to regenerate cortical tissue from the capsule but which no longer contained medullary tissue) (Fig. 2). Three weeks after adrenal enucleation the plasma corticosterone in the rats was 36 ± 4.5 compared with 50 ± 4 μg/100 ml in sham-operated controls.

Although on any single occasion on which an effect of adrenalectomy on cardiac MAO has been observed, the differences between control rats and

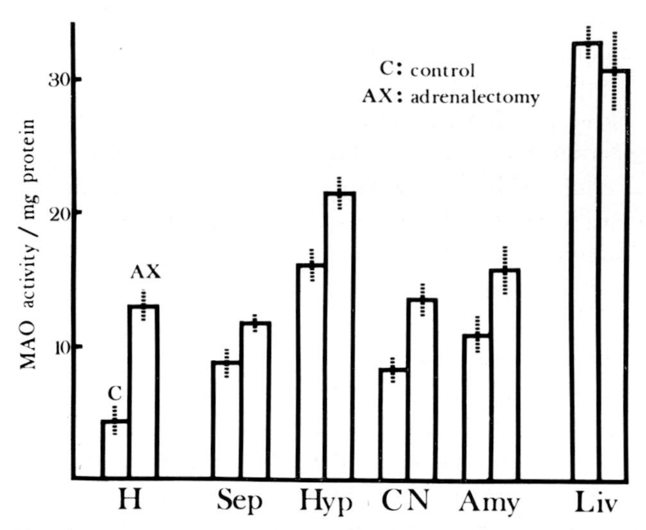

FIG. 1. Monoamine oxidase activity in several organs of the rat 10 days after bilateral adrenalectomy. C: sham-operated controls (first columns); Ax: adrenalectomized rats (second columns); H: heart; Sep.: septum (brain); Hyp.: hypothalamus; CN: caudate nucleus; Amy: part of the amygdala complex; Liv: liver. Enzyme activity expressed in arbitrary units, kynuramine used as substrate (mean values ± S.E.M.).

adrenalectomized rats were highly significant (see e.g. Fig. 1), there were occasions when no differences could be seen. This has also been experienced by other workers (e.g. B. A. Callingham & L. de la Corte, personal communication). So far these failures remain unexplained.

The mode of action by which steroids influence MAO activity is still a matter of conjecture. Some possibilities were discussed by Youdim *et al.* (1974). It is now generally accepted for most steroid hormones that on entering the target cell they associate with an extranuclear receptor protein to form a steroid–protein complex which migrates to the nucleus where it binds to a specific acceptor site in the chromatin and initiates the 'hormonal response' (for reference see O'Malley & Schrader 1972). In this way some steroids may affect MAO activity by inducing new enzyme synthesis at the transcriptional or post-transcriptional level. We intend to examine this problem with a recently developed method (Youdim 1975) which allows the titration of the exact amount of enzyme present in a tissue, using radioactive MAO inhibitors.

An involvement of monoamine oxidase in the mechanism of insulin release is also being discussed (Aleyassine & Gardiner 1975). Whether this effect is mediated by monoamines or by the products of their enzymic deamination as in the thyroid (Huang & Schulz 1972) is not known.

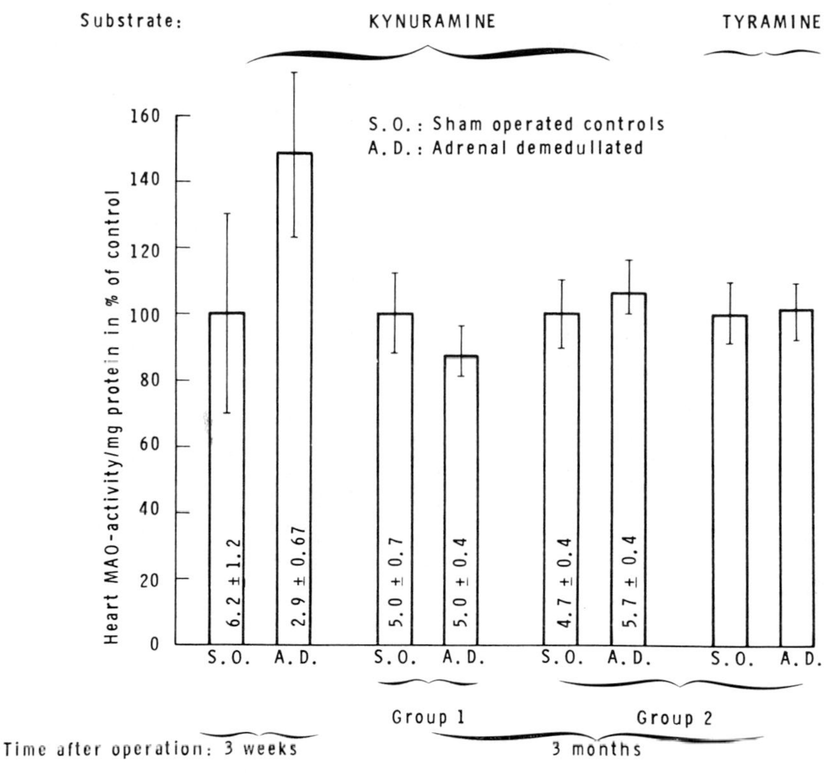

FIG. 2. Monoamine oxidase activity in the hearts of adrenal demedullated rats 3 weeks or 3 months after the operation. Activity expressed as % of that in sham-operated controls (mean values ± S.E.M.). Figures inside the columns: mg adrenal protein/rat.

Furthermore, it has still to be established whether the changes in MAO activity caused by hormones are large enough to affect the rate at which monoamines are being metabolized in the body.

MAO IN THE ADRENAL GLAND

The adrenal gland is of special interest in this context, not only because its hormones exert an influence on the enzyme activity of other organs; it also contains the highest concentrations of monoamines in the body. MAO activity can be found both in the adrenal medulla and in the adrenal cortex. In homogenates from ox adrenal medulla Blaschko *et al.* (1955) found it predominantly in 'large granules'.

In a study on adrenal glands from pigs we found more MAO in the cortex

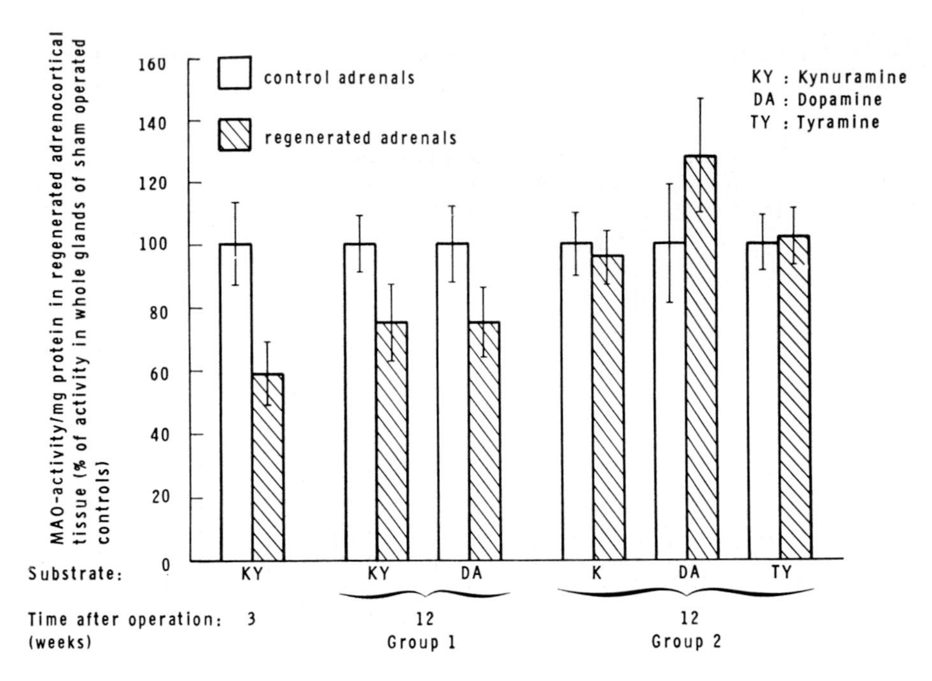

FIG. 3. Specific activity of monoamine oxidase in regenerated adrenocortical tissue of rats 3 or 12 weeks after adrenal demedullation. Results expressed as % of values in sham-operated controls (mean values ± S.E.M.).

than in the medulla. When dopamine was used as substrate cortical MAO activity per mg protein was seven times higher than medullary; with kynuramine it was six times higher and with tyramine four times higher. In the dog about twice as much MAO activity per mg protein could be found in the isolated cortex as in the whole adrenal gland. In the rat pure cortical tissue can be obtained after enucleation of the gland and regeneration of the cortex from the capsule. Three weeks after the operation the regenerated tissues had 42% less MAO activity towards kynuramine than the intact glands ($P < 0.05$). After three months of regeneration the MAO activity was the same as that of undisturbed adrenal glands containing medullary tissue (Fig. 3).

Adrenal MAO can be influenced by pretreating rats with ovarian hormones. As shown in Fig. 4 the extent of the changes in MAO activity was found to be dependent on the substrate used, which indicates the possible presence of multiple forms of MAO in the adrenal gland (Tipton *et al.* 1972; Youdim *et al.* 1974).

Progesterone occurs in the adrenal gland of the rat in quantities similar to those in the ovary (Fajer *et al.* 1971). In the adrenal gland it is an essential

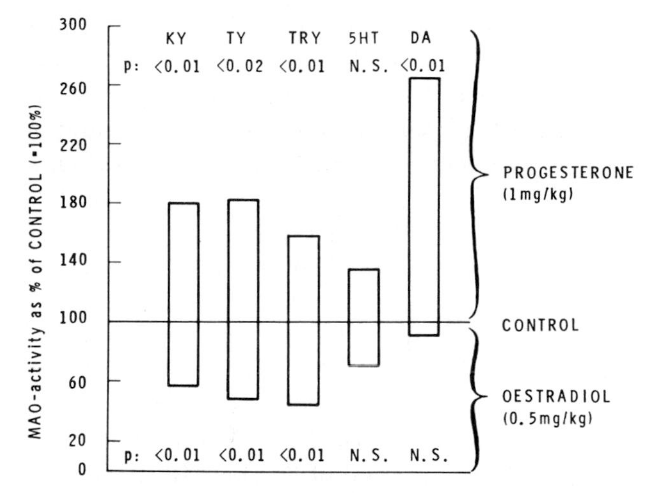

FIG. 4. Effect of 8 daily intraperitoneal injections of either progesterone (1 mg/kg/day) or oestradiol (0.5 mg/kg/day) on the monoamine oxidase activity of the rat adrenal gland. Activity expressed in % of control values (injected with 0.9% NaCl). Substrates: KY: kynuramine; TY: tyramine; TRY: tryptamine; 5HT: 5-hydroxytryptamine; DA: dopamine. (N.S.: not statistically significant).

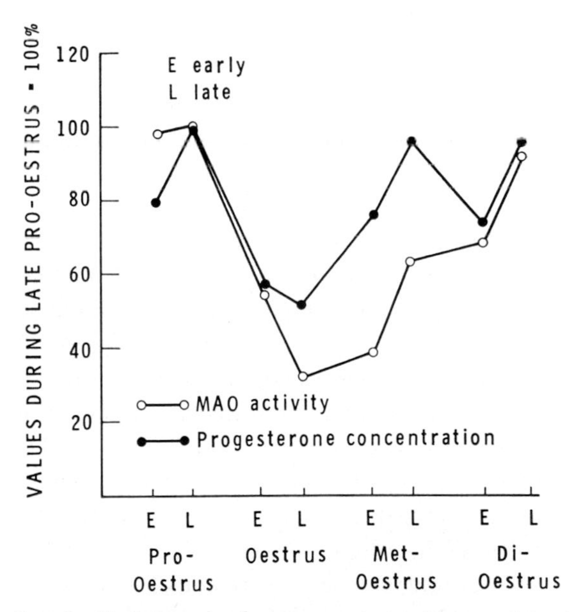

FIG. 5. Variations in the monoamine oxidase activity (per mg protein, ○—○, substrate kynuramine) and progesterone concentration (●—●) of the rat adrenal gland during the oestrous cycle. Values expressed as % of those found in late pro-oestrus. Rats kept under controlled lighting conditions (12 h white light, 12 h red light). E: early (8 h after onset of white light); L: late (3 h after onset of red light).

precursor steroid for the synthesis of the gluco- and mineralocorticoids. Under certain conditions the intrinsic progesterone in the adrenal cortex seems to be able to stimulate adrenal MAO activity. This is indicated during the oestrous cycle of the rat where both the MAO activity and progesterone concentration reach a maximum in late pro-oestrus and a minimum in late oestrus (Fig. 5) (Holzbauer & Youdim 1973).

In the adrenal gland progesterone and MAO are present in highest concentrations in the mitochondrial fraction. MAO is known to be associated with the outer mitochondrial membrane (Schnaitman *et al.* 1967). The exact location of progesterone has not yet been defined. It is however known that the enzyme 11β-hydroxylase, which is required for the transformation of progesterone into the major glucocorticoids, is associated with the inner mitochondrial membrane (Yago & Ichii 1969). The first intermediate, 11βOH-progesterone, is also present in the adrenal gland of the dog. In studies concerned with the subcellular distribution of steroids in the adrenal cortex of the dog we have been using MAO as 'mitochondrial marker'. In the course of this work it was observed that the rise in progesterone in purified adrenocortical mitochondria two hours after the intravenous injection of ACTH was accompanied by a rise in the specific activity of MAO in the same fraction (Holzbauer *et al.* 1973) (Fig. 6). When ACTH was added to dog adrenal slices under *in vitro* conditions for two hours there was also a significant rise in mitochondrial progesterone; however, no rise in MAO activity could be observed. It is known that progesterone added to tissue homogenates *in vitro* does not stimulate MAO activity. The reason for this failure may also be the reason why intracellularly formed progesterone does not stimulate MAO activity *in vitro*.

The effect of stimulation or inhibition of corticosterone production by hypophysectomy and subsequent treatment with ACTH of rats on adrenal MAO activity has been investigated on several occasions. Wurtman & Axelrod (1966) found in Sprague-Dawley rats 30 days after hypophysectomy MAO activity towards tryptamine decreased when expressed per pair of adrenal glands. As the adrenal weight was also considerably reduced the decrease in enzyme activity expressed per mg gland was, however, only 8%. Similar results were obtained by Bhagat *et al.* (1973). In our experiments on Wistar rats, using kynuramine as substrate and expressing MAO activity per mg adrenal protein, a significant decrease in adrenal MAO (52%, $P < 0.02$) occurred eight days after hypophysectomy and the treatment of hypophysectomized rats for eight days with ACTH (20 m-units/rat/day, i.p.) caused a rise in specific MAO activity by 83% ($P < 0.05$) (Fig. 7). The discrepancy between our observations and the previous reports may be due to the different rat strains and/or substrates used.

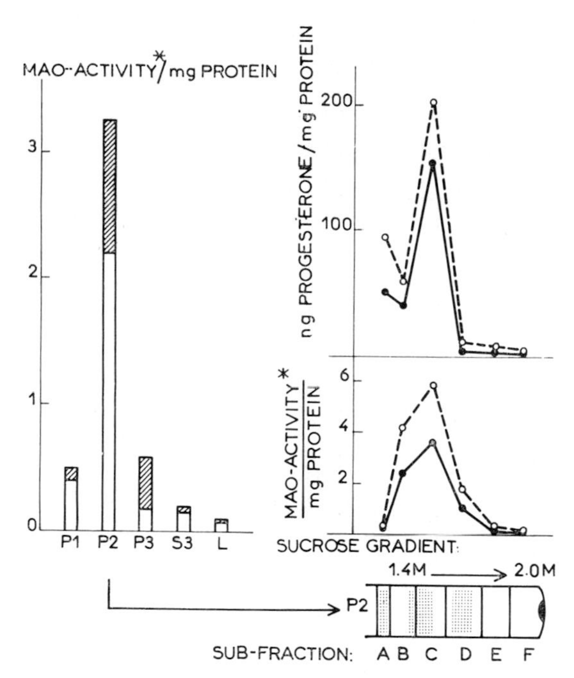

FIG. 6. Monoamine oxidase activity (*: arbitrary units, kynuramine as substrate) and progesterone concentration in subcellular fractions of dog adrenal cortex before (open part of columns and ●—●) and 2 h after an i.v. injection of 2 i.u. ACTH/kg (open plus shaded part of columns and ○- - -○). P1: nuclear fraction; P2: mitochondrial fraction; P3: microsomal fraction; S3: high-speed supernatant; L: lipids.

ONTOGENESIS OF MAO

Another physiological event which causes changes in MAO activity is development and growth (e.g. Karki *et al.* 1962; Bennet & Gairman 1965; Vaccari *et al.* 1972). In most of the studies so far concerned with this problem either only one organ or only one substrate has been tested.

We have recently made a more detailed investigation in which enzyme activity was measured in littermate male rats from five different colonies between the age of 5 and 80 days. The tissues studied were the liver, heart, adrenal glands and four discrete brain regions: hypothalamus, striatum, septum and cerebellum. Five different substrates were used. Particular emphasis was laid on the possibility that different forms of the MAO enzyme system might develop at different rates. In the following some examples of results obtained in this work will be discussed.

Fig. 8 illustrates the development of MAO in the adrenal glands of littermate

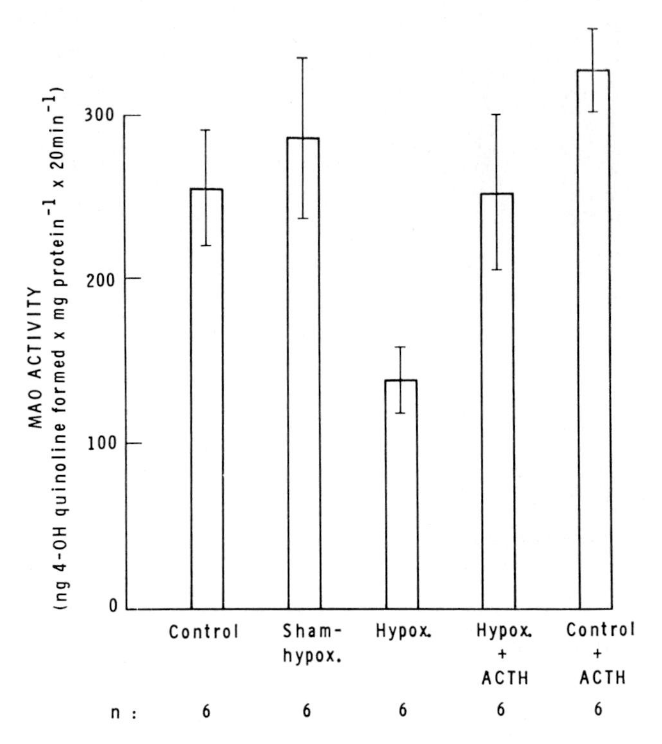

FIG. 7. Effect of hypophysectomy (14 days) and ACTH (20 m-units/rat/day i.p. from day 7-14 after hypophysectomy) on the specific activity of rat adrenal monoamine oxidase. (Mean values $\pm$ S.E.M.)

male rats of two Wistar colonies. In the first colony ('1') rats were killed between 5 and 80 days after birth, in the second colony ('2') 5 and 20 days after birth. In colony '1' there was a progressive rise in MAO activity with age when tyramine was used as substrate. MAO activity towards dopamine in the 5-, 10- and 20-day-old rats was either very low or not detectable. When 5-hydroxytryptamine (5HT) was used as substrate, in colony '1' the glands of 5- and 10-day-old rats showed higher activities than those from the 20- and 40-day-old rats. The large differences which can exist between different rat colonies are indicated by the findings in the Wistar '2' colony (Fig. 8), in which the 5-day-old rats metabolized less 5HT than the 20-day-old rats.

MAO activity was also measured in the hearts from the same rats (Fig. 9). As in the adrenal gland, the activity towards kynuramine and tyramine showed a steep rise after day 20, the activity towards 5HT a fall. Towards dopamine, very little or no activity could be detected between days 5 and 20. The ratio

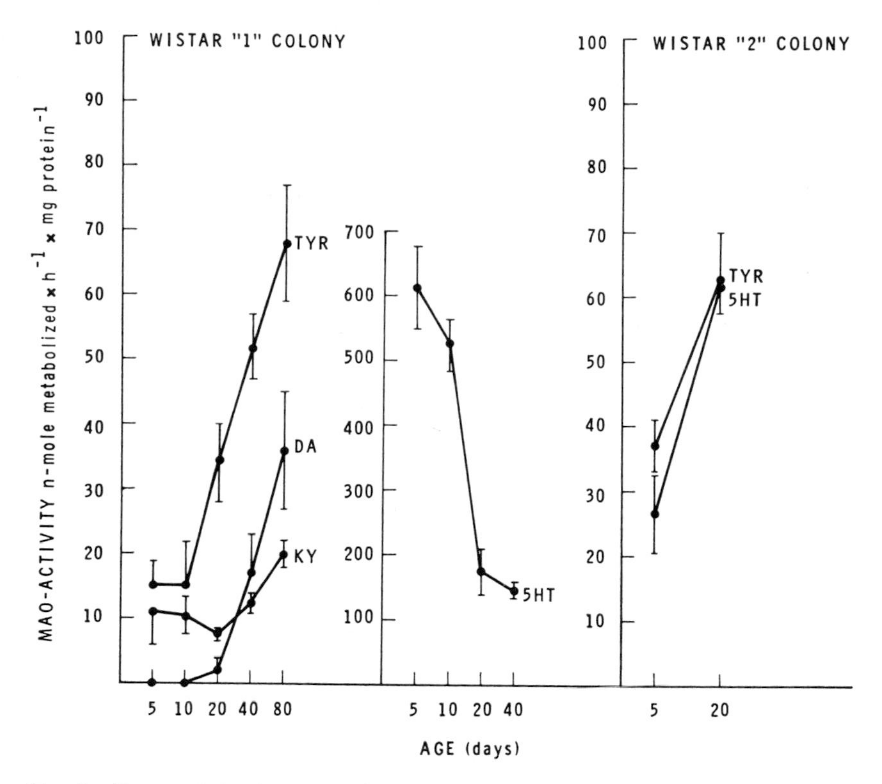

FIG. 8. Postnatal development of monoamine oxidase activity in the adrenal glands of littermate rats from two Wistar colonies. Substrates: TYR: tyramine; DA: dopamine; KY: kynuramine; 5HT: 5-hydroxytryptamine. (Mean values ± S.E.M.) No MAO activity towards dopamine could be detected in the rats of the Wistar '2' colony on days 5 and 20.

of tyramine to dopamine metabolized decreased with age, as was also the case for the adrenal gland, indicating the possibility of an independent development of different forms of MAO in this tissue. Callingham & Lyles (1975) studied MAO activity in the rat heart during development using specific enzyme inhibitors to differentiate between the A and B form of MAO (Johnston 1968). They found mainly form B (up to 100%) in the young rat (30–60 g) whereas in the adult rat (300-400 g) form A was predominant (up to 70%).

Results obtained with tyramine as substrate on the brain and liver homogenates from the rats of the Wistar '1' colony are shown in Fig. 10. The enzyme activity increased in all four brain regions and in the liver until about day 40. On day 80 the values were either similar to or lower than those on day 40, which stands in sharp contrast to the heart and adrenal glands. Very little

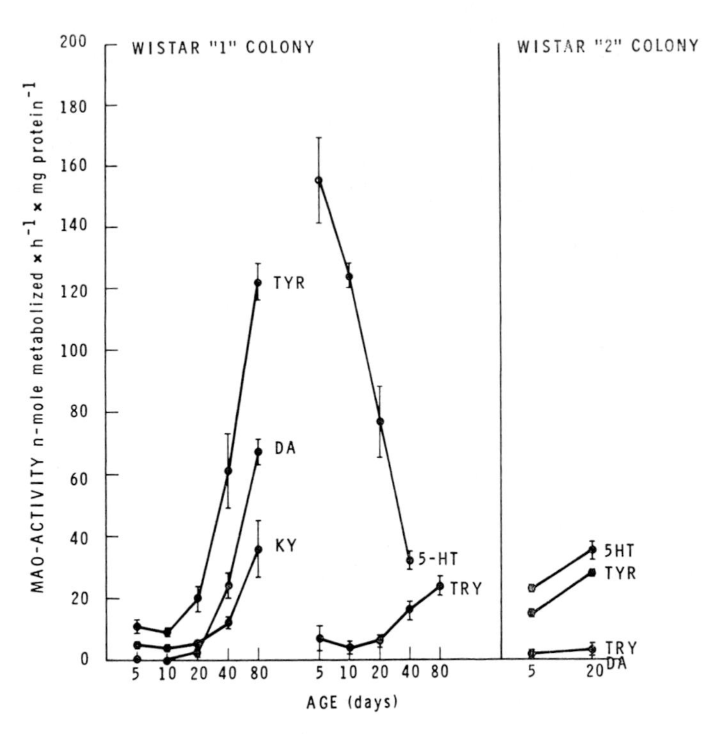

FIG. 9. Postnatal development of cardiac monoamine oxidase activity; same rats as in Fig. 8. Abbreviations as in Fig. 8. No activity towards DA in the Wistar '2' colony was detectable.

dopamine was metabolized by the brain homogenates up to day 20. However, the liver homogenates deaminated dopamine at all ages about half as efficiently as tyramine. Similar observations were made on other rat colonies (hooded rats, Porton rats and two other Wistar colonies).

In Fig. 11 the MAO activity in the hypothalamus of the Wistar '1' rats is expressed as a percentage of the highest activity on day 40. MAO activity towards tyramine and kynuramine was doubled between days 10 and 40. Dopamine activity increased fourfold during this period. The figure also includes observations made with tryptamine and 5HT as substrates. In contrast to previous reports (Karki *et al.* 1962; Bennet & Gairman 1965) the hypothalami of the 5-day-old rats of the Wistar '1' strain exhibited a very high activity which fell with age towards both indolalkylamines. This phenomenon was seen in some but not all the other colonies studied and requires further investigation.

Considerable strain differences were also seen in the development of MAO activity towards dopamine. Although very low activity was seen in the 5- and 10-day-old rats of most strains, the rate of the subsequent rise varied between

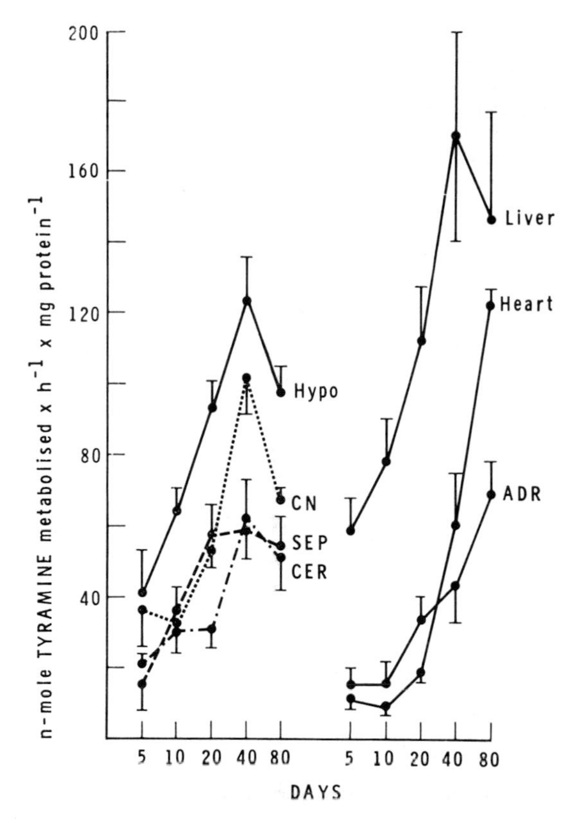

FIG. 10. Development of cerebral and hepatic monoamine oxidase activity towards tyramine in the Wistar '1' colony (same rats as Fig. 8 and 9). Hypo: hypothalamus; CN: caudate nucleus; SEP: septum; CER: cerebellum; ADR: adrenal gland.

strains. This emphasizes the danger of considering findings in a single rat colony to be representative for the whole species.

The possibility of an independent development in the brain of a form of MAO preferentially deaminating dopamine is suggested by the results in Fig. 11. If tyramine and dopamine were deaminated by the same enzyme system, one would expect the ratios of the absolute activities towards these two substrates to remain constant during development. This is in fact the case for the liver, in which it was about two at all ages and all colonies. In contrast, in the hypothalami and striata this ratio was much higher in the 5-day-old rats (between 4 and $>$ 10) than in the 20- to 80-day-old rats (approximately 2).

Evidence for differences in the localization of multiple forms of MAO in the brain of the adult rat was obtained in studies on isolated brain mitochondria.

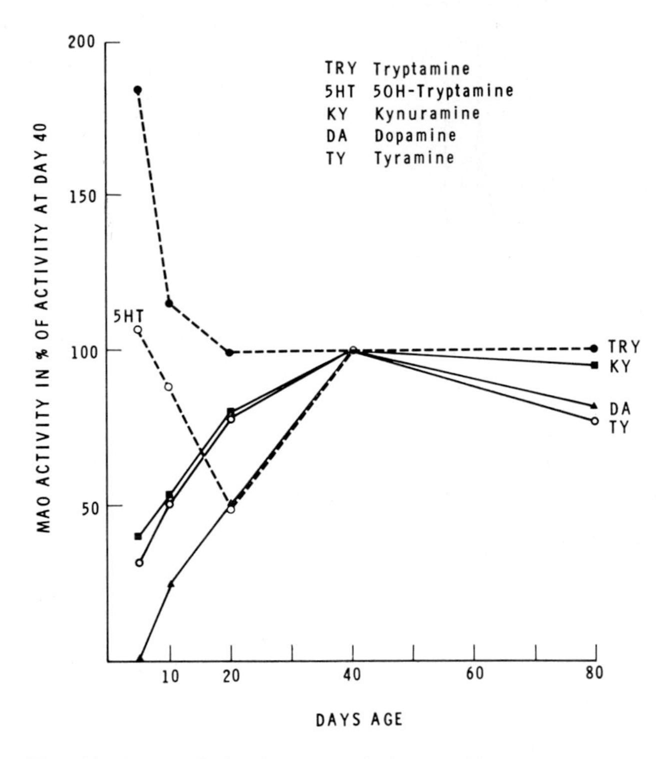

FIG. 11. Rate of development of the specific activity of monoamine oxidase towards different substrates in the hypothalamus of the rats from the Wistar '1' colony. Results expressed as % of the activity on day 40.

After their distribution on a sucrose-Ficoll gradient the peak of activity (per mg protein) towards dopamine was separated from that towards tyramine (Youdim 1974). The highest activity towards dopamine was associated with synaptosomes. The existence of distinct populations of brain mitochondria deaminating different monoamines was also shown by Kroon & Veldstra (1972).

It is tempting to suggest the existence of a specific dopamine-metabolizing enzyme in certain brain regions whose development may reflect the maturation of the neuronal system employing dopamine as the transmitter substance. No evidence has been obtained for the independent development of a dopamine-metabolizing enzyme system in the liver. There is no reason to assume that hepatic MAO is involved in neuronal functions. Its main physiological role is more likely the deamination of dietary monoamines and the surplus of biogenic amines in the blood.

Our experiments tested only the ability of tissue homogenates to deaminate dopamine *in vitro*. Whether they give a true picture of the conditions prevailing *in vivo* remains to be proved. Keller *et al.* (1973) have measured striatal homovanillic acid (HVA) during the development of rats (Wistar, Füllinsdorf colony) and found low concentrations in 4-day-old rats. However, on day 12 the striatal MAO concentration was already similar to that found in the adult rat. Because of the large strain differences seen in our experiments and the strain differences observed by Loizou (1972) in the early anatomical development of dopamine-containing fibres in the brain, we cannot make a direct comparison between our results and those of Keller *et al.* (1973). Experiments are in progress in collaboration with Dr D. F. Sharman in which HVA, dihydroxyphenylacetic acid and MAO are measured simultaneously in striata of littermate rats during development.

The increase in tissue MAO activity with age occurs concomitantly with the development of other tissue components which reach their adult composition and function only after several weeks or months. It may be that in the newborn rat relatively less enzyme protein is present. However, developmental changes of factors in the immediate environment of the enzyme which are required for its full activity, like cofactors or allotopic and allosteric effectors, will certainly contribute to the progressive rise in MAO activity in the young rat.

For example, flavin is an essential cofactor of MAO (for review see Youdim 1976). The enzyme effecting the synthesis of flavin is known to increase with age and is thought to be regulated by thyroid hormones (Rivlin *et al.* 1976). The low tissue content of flavin in the newborn rat (Kuzuya & Nagatsu 1969) may be the consequence of the lack of its synthesizing enzyme.

Of great importance for MAO activity are also the phospholipids found in juxtaposition to MAO in the outer mitochondrial membrane. Variations in phospholipids are, for example, believed to be responsible for the multiple forms of MAO, for substrate-dependent differences in the rate of its heat inactivation, and also for inhibitor specificity (Houslay & Tipton 1973; Tipton *et al.* 1973). Brain lipid concentrations are low at birth and increase gradually during the first three months. In Fig. 12 the development of MAO activity towards dopamine in the hypothalamus (Wistar '1' colony) is compared with that of brain lipids and myelin as observed by Crawford & Sinclair (1972). The lower lipid content of the tissues of the young rats may also be responsible for the lower heat stability of MAO in adrenal glands of 5-day-old rats when compared with 40-day-old rats (Fig. 13). Similar differences were seen with brain and liver.

Studies on the turnover rates of MAO in the mitochondria of different organs during development are essential for the understanding of the processes involved in the increase of the specific activity of MAO with age. However,

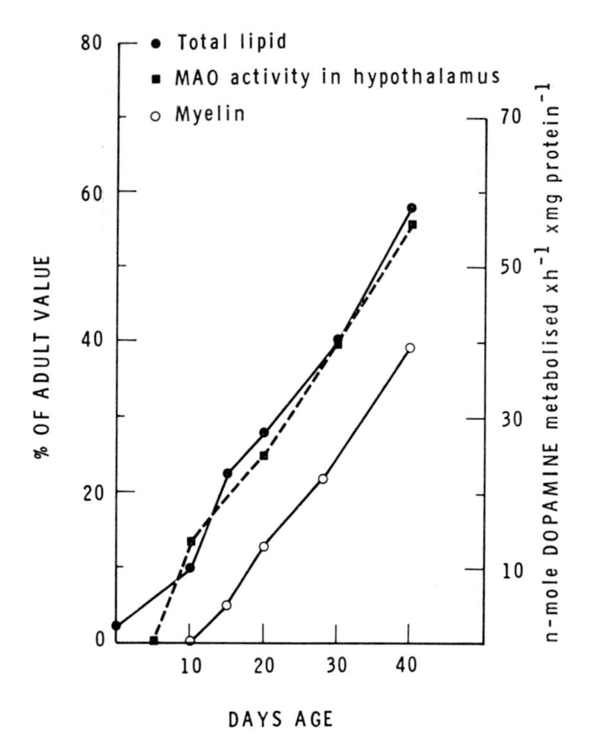

FIG. 12. Comparison between the postnatal development of brain lipids (●—●) and myelin (○—○) as observed by Crawford & Sinclair (1972) and monoamine oxidase activity towards dopamine in the Wistar '1' colony (■ - - - ■).

no information on this point is yet available. Gregson & Williams (1969) found that in the brain the number of mitochondria per g wet weight does not increase with age; however, succinate dehydrogenase activity increased fourfold and cytochrome oxidase activity sixfold. From these observations it appears that after birth the MAO activity per mitochondrion increases.

Age-dependent changes in MAO activity in the guinea pig tissues are at present disputed (Karki *et al.* 1962; Ghosh & Guha 1972). In the human a rise in brain and blood platelet MAO between 35 and 70 years has been reported by Robinson (1975).

CONCLUSIONS

Our knowledge of the physiology of monoamine oxidase and the regulation of its activity is still very sparse. There is little doubt that this enzyme plays an important role in the metabolism of biogenic amines which are transmitter

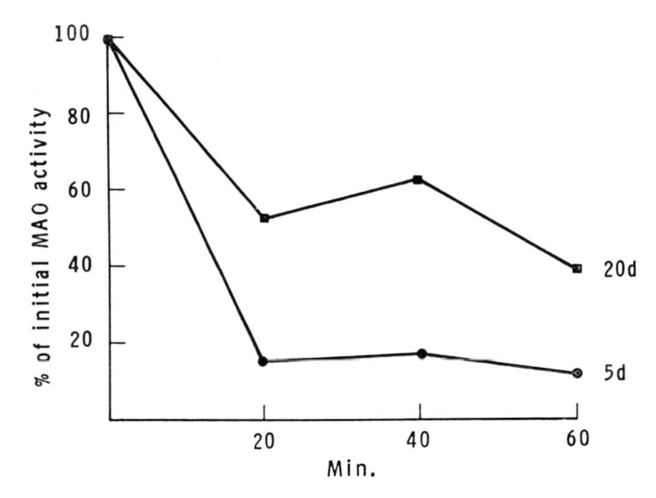

FIG. 13. Heat inactivation of adrenal monoamine oxidase in 5- and 20-day-old rats. Substrate: kynuramine. *Abscissa:* incubation time in minutes at 48 °C in Na phosphate buffer (0.05 M, pH 7.4).

substances. The behavioural effects seen after MAO inhibition show, for example, that the enzyme is involved in the regulation of nervous function (Green & Youdim, this volume, pp. 231-240). Whether the beneficial effects achieved in certain mental conditions by substances which inhibit MAO *in vitro* can be attributed to the inhibition of the enzyme *in vivo* and thus to a slower catabolism of biogenic amines in the brain is still a matter of conjecture. The crucial question whether changes in the normal rate of amine metabolism can lead to mental disturbances is still far from settled (see e.g. Murphy *et al.* 1974). The fact that some patients treated with MAO inhibitors respond to the ingestion of certain dietary amines with a large rise in blood pressure suggests that MAO plays an essential role in the disposition of these compounds.

Several physiological factors and conditions which can affect MAO activity have by now been described in experimental animals and also in humans. It remains to be seen if the changes in enzyme activity induced by hormones or those occurring during development are large enough to affect amine catabolism.

For further experimental work on the function of brain MAO it will be important to concentrate on smaller regions than those investigated in the present work. This is indicated by the fact that developmental variations in MAO activity are much more obvious in homogenates of discrete regions of the brain than in homogenates of whole brain.

Whether 'multiple forms' of MAO are due to the existence of different molecular forms of the enzyme or to the presence of different allotopic or allosteric effectors in the environment of the enzyme seems to be of minor physiological importance. The fact is that the so-called multiple forms exert independent functions under certain biological conditions. For example, their activity can be affected independently by steroid hormones (see Fig. 4, p. 111) and there are differences in the rate of their postnatal development.

Recent experiments on the catabolism of biogenic amines (5-hydroxytryptamine, dopamine and phenylethylamine) in isolated perfused rat lungs have also shown that the differentiation of MAO into 'type A' and 'type B' may be of physiological significance (Bakhle & Youdim 1975, 1976). The demonstration that types A and B MAO can operate independently in an organized tissue makes it likely that these two enzyme activities play a real and independent role *in vivo* not only in the lungs but also in other tissues, including the brain.

ACKNOWLEDGEMENT

We would like to thank Mr D. Blatchford for his skilful technical assistance.

References

ALEYASSINE, H. & GARDINER, R. J. (1975) Dual action of antidepressant drugs (MAO inhibitors) on insulin release. *Endocrinology 96*, 702-710

AVAKIAN, V. M. & CALLINGHAM, B. A. (1968) An effect of adrenalectomy upon catecholamine metabolism. *Br. J. Pharmacol. Chemother. 33*, 211*P*

BAILEY, A. R., CROW, T. J., JOHNSTONE, E. C. & OWEN, F. (1975) Platelet monoamine oxidase activity in patients with chronic schizophrenic illnesses untreated by neuroleptic drugs. *Br. J. Clin. Pharmacol. 2*, 380-381*P*

BAKHLE, Y. S. & YOUDIM, M. B. H. (1975) Inactivation of phenylethylamine and 5-hydroxytryptamine in rat isolated lungs: evidence for monoamine oxidase A and B in lung. *J. Physiol. (Lond.) 248*, 23-25*P*

BAKHLE, Y. S. & YOUDIM, M. B. H. (1976) Metabolism of phenylethylamine in rat perfused lung: evidence for monoamine oxidase 'type' B in lung. *Br. J. Pharmacol. 56*, 125-127

BELMAKER, R. H., MURPHY, D. L., WYATT, R. J. & LORIAUX, D. L. (1974) Human platelet monoamine oxidase changes during the menstrual cycle. *Arch. Gen. Psychiatry 31*, 553-556

BENNET, D. S. & GAIRMAN, N. J. (1965) Schedule of appearance of 5-hydroxytryptamine (serotonin) and associated enzymes in the developing rat brain. *J. Neurochem. 11*, 911-918

BHAGAT, B., BRYAN, R. J. & LEE, Y. C. (1973) Increase in adrenal monoamine oxidase activity in hypophysectomized rats after ACTH. *Neuropharmacology 12*, 1199-1202

BLASCHKO, H. (1974) The natural history of amine oxidases. *Rev. Physiol. Biochem. Pharmacol. 70*, 83-148

BLASCHKO, H., HAGEN, P. & WELCH, A. D. (1955) Observations on the intracellular granules of the adrenal medulla. *J. Physiol. (Lond.) 129*, 27-49

CALLINGHAM, B. A. & LYLES, G. A. (1975) Some effects of age upon irreversible inhibition of cardiac MAO. *Br. J. Pharmacol. 53*, 258-259*P*

CEASAR, P. M., COLLINS, G. G. S. & SANDLER, M. (1970) Catecholamine metabolism and monoamine oxidase activity in adrenalectomized rats. *Biochem. Pharmacol. 19*, 921-926

COLLINS, G. G. S., PRYSE-DAVIES, J., SANDLER, M. & SOUTHGATE, J. (1970) Effect of pretreatment with estradiol, progesterone and DOPA on monoamine oxidase activity in the rat. *Nature (Lond.) 226*, 642-643

CRAWFORD, M. A. & SINCLAIR, A. J. (1972) Nutritional influences in the evolution of mammalian brain, in *Lipids, Malnutrition and the Developing Brain (Ciba Found. Symp. 3)*, pp. 267-292, Associated Scientific Publishers, Amsterdam

FAJER, A. B., HOLZBAUER, M. & NEWPORT, H. M. (1971) The contribution of the adrenal gland to the total amount of progesterone produced in the female rat. *J. Physiol. (Lond.) 214*, 115-126

GHOSH, S. K. & GUHA, S. R. (1972) Oxidation of monoamines in developing rat and guinea pig brain. *J. Neurochem. 19*, 229-231

GRAHAM, C. W., GREEN, A. R., WOODS, H. F. & YOUDIM, M. B. H. (1975) 5-Hydroxytryptamine synthesis in the isolated, perfused rat brain. *Br. J. Pharmacol. 53*, 250-251*P*

GREEN, A. R. & YOUDIM, M. B. H. (1975) This volume, pp. 231-240

GREGSON, N. A. & WILLIAMS, P. L. (1969) A comparative study of brain and liver mitochondria from new born and adult rats. *J. Neurochem. 16*, 617-626

HARE, M. L. C. (1928) Tyramine oxidase. I. A new enzyme system in liver. *Biochem. J. 22*, 968-979

HOLZBAUER, M., BULL, G., YOUDIM, M. B. H., WOODING, F. B. P. & GODDEN, U. (1973) Subcellular distribution of steroids in the adrenal gland. *Nature New Biol. 242*, 117-119

HOLZBAUER, M. & YOUDIM, M. B. H. (1972) The influence of endocrine glands on central and peripheral monoamine oxidase activity. *Br. J. Pharmacol. 44*, 355-356*P*

HOLZBAUER, M. & YOUDIM, M. B. H. (1973) The oestrous cycle and monoamine oxidase activity. *Br. J. Pharmacol. 48*, 600-608

HOUSLAY, M. D. & TIPTON, K. F. (1973) The nature of the electrophoretically separable multiple forms of rat liver monoamine oxidase. *Biochem. J. 135*, 173-186

HUANG, C. L. & SCHULZ, A. R. (1972) The effect of inhibitors of thyroid MAO on the incorporation of iodide into thyroid slice protein. *Life Sci. (Oxford) 11*, 975-982

JOHNSTON, J. P. (1968) Some observations upon a new inhibitor of monoamine oxidase in brain tissue. *Biochem. Pharmacol. 17*, 1285-1297

KARKI, N., KUNTZMAN, R. & BRODIE, B. B. (1962) Storage synthesis and metabolism of monoamines in the developing brain. *J. Neurochem. 9*, 53-58

KELLER, H. H., BARTHOLINI, G. & PLETSCHER, A. (1973) Spontaneous and drug-induced changes of cerebral dopamine turnover during postnatal development of rats. *Brain Res. 64*, 371-378

KROON, M. C. & VELDSTRA, H. (1972) Multiple forms of rat brain mitochondrial monoamine oxidase. Subcellular localization. *FEBS Lett. 24*, 173-176

KUZUYA, H. & NAGATSU, T. (1969) Flavins and monoamine oxidase activity in the brain, liver and kidney of the developing rat. *J. Neurochem. 16*, 123-126

LEMAY, A. & BERLINGUET, L. (1967) Influence de la castration et des hormones sexuelles sur l'activité de la monoamine oxydase chez le rat. *Rev. Can. Biol. 26*, 295-303

LOIZOU, L. A. (1972) The postnatal ontogeny of monoamine containing neurones in the central nervous system of the albino rat. *Brain Res. 40*, 395-418

LUINE, V. N., KHYLCHEVSKAYA, R. I. & MCEWEN, B. S. (1975) Effect of gonadal steroids on activities of monoamine oxidase and choline acetylase in rat brain. *Brain Res. 86*, 293-306

LYLES, G. A. & CALLINGHAM, B. A. (1974) The effect of thyroid hormones on monoamine oxidase in the rat heart. *J. Pharm. Pharmacol. 26*, 921-930

MURPHY, D. L., BELMAKER, R. & WYATT, R. J. (1974) MAO in schizophrenia and other behavioral disorders. *J. Psychiatr. Res. 11*, 221-247

NOVICK, W. J., JR (1961) The effect of age and thyroid hormones on the monoamine oxidase of rat heart. *Endocrinology 69*, 55-59

O'MALLEY, B. W. & SCHRADER, W. T. (1972) Progesterone receptor components: identification of subunits binding to the target-cell genome. *J. Steroid Biochem. 3*, 617-629

PARVEZ, H. & PARVEZ, S. (1973) The effects of metopirone and adrenalectomy on the regulation of the enzymes monoamine oxidase and catechol-*O*-methyl transferase in different brain regions. *J. Neurochem. 20*, 1011-1020

PARVEZ, S., PARVEZ, S. H. & YOUDIM, M. B. H. (1975) Variation in activity of monoamine metabolizing enzymes in rat liver during pregnancy. *Br. J. Pharmacol. 53*, 241-246

RIVLIN, R. S., FAZEKAS, A. G., HUANG, Y. P. & CHAUDHURI, R. (1976) Hormonal control of flavin metabolism, in *Flavins and Flavoproteins* (Singer, T. P., ed.), Elsevier Scientific Publishing Company, Amsterdam, in press

ROBINSON, D. S. (1975) Changes in monoamine oxidase and monoamines with human development and aging. *Fed. Proc. 34*, 103-107

SANDLER, M. & YOUDIM, M. B. H. (1972) Multiple forms of monoamine oxidase: functional significance. *Pharmacol. Rev. 24*, 331-349

SCHNAITMAN, C., ERWIN, V. G. & GREENAWALT, J. W. (1967) Submitochondrial localization of monoamine oxidase. *J. Cell Biol. 34*, 719-735

SOUTHGATE, J., GRANT, E. C. G., POLLARD, W., PRYSE-DAVIES, J. & SANDLER, M. (1968) Cyclic variation in endometrial monoamine oxidase: correlation of histochemical and quantitative biochemical assays. *Biochem. Pharmacol. 17*, 721-726

TIPTON, K. F., YOUDIM, M. B. H. & SPIRES, I. P. C. (1972) Beef adrenal medulla monoamine oxidase. *Biochem. Pharmacol. 21*, 2197-2204

TIPTON, K. F., HOUSLAY, M. D. & GARRETT, N. J. (1973) Allotopic properties of human brain monoamine oxidase. *Nature New Biol. 246*, 213-214

VACCARI, A., MAURA, M., MARCHI, M. & CUGURRA, A. F. (1972) Development of monoamine oxidase in several tissues in the rat. *J. Neurochem. 19*, 2453-2457

WURTMAN, R. J. & AXELROD, J. (1966) Control of enzymatic synthesis of adrenaline in the adrenal medulla by adrenal cortical steroids. *J. Biol. Chem. 241*, 2301-2305

YAGO, N. & ICHII, S. (1969) Submitochondrial distribution of components of the steroid-11β-hydroxylase and cholesterol sidechain-cleaving enzyme system in hog adrenal cortex. *J. Biochem. (Tokyo) 65*, 215-224

YOUDIM, M. B. H. (1974) Heterogeneity of rat brain mitochondrial monoamine oxidase. *Adv. Biochem. Psychopharmacol. 11*, 59-63

YOUDIM, M. B. H. (1975) Assay and purification of brain monoamine oxidase, in *Research Methods in Neurochemistry* (Marks, N. & Rodnight, R., eds.), pp. 167-200, Plenum Press, New York

YOUDIM, M. B. H. (1976) Rat liver mitochondrial monoamine oxidase – an iron requiring flavoprotein, in *Flavins and Flavoproteins* (Singer, T. P., ed.), Elsevier Scientific Publishing Company, Amsterdam, in press

YOUDIM, M. B. H. & WOODS, H. F. (1975) The influence of tissue environment on the rates of metabolic processes and the properties of enzymes. *Biochem. Pharmacol. 24*, 317-323

YOUDIM, M. B. H., HOLZBAUER, M. & WOODS, H. F. (1974) Physicochemical properties, development and regulation of central and peripheral monoamine oxidase activity, in *Neuropsychopharmacology of Monoamines and their Regulatory Enzymes* (Usdin, E., ed.), pp. 11-28, Raven Press, New York

YOUDIM, M. B. H., WOODS, H. F., MITCHELL, B., GRAHAME-SMITH, D. G. & CALLENDER, S. (1975) Human platelet monoamine oxidase in iron-deficiency anaemia. *Clin. Sci. Mol. Med. 48*, 289-295

ZILE, M. H. (1960) The effect of thyroxine and related compounds on monoamine oxidase activity. *Endocrinology 66*, 311-312

Discussion

Blaschko: You discussed the effects of hormones on MAO activity, Dr Youdim. One point seems to be quite important and has to be considered when one thinks of nervous tissue, especially in connection with the stimulation of enzyme formation. It concerns the dimensions of the cell. If you assume that a hormonal effect is one which turns on enzyme formation, you would pick up the increased activity in the neighbourhood of the nerve *endings* after a time lag. I was interested, therefore, that there were some areas in the brain in which there was evidence of a lag period between the application of progesterone and MAO activity. However, the times involved were rather larger than one would expect from what is known of the rate of mitochondrial transport. Nevertheless, one should keep this point in mind when comparing say, the adrenal cortical cell, which has no particular dimensions, and an area of the brain where any increase in activity could only show upon arrival of the mitochondria which carry the enzyme, the synthesis of which may have been stimulated some days before.

Youdim: In our studies of the adrenal gland, there was a lag between the adrenal content of progesterone and the MAO activity.

Sandler: May I ask about your confirmatory experiments done to check MAO activity in pregnancy, in which you injected catecholamines? You measured the excretion of 4-hydroxy-3-methoxymandelic acid (vanillylmandelic acid), but this is only a minor metabolite of noradrenaline in the rat (Ceasar *et al.* 1969); 4-hydroxy-3-methoxyphenylglycol is the major metabolite. Your findings may therefore not have reflected the true changes. I was also worried about this because when one gives noradrenaline and also MAO inhibitors, in man, for instance, there is no increase in pressor activity (Horwitz *et al.* 1960). Only when you give tricyclic drugs and noradrenaline do you get potentiation (Svedmyr 1968). MAO inhibitors as such don't seem to affect the metabolism of intravenously administered catecholamines very much, to judge indirectly from these effects.

Youdim: Yes. After these initial experiments we felt that some other substrate, such as 5-hydroxytryptamine (5HT), should have been used.

Fuller: When you looked at MAO in various brain regions during the oestrous cycle in the rat, you used kynuramine as substrate, and you found the lowest MAO activity during late oestrus. In an early study, Zolovick *et al.* (1966) reported high MAO activity on the day of oestrus using 5HT as substrate. Dick Wurtman and I (unpublished studies) have independently tried to repeat that, and we never found any change in MAO in any of the three brain regions that they studied. You are now finding directly opposite changes, but with a different substrate, and I wonder if you have looked at any other substrates?

Youdim: No, we have not. This is an important point. In fact Kamberi & Kobayashi (1970) have reported cyclic variations using tyramine as a substrate. The differences may have arisen because they did not measure monoamine oxidase activity during the dark period, where we find a fall in MAO activity. They measured MAO activity during the light periods only. This, together with the use of different substrates, may explain some of the differences.

I have always stressed the importance of using more than one substrate for this type of study.

Murphy: You used five substrates when you directly studied progesterone and oestradiol, but you didn't use phenylethylamine or benzylamine.

Youdim: This was simply because we have never really believed that one should use the criteria of MAO type A and B. But we haven't used phenylethylamine, and maybe we should.

Murphy: Those observations were all made in the adrenal gland. Do you have observations in other tissues?

Youdim: No. Professor Sandler's group has reported similar results for the uterus.

Neff: The adrenal gland releases catecholamines into the circulation. What would you imagine the change in MAO is related to in this organ?

Youdim: I wouldn't like to say, because we haven't looked at the catecholamines. This type of observation is badly needed, however. The problem is of being able to study the enzyme, the various amines and their metabolites in the same tissue homogenate. You can imagine how difficult it is if a small brain area such as the striatum is used.

Neff: You stress MAO; however, it is just a marker for mitochondria. Did you look at other enzymes associated with mitochondria?

Youdim: We did not look at other enzymes but it is known that succinate dehydrogenase also changes with age and during the oestrous cycle.

Neff: Do the changes run parallel?

Youdim: I don't know.

Neff: So the changes you measured would not be directly related to MAO, but rather to the development of mitochondria.

Youdim: We have never said that they are *directly* related to MAO.

Fuller: Dr Youdim, you observed a decrease in tryptamine oxidation at early times which coincided somewhat with the results for 5HT but not with any of the other substrates that you looked at. We discussed earlier (p. 27) whether tryptamine is a type A or type B substrate. Could you comment on those data in regard to that?

Youdim: In our hands tryptamine behaves as a substrate for MAO type B. We have studied the development of MAO activity in five different strains of rats and we wanted to report on the one strain which was different to the previously reported results. We have a strain of rats in which 5HT and tryptamine deamination increased with age.

Gorkin: Did I understand you to say that progesterone doesn't affect the activity of purified MAO?

Youdim: That is correct. Some years ago Dr Collins was studying the effect

Neff: We have shown that sonication does not affect the relative activities of A and B enzyme. That is, the ratio of A to B activity is similar in a crude homogenate and in a sonicated preparation (Goridis & Neff 1971).

Iversen: So there appears to be a differential distribution of the A and B forms between mitochondria inside synaptosomes and free mitochondria?

Youdim: There appears to be a difference in substrate specificities between mitochondria inside synaptosomes and free mitochondria (Youdim 1973). I should mention that our preliminary results have shown that A and B type enzymes are present in all the particulate fractions tested. Our studies have shown that mitochondria are heterogeneous with regard to MAO, succinate dehydrogenase and NADH dehydrogenase. Some mitochondria may have more of one enzyme than another but I don't know whether it has anything to do with the synaptosomes.

Neff: All I can say from our studies is that there is a heterogeneous population of mitochondria, and that some mitochondria have more of one type of enzyme than another type. But I could not say that the distribution is related to free mitochondria or mitochondria in synaptosomes.

van Praag: Hendley & Snyder in 1968 reported that certain MAO inhibitors, apart from being capable of inhibiting MAO, were inhibitors of the re-uptake of monoamines. They said that there was an even closer correlation between re-uptake inhibition by MAO inhibitors and clinical efficacy, than between MAO inhibition and clinical effectiveness.

Sandler: Iproniazid didn't fit in with the data at all, however.

van Praag: No, but I wonder if their observations are still valid, and could they have influenced your findings on the effect of MAO inhibitors on the metabolism of monoamines, Dr Maître?

Maître: Dr Iversen showed that various MAO inhibitors do inhibit uptake, not as powerfully as the tricyclic antidepressants, but they do, particularly tranylcypromine (Iversen 1965). We have tried to look at the inhibition of noradrenaline uptake in the rat brain *in vivo*. We could not demonstrate potent inhibition of noradrenaline uptake after subcutaneous injection of various MAO inhibitors of different chemical classes, like iproniazid, tranylcypromine and pargyline. In no case was there a significant and dose-dependent inhibition of noradrenaline uptake.

Knoll: (+)-Deprenyl inhibits noradrenaline uptake very efficiently and tranylcypromine also does so quite strongly. We are therefore using (—)-deprenyl, because it does not do this. All MAO inhibitors of course have many other effects beside MAO inhibition; if pharmacologists are aware of them, they have to take them into consideration.

Sandler: Since we are broadening the discussion to include inhibition of uptake, we might also discuss an even more fascinating property of MAO and its inhibitors. I am thinking of Cotzias's work (Cotzias *et al.* 1974) on a possible transport function of MAO. He claims that giving MAO inhibitors to an experimental animal facilitates the transport of amines, not only of dopamine and methylene blue but also of apomorphine, which may be bound but not metabolized by MAO.

Rafaelsen: Cotzias *et al.* (1974) investigated whether methylene blue, which is bound by monoamine oxidase, would be transported into the brain, if an MAO inhibitor (nialamide) had been given beforehand (in contrast to Evans blue, which is bound to albumin and would not be expected to be transported into the brain). Cotzias showed histologically that there was uptake after treatment with nialamide. We were interested in this question but when we tried to repeat the experiment there was no staining of the brain. Moreover, if you calculate the dose of nialamide per kilogram weight, it is around 100 times greater than that given to a human being on a daily basis, and even related to metabolism it is a ten-fold higher dose. Our mice did not like it, and indeed some of them died. So I think the transfer of methylene blue from blood to brain may be a terminal phenomenon at a time when the blood–brain barrier is broken down.

Coppen: We are interested in the possible relevance of your observations to the human situation, Dr Youdim. There is a dramatic change in progesterone and oestrogens around the time of the menopause in women. There is a sudden decline in both steroids. Is there any comparable change before and after the menopause in MAO in any tissues?

Murphy: Robinson *et al.* (1971) reported that MAO keeps increasing with age in man, but in our data we haven't seen much of an increase in platelet MAO with age in either sex. We looked specifically at pre-menopausal and post-menopausal women, about 25 in each group, and found no differences. But I was surprised at our finding, because D. E. Redmond has worked with us, attempting to follow up some of the studies of platelet MAO in the rhesus monkey (Redmond *et al.* 1975, 1976). In female animals, bilateral ovariectomy was associated with a significant increase in platelet MAO activity but no change in the plasma MAO activity. In male animals, castration was also associated with some increase in platelet MAO activity, but the comparisons with controls were not statistically significant. However, changes in platelet MAO activity did seem to follow changes in testosterone levels in a group of animals compared during the mating season and the non-mating season. Testosterone levels dropped approximately 50% during the non-mating season, while platelet MAO activity was significantly increased. There may also be an interaction between age and development with the sex steroids, because a moderately sized positive correlation between sexual maturity and platelet MAO activity was found in these studies.

In regard to the menstrual cycle changes discussed by Dr Youdim, platelet MAO activity in both man and the rhesus monkey exhibits a pre-ovulatory increase in activity. The trough comes during the post-ovulation period, and this is in inverse relation to the progesterone levels. There are many possible explanations: an effect of light, an effect of the hormone early during the development of the platelet in the bone marrow, which is only seen a week or so later in our measurements on circulating platelets, or an effect of the B enzyme in platelets on the substrates we studied.

Youdim: We have reported that ovariectomy in rats results in the reduction of MAO activity in the brain areas tested.

Murphy: That is why I was very interested in the B substrates, because of course here we are working with platelets and their MAO B activity.

Youdim: However, the platelet MAO may not be the same as the brain enzyme. It might only be a B type enzyme.

Murphy: I agree. We can't extrapolate from one tissue to another.

Kety: Do you have enough data on platelet MAO in pre-puberal and post-puberal individuals, and is there any significant difference?

Murphy: We find a significant sex difference before puberty, with higher MAO levels in platelets in females. It does look as if there is some interaction of sex steroids with platelet MAO.

Sandler: We got results opposite to yours, Dr Murphy—a slight but significant rise in platelet MAO in the latter half of the menstrual cycle. In fact, we found a pattern fitting rather well with our earlier endometrial work (Southgate *et al.* 1968). Our results also differed from yours, Dr Coppen (Robinson *et al.* 1972), in that with an indirect method of measuring MAO activity we were unable to detect any age changes (Sandler *et al.* 1975).

Coppen: Have you any correlative data on the effect of steroids? Does the platelet MAO follow the brain or other organs as far as steroid-induced changes are concerned, Dr Youdim?

Youdim: We haven't looked at this.

Coppen: It would be very useful for the clinical field if the platelets could be used as an indicator of changes in the brain enzyme.

Youdim: I have already mentioned that the platelet MAO may be different from the brain or liver enzyme in substrate and inhibitor specificity. Some years ago Collins & Sandler (1971) found that, unlike the brain and liver enzyme, the platelet MAO migrated as a single band electrophoretically.

Sandler: I should add now that there was a second very labile band which made its appearance if we solubilized the platelet enzyme and ran it immediately by electrophoresis (Henley 1973). If we kept the preparation overnight, however, there was only the single band.

Neff: Has anyone studied platelet MAO activity in relation to platelet number, rather than per mg of platelet protein? Do results expressed in other ways yield similar values?

Murphy: We have been doing that for the last four years, as well as measuring protein, with very similar results. The MAO assay method is slightly more replicable if we use the platelet number, because there are several ways in which 'platelet' protein values can be contaminated by leucocytes, erythrocytes, fibrinogen and other proteins in blood.

Youdim: We have tried to measure MAO in terms of platelet count in our iron deficiency studies, and it made our results even more significant, because in iron deficiency the number of platelets increases!

Neff: Is there a change of platelet size?

Youdim: We are now measuring the platelets' size and studying their distribution on sucrose density gradients.

Fuller: Did you measure platelet 5HT levels in your iron deficiency studies?

Youdim: Not yet; we are doing it now.

References

ANDEN, N.-E., ROOS, B. E. & WERDINIUS, B. (1963) On the occurrence of homovanillic acid in brain and cerebrospinal fluid and its determination by a fluorimetric method. *Life Sci. 7*, 448-458

CEASAR, P. M., RUTHVEN, C. R. J. & SANDLER, M. (1969) Catecholamine and 5-hydroxyindole metabolism in immunosympathectomized rats. *Br. J. Pharmacol. 36*, 70-78

COLLINS, G. G. S. & SANDLER, M. (1971) Human blood platelet monoamine oxidase. *Biochem. Pharmacol. 20*, 289-296

COTZIAS, G. C., TANG, L. C. & GINOS, J. Z. (1974) Monoamine oxidase and cerebral uptake of dopaminergic drugs. *Proc. Natl. Acad. Sci. U.S.A. 71*, 2715-2719

GORIDIS, C. & NEFF, N. H. (1971) Evidence for a specific monoamine oxidase associated with sympathetic neurons. *Neuropharmacology 10*, 557-564

HENDLEY, E. & SNYDER, S. H. (1968) Relationship between the action of monoamine oxidase inhibitors on the noradrenaline uptake system and their antidepressant efficacy. *Nature (Lond.) 220*, 1330-1331

HENLEY, C. (1973) Studies on monoamine oxidase and its multiple forms in human and animal tissues, Ph.D. Thesis, University of London

HORWITZ, D., GOLDBERG, L. I. & SJOERDSMA, A. (1960) Increased blood pressure responses to dopamine and norepinephrine produced by monoamine oxidase inhibitors in man. *J. Lab. Clin. Med. 56*, 747-753

IVERSEN, L. L. (1965) The inhibition of noradrenaline uptake by drugs. *Adv. Drug Res. 2*, 5-23

KAMBERI, I. A. & KOBAYASHI, Y. (1970) Monoamine oxidase activity in the hypothalamus and various other brain areas and in some endocrine glands of the rat during the estrus cycle. *J. Neurochem. 17*, 261-268

MURPHY, G. F., ROBINSON, D. & SHARMAN, D. F. (1969) The effect of tropolone on the formation of 3,4-dihydroxyphenylacetic acid and 4-hydroxy-3-methoxyphenylacetic acid in the brain of the mouse. *Br. J. Pharmacol. 36*, 107-115

REDMOND, D. E., MURPHY, D. L., BAULU, J., ZIEGLER, M. G. & LAKE, C. R. (1975) Menstrual cycle and ovarian hormone effects on plasma and platelet monoamine oxidase (MAO) and plasma dopamine-beta-hydroxylase (DBH)) activities in the Rhesus monkey. *Psychosom. Med. 37*, 417

REDMOND, D. E., BAULU, J., MURPHY, D. L., LORIAUX, D. L., ZIEGLER, M. G. & LAKE, C. R. (1976) The effects of testosterone on plasma and platelet monoamine oxidase (MAO) and plasma dopamine-B-hydroxylase activities in the male Rhesus monkey. *Psychosom. Med.* in press

ROBINSON, D. S., DAVIS, J. M., NIES, A., RAVARIS, C. L. & SYLWESTER, D. (1971) Relation of sex and aging to monoamine oxidase activity of human brain, plasma, and platelets. *Arch. Gen. Psychiatry 24*, 536-539

ROBINSON, D. S., DAVIS, J. M., NIES, A., COLBURN, R. W., DAVIS, J. N., BOURNE, H. R., BUNNEY, W. E., SHAW, D. M. & COPPEN, A. J. (1972) Ageing, monoamines and monoamine oxidase levels. *Lancet 1*, 290-291

SHARMAN, D. F. (1971) Methods of determination of catecholamines and their metabolites, in *Methods of Neurochemistry* (Fried, R., ed.), vol 1, pp. 83-127, Dekker, New York

SANDLER, M., BONHAM CARTER, S., CUTHBERT, M. F. & PARE, C. M. B. (1975) Is there an increase in monoamine oxidase activity in depressive illness? *Lancet 1*, 1045-1049

SOUTHGATE, J., GRANT, E. C., POLLARD, W., PRYSE-DAVIES, J. & SANDLER, M. (1968) Cyclical variations in endometrial monoamine oxidase: correlation of histochemical and quantitative biochemical assays. *Biochem. Pharmacol. 17*, 721-726

SVEDMYR, N. (1968) The influence of a tricyclic antidepressant agent (protriptyline) on some of the circulatory effects of noradrenaline and adrenaline in man. *Life Sci. 7*, 77-84

WALDMEIER, P. C., DELINI-STULA, A. & MAÎTRE, L. (1976) Preferential deamination of dopamine by an A type monoamine oxidase in rat brain. *Naunyn-Schmiedeberg's Arch. Pharmacol. 292*, 9-14

YOUDIM, M. B. H. (1973) Heterogeneity of rat brain and liver mitochondrial monoamine oxidase: subcellular fractionation. *Biochem. Soc. Trans. 1*, 1126-1127

ZOLOVICK, A. J., PEARSE, R., BOEHLKE, K. W. & ELEFTHERIOU, B. E. (1966) Monoamine oxidase activity in various parts of the rat brain during the estrous cycle. *Science (Wash. D.C.) 154*, 649

Analysis of the pharmacological effects of selective monoamine oxidase inhibitors

J. KNOLL

Department of Pharmacology, Semmelweis University of Medicine, Budapest

Abstract The simultaneous oxidation of labelled monoamines in the presence and absence of selective monoamine oxidase (MAO) inhibitors (deprenyl and clorgyline) by rat liver and brain mitochondrial MAO was studied with the double-labelling technique. 5-Hydroxytryptamine (5HT), phenylethylamine (PEA), benzylamine (BZA) and tyramine (TYR) inhibit the oxidation of each other. According to the Lineweaver-Burk analysis the inhibition was found to be non-competitive between 5HT and BZA, and competitive between BZA and PEA, TYR and BZA, and 5HT and TYR.

The nictitating membrane contracting effects of intravenous 5HT, tryptamine, PEA, TYR and BZA in cats before and after exclusion of the liver from the circulation and under the influence of MAO inhibitors (tranylcypromine, nialamide, pargyline, (—)-deprenyl and clorgyline) were measured. Taking the dose-related increase of the effects of an amine as an index of the *in vivo* rate of biotransformation, the existence of two different enzymes, namely phenylethylamine oxidase, selectively inhibited by (—)-deprenyl, and 5-hydroxytryptamine oxidase, selectively blocked by clorgyline, was deduced.

Striking differences in the effects of (–)-deprenyl and clorgyline on isolated organs (guinea pig ileum and vas deferens, cat papillary muscle) supported this view. Clorgyline was found to possess a high affinity for the 5HT receptor, blocking it in a non-competitive manner.

Characteristic differences in the responsiveness of the field-stimulated isolated guinea pig and rat vas deferens were demonstrated. Noradrenaline appears to be the transmitter in both organs but profound differences in the transmission machinery seem to exist. In the 'pure' system of the rat an unrestricted noradrenaline release from the granules follows nerve stimulation and in the 'mixed' system of the guinea pig vas deferens stimulation is supposed to release both noradrenaline and PEA from the granules, and the latter may increase the outflow of noradrenaline from a cytoplasmic pool. Prostaglandin restricts the release of noradrenaline in the 'mixed' system only. There is a cholinergic modulation of noradrenergic transmission in the rat vas deferens, and in the guinea pig vas 5HT also serves as a modulator.

The therapeutic aspects of the selective inhibitors were evaluated. (—)-Deprenyl combined with phenylalanine is hoped to be efficient in certain types of depression and combined with L-dopa in Parkinson's disease. In contrast to clorgyline and other MAO inhibitors, (−)-deprenyl inhibits MAO activity slightly in the intestine, antagonizes the effects of tyramine and inhibits the outflow of [^{3}H]noradrenaline from brain synaptosomes; consequently the absence of hypertonic crises in patients treated with (—)-deprenyl can be expected.

Since the first demonstration of the efficacy of iproniazid in psychotic depression (Kline 1958) and the observation that when this drug is used in psychiatry postural hypotension is a frequent side-effect (cf. Goldberg 1964), the therapeutic uses of monoamine oxidase inhibitors have been related to their effects on the central nervous and cardiovascular systems (monoamine oxidase—amine: oxygen oxidoreductase [deaminating] [flavin-containing]; EC 1.4.3.4) (MAO). It now seems justified to attribute the therapeutic effects of the MAO inhibitors to multiple effects of these drugs on the action and economy of biogenic amines (Knoll & Magyar 1972).

The sometimes fatal acute hypertensive reactions following the interaction of MAO inhibitors with a variety of pressor drugs and foods (Blackwell *et al.* 1967) undermined confidence in these drugs and in most countries there has been a great deal of flux in the introduction and withdrawal of different MAO inhibitors in the last decade. The facts that multiple forms of MAO exist (cf. Youdim 1972, 1974) and that we possess such highly selective MAO inhibitors as deprenyl (Knoll *et al.* 1965) and clorgyline (Johnston 1968) offer unexploited potentialities which hold out new hopes for experimental and clinical research. This study is an example of this approach.

THE FEASIBILITY OF TWO FUNCTIONALLY DIFFERENT MONOAMINE OXIDASES: PHENYLETHYLAMINE AND 5-HYDROXYTRYPTAMINE OXIDASES

It now seems well-established that structurally different MAO inhibitors differ from each other in their ability to inhibit the oxidation of various MAO substrates (cf. Costa & Sandler 1972). To analyse this phenomenon we used two selective inhibitors: (−)-deprenyl and clorgyline. Deprenyl, especially the (—)isomer, developed by us (Knoll *et al.* 1964, 1965, 1968; Magyar *et al.* 1967; Knoll & Magyar 1972), was found to be the *only* potent selective inhibitor of the so called B type MAO, and clorgyline, introduced by Johnston (1968), was selected as one of the highly specific inhibitors of the A type MAO.

The oxidation of [^{3}H]5-hydroxytryptamine or [^{14}C]benzylamine or [^{14}C] tyramine by rat liver and brain mitochondrial MAO in the presence of another

TABLE 1

The oxidation of [^{3}H]5-hydroxytryptamine ([^{3}H]5HT) by rat liver and rat brain mitochondrial MAO under different conditions

Conditions	Liver		Brain	
	μmol [^{3}H]5HT/ 100 mg protein/ 10 min	*Activity in %*	*μmol [^{3}H]5HT/ 100 mg protein/ 10 min*	*Activity in %*
[^{3}H]5HT	0.46	100	0.187	100
[^{3}H]5HT + clorgyline	0.04	8.7	0	0
[^{3}H]5HT + (−)-deprenyl	0.38	82.6	0.170	92.1
[^{3}H]5HT + [^{14}C]BZA[a]	0.39	84.7	0.061	32.3
[^{3}H]5HT + [^{14}C]BZA + clorgyline	0.16	34.8	0.043	23.1
[^{3}H]5HT + [^{14}C]BZA + (−)-deprenyl	0.25	55.8	0.048	26.7

[a] [^{14}C]BZA, [^{14}C]benzylamine.
Concentrations of [^{3}H]5HT and [^{14}C]BZA: 2.5 mM, and of (−)-deprenyl and clorgyline: 10^{-5}M. Radiometric assay, using the double-labelling technique.

TABLE 2

The oxidation of [^{14}C]benzylamine ([^{14}C]BZA) by rat liver and rat brain mitochondrial MAO under different conditions

Conditions	Liver		Brain	
	μmol [^{14}C]BZA/ 100 mg protein/ 10 min	*Activity in %*	*μmol [^{14}C]BZA/ 100 mg protein/ 10 min*	*Activity in %*
[^{14}C]BZA	9.99	100	1.790	100
[^{14}C]BZA + clorgyline	9.40	94.1	1.716	95.8
[^{14}C]BZA + (−)-deprenyl	1.20	12.0	0	0
[^{14}C]BZA + [^{3}H]5HT[a]	9.54	95.5	1.179	66.0
[^{14}C]BZA + [^{3}H]5HT + clorgyline	8.83	88.4	1.240	69.4
[^{14}C]BZA + [^{3}H]5HT + (−)-deprenyl	1.21	12.1	0	0

[a] [^{3}H]5HT, [^{3}H]5-hydroxytryptamine.
Concentrations of [^{3}H]5HT and [^{14}C]BZA: 2.5 mM, and of (−)-deprenyl and clorgyline: 10^{-5}M. Radiometric assay, using the double-labelling technique.

amine or a selective inhibitor, or different combinations of them, was studied, and the results are shown in Tables 1–5. The data show that 5-hydroxytryptamine (5HT), phenylethylamine (PEA), benzylamine and tyramine inhibit the oxidation of each other. Table 6 demonstrates that according to the Lineweaver–

TABLE 3

The oxidation of [^{3}H]5-hydroxytryptamine ([^{3}H]5HT) and [^{14}C]tyramine ([^{14}C]TYR) by rat liver mitochondrial MAO under different conditions

Conditions	*μmol [^{3}H]5HT/ 100 mg protein/ 10 min*	*Activity in %*
[^{3}H]5HT	0.538	100
[^{3}H]5HT + clorgyline	0.084	15.61
[^{3}H]5HT + (−)-deprenyl	0.445	82.71
[^{3}H]5HT + [^{14}C]TYR	0.268	49.81
[^{3}H]5HT + [^{14}C]TYR + clorgyline	0.126	23.42
[^{3}H]5HT + [^{14}C]TYR + (−)-deprenyl	0.235	43.68
	μmol [^{14}C]TYR/ 100 mg protein/ 10 min	
[^{14}C]TYR	15.831	100
[^{14}C]TYR + clorgyline	7.899	49.90
[^{14}C]TYR + (−)-deprenyl	5.075	32.06
[^{14}C]TYR + [^{3}H]5HT	12.025	75.96
[^{14}C]TYR + [^{3}H]5HT + clorgyline	7.420	46.87
[^{14}C]TYR + [^{3}H]5HT + (−)-deprenyl	2.714	17.14
[^{14}C]TYR + BZA[a]	4.866	30.74
[^{14}C]TYR + BZA + clorgyline	3.386	21.39
[^{14}C]TYR + BZA + (−)-deprenyl	2.588	16.35

[a] BZA, benzylamine.
Concentrations of [^{14}C]TYR, [^{3}H]5HT and BZA: 2.5 mM, and of (−)-deprenyl and clorgyline: 10^{-5} M. Radiometric assay, using the double-labelling technique.

Burk analysis the inhibition is non-competitive between 5HT and benzylamine and competitive between benzylamine and PEA, 5HT and tyramine, and benzylamine and tyramine, respectively.

These data demonstrate that from a physiological and pharmacological point of view we definitely have to reckon with functionally different monoamine oxidases in the organism. It seems reasonable to assume that in this series of experiments we were concerned with two enzymes; one of them recognizes PEA and the other 5HT. Let us call them phenylethylamine

TABLE 4

The oxidation of [^{14}C]tyramine ([^{14}C]TYR) by rat brain mitochondrial MAO under different conditions

Conditions	*µmol [^{14}C]TYR/ 100 mg protein/ 10 min*	*Activity in %*
[^{14}C]TYR	2.668	100
[^{14}C]TYR + clorgyline	1.028	38.54
[^{14}C]TYR + (−)-deprenyl	1.227	46.00
[^{14}C]TYR + [^{3}H]5HT[a]	1.763	66.07
[^{14}C]TYR + [^{3}H]5HT + clorgyline	0.981	36.77
[^{14}C]TYR + [^{3}H]5HT + (−)-deprenyl	0.772	28.95
[^{14}C]TYR + BZA[b]	0.8436	31.62
[^{14}C]TYR + BZA + clorgyline	0.2038	7.64
[^{14}C]TYR + BZA + (−)-deprenyl	0.5403	20.25

[a] [^{3}H]5HT, [^{3}H]5-hydroxytryptamine.
[b] BZA, benzylamine.
Concentrations of [^{14}C]TYR, [^{3}H]5HT and BZA: 2.5 mM, and of (—)-deprenyl and clorgyline: 10^{-5} M. Radiometric assay, using the double-labelling technique.

TABLE 5

The influence of tyramine (TYR) on the oxidation of [^{3}H]5-hydroxytryptamine ([^{3}H]5HT) and [^{14}C]benzylamine ([^{14}C]BZA) by rat brain mitochondrial MAO

Conditions	*µmol [^{3}H]5HT/ 100 mg protein/ 10 min*	*Activity in %*
[^{3}H]5HT	0.279	100
[^{3}H]5HT + [^{14}C]BZA	0.222	78.67
[^{3}H]5HT + TYR	0.165	59.85
[^{3}H]5HT + [^{14}C]BZA + TYR	0.179	63.65
	µmol [^{14}C]BZA/ 100 mg protein/ 10 min	
[^{14}C]BZA	3.301	100
[^{14}C]BZA + TYR	2.310	70.03
[^{14}C]BZA + [^{3}H]5HT	2.235	67.81
[^{14}C]BZA + [^{3}H]5HT + TYR	1.903	57.73

Monoamine concentrations: 2.5 mM.

TABLE 6

Interactions between different MAO substrates (kinetic analysis according to the Lineweaver–Burk method)

Substrate	*Inhibitor*	*Concentration of inhibitor (mM)*	*Nature of inhibition*	K_m (mM)	K_1 (mM)
[^{3}H]5HT	BZA	2.5	Non-competitive	1.0	0.956
[^{14}C]BZA	5HT	2.5	Non-competitive	1.44	4.361
[^{14}C]TYR	BZA	2.5	Competitive	0.909	0.394
[^{14}C]BZA	PEA	0.25	Competitive	1.44	0.255
[^{14}C]BZA	PEA	0.125	Competitive	1.44	0.291
[^{3}H]5HT	TYR	2.5	Competitive	1.0	1.281

5HT, 5-hydroxytryptamine; BZA, benzylamine; PEA, phenylethylamine; TYR, tyramine.

oxidase (PEAO) and 5-hydroxytryptamine oxidase (5HTO). Johnston (1968) concluded that two types of MAO (A and B) exist. They probably represent two families of enzymes. If so, 5HTO belongs to the A type and PEAO to the B type MAO.

PEA, benzylamine and tyramine are substrates of PEAO; 5HT is a non-competitive inhibitor of this enzyme. PEA has a much higher affinity for PEAO than either benzylamine or tyramine. In a recent study Williams (1974), using porcine brain mitochondria, found the K_m of PEA to be 625 μM and that of benzylamine, 11 μM. On the other hand, 5HT and tyramine are substrates of 5HTO; PEA and benzylamine are non-competitive blockers of this enzyme.

If we take the calculated conformation of 5HT (Kier 1971), the interaction of the amine and the enzyme can be visualized as shown in Fig. 1. The phenol alkylamines and catecholamines may comply with the requirements of a substrate of this enzyme if we assume that the presence of one of the two characteristic features of 5HT (OH to nitrogen distance of 6.98 Å and inter-nitrogen distance of 5.84 Å) indicated in Fig. 1 is sufficient for the formation of a complex between substrate and receptor. On the other hand, the absence of the favoured patterns might be responsible for the non-competitive inhibitory effect of the phenylalkylamines, such as benzylamine.

Regarding the interaction between PEAO and its substrates, we may accept the 'three-point accommodation' proposed and visualized by Belleau & Moran (1963). Phenyl and phenolalkylamines are suitable substrates of this enzyme; indole monoamines are non-competitive inhibitors.

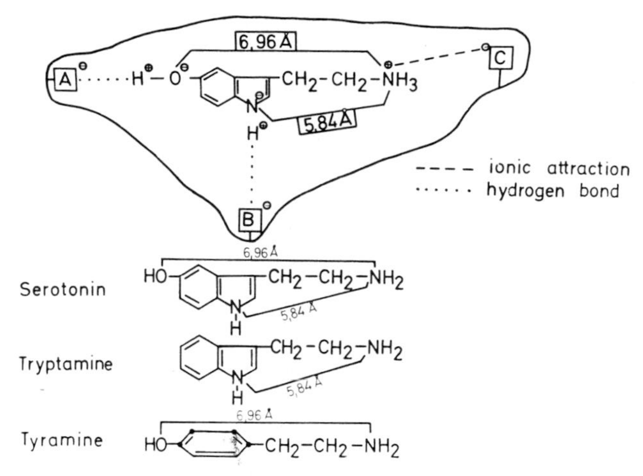

FIG. 1. Model of binding of 5-hydroxytryptamine (5HT) (serotonin) and structurally similar monoamines to 5-hydroxytryptamine oxidase. Three points of accommodation (amine nitrogen, indole nitrogen and oxygen) and the interatomic distances calculated by Kier (1971) are proposed to determine the attachment of 5HT to the enzyme.

DETECTION OF IN VIVO ACTIVITY OF PEAO AND 5HTO IN THE CAT

The cat's nictitating membrane is very sensitive and responds dose dependently to catecholamines, PEA, tyramine, 5HT and tryptamine; that is, to physiological MAO substrates. Catecholamines have direct effects on the α-adrenoceptive receptors of the smooth muscle; phentolamine blocks their effects. The phenyl and phenolalkylamines (e.g. PEA and tyramine) act through the release of noradrenaline; pretreatment with reserpine prevents their effects. 5HT and tryptamine act on an indole receptor; methysergid antagonizes the effects.

We analysed in 55 cats the dose-dependent contractions of the nictitating membrane in response to the intravenous administration of the following monoamines: 5HT, tryptamine, tyramine, PEA and benzylamine. The lowest doses which usually evoked contractions of the nictitating membrane were found to be as follows:

5HT	10– 15 μg/kg
Tryptamine	20– 30 μg/kg
Tyramine	50–100 μg/kg
PEA	150–250 μg/kg
Benzylamine	5– 10 mg/kg

In 12 cats the dose–response relationships were analysed before and after exclusion of the liver from the circulation. The dose-related increase in the effects of an amine was an index of the *in vivo* rate of biotransformation of this

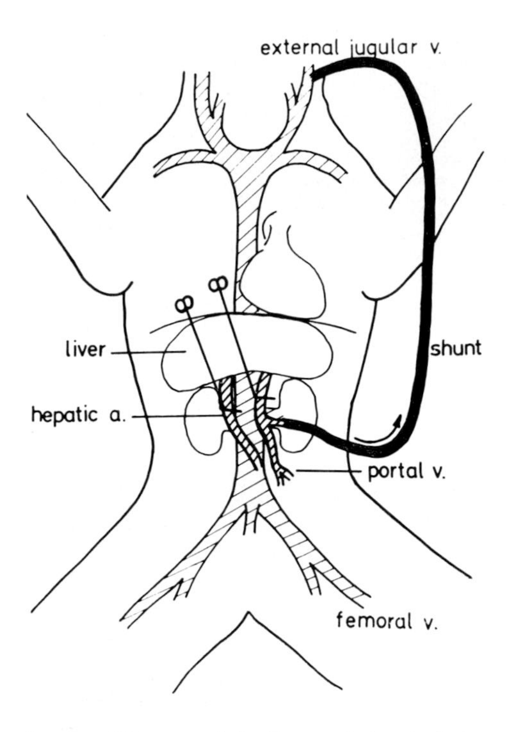

FIG. 2A. Diagram of the liver exclusion technique used.

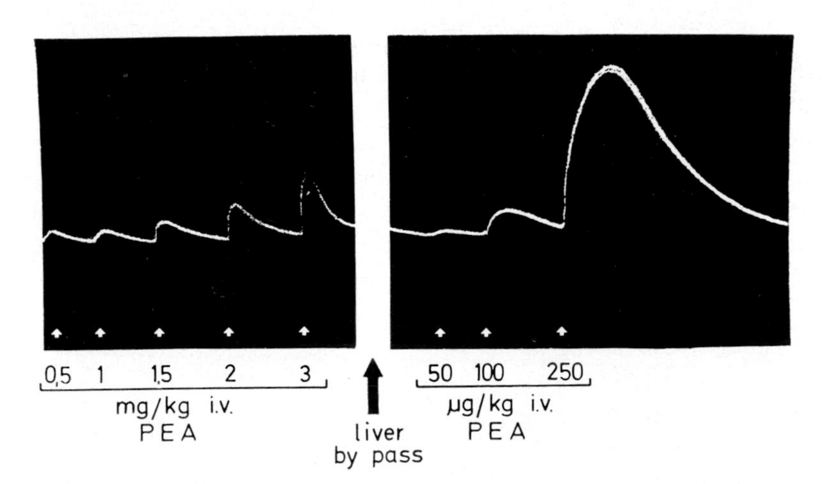

FIG. 2B. The dose-related increase in the effects of phenylethylamine (PEA), after exclusion of the liver from the circulation, on contractions of the nictitating membrane of the cat. Weight of animal, 3 kg. Chloralose–urethane anaesthesia. Artificial respiration.

monoamine. Fig. 2 shows the scheme of liver by-pass and illustrates (Fig. 2B) how the *in vivo* metabolism of the biogenic amines can be checked by the method described. Liver by-pass strongly potentiated the effects of the five monoamines, showing that all are rapidly metabolized by the liver. We usually observed with PEA and tyramine the highest and with benzylamine the lowest grade of potentiation.

In 37 cats the effects of selective MAO inhibitors (deprenyl and clorgyline) and of non-selective ones (pargyline, tranylcypromine and nialamide) were compared.

The dose–response relationships with the five monoamines were determined and were repeated 15 minutes after the intravenous administration of a selected dose of one of the MAO inhibitors. The inhibitors potentiated the effects of the monoamines with great qualitative and quantitative variations. (—)-Deprenyl was found to be the most potent MAO inhibitor in the test. Its effect on PEA was particularly striking. Fig. 3 gives an example and, by contrasting the effect of 0.25 mg/kg of (—)-deprenyl with that of 1 mg/kg of clorgyline, it summarizes the results. Tranylcypromine and nialamide were less and pargyline only slightly more effective than clorgyline in potentiating the effects of PEA.

In some cats even very high doses of clorgyline (5–10 mg/kg) only moderately potentiated the effects of PEA (Fig. 4A). On the other hand, Fig. 4B illustrates that the potentiating effect of clorgyline was much greater on the action of 5HT.

Thus the *in vivo* experiments are in agreement with the conclusions of the biochemical analysis and support the assumption that there are two different enzymes, namely PEAO and 5HTO, in the liver.

COMPARISON OF EFFECTS OF SELECTIVE MAO INHIBITORS ON ISOLATED ORGANS

The assumption that functionally different MAOs exist in the organism and the demonstration of the presence of PEAO, specific for PEA and inhibited selectively by (—)-deprenyl, and of 5HTO, specific for 5HT and selectively inhibited by clorgyline, suggested a comparative analysis of these inhibitors on isolated organs which contain specific receptors for different monoamines. Three organs were chosen for the analysis:

(1) The guinea pig ileum, a cholinergic organ, supplied with both neural and smooth muscle 5HT receptors,
(2) The guinea pig vas deferens, an essentially noradrenergic organ, containing 5HT receptors, and
(3) The papillary muscle of the cat, an organ profoundly modulated by noradrenaline and 5HT.

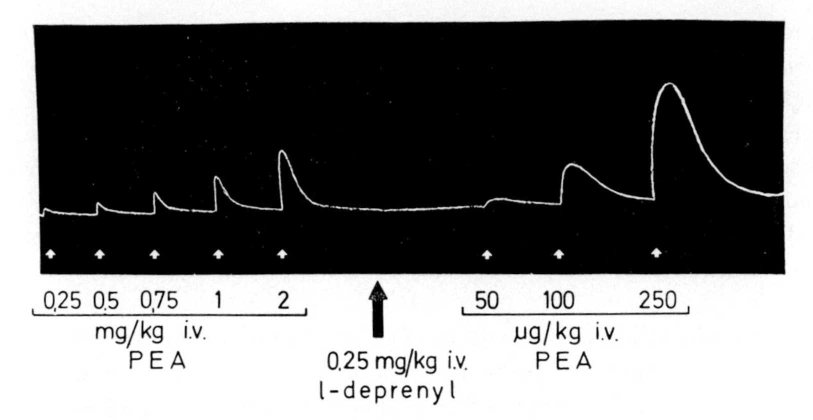

FIG. 3A. The dose-dependency of the phenylethylamine (PEA)-induced contractions of the nictitating membrane and the increase of the effects of PEA 15 minutes after the intravenous administration of 0.25 mg/kg (—)-deprenyl. Weight of the cat, 2.8 kg. Chloralose–urethane anaesthesia. Artificial respiration.

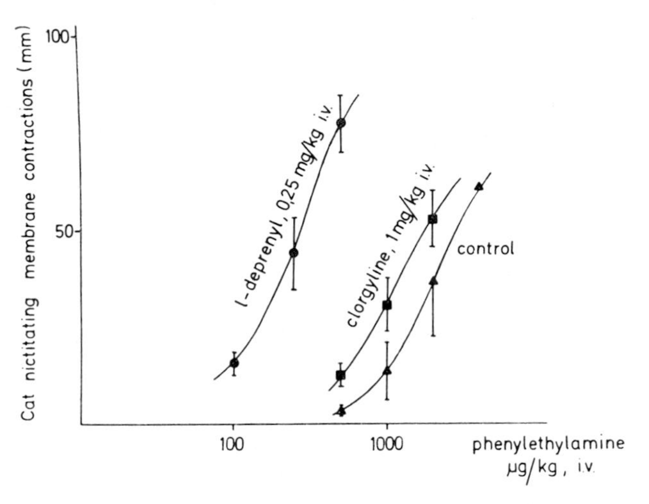

FIG. 3B. PEA-induced dose-related nictitating membrane contractions and the potentiation of the effects by (—)-deprenyl (0.25 mg/kg, i.v.) and clorgyline (1 mg/kg, i.v.) Each point indicates the average of experiments on four cats.

(1) Guinea pig ileum

Acetylcholine, histamine and 5HT produce dose-dependent contractions in this test. Fig. 5 shows that in striking contrast to deprenyl, clorgyline is a very potent, selective, non-competitive inhibitor of the 5HT receptor, leaving the

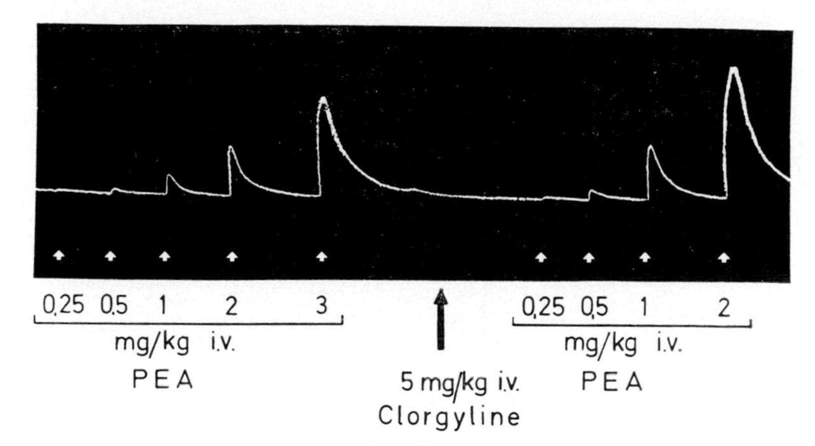

FIG. 4A. The influence of a high dose of clorgyline on phenylethylamine (PEA)-induced nictitating membrane contractions. Weight of the cat, 2.5 kg. Chloralose–urethane anaesthesia. Artificial respiration.

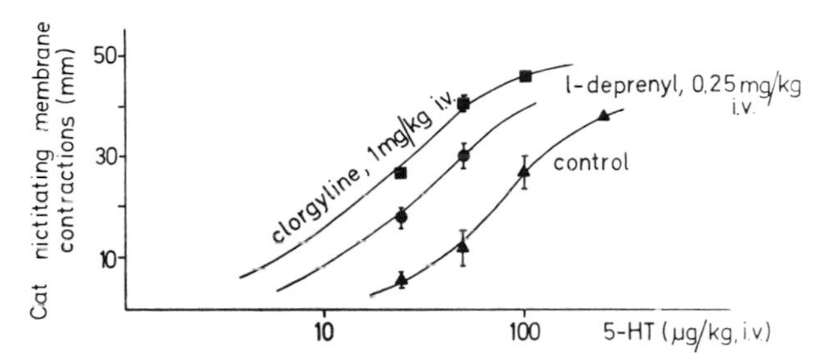

FIG. 4B. 5-hydroxytryptamine (5HT)-induced dose-related nictitating membrane contractions and the potentiation of the effects by (—)-deprenyl (0.25 mg/kg, i.v.) and clorgyline (1 mg/kg, i.v.). Each point indicates the average of experiments on four cats.

acetylcholine and histamine receptors unaltered. In some preparations clorgyline itself stimulated the 5HT receptor before blocking it.

The effect of clorgyline on the 5HT receptor of the ileum, together with the ineffectiveness of (—)-deprenyl, seems to be in good agreement with the working hypothesis that clorgyline is selectively related to the serotonergic system.

The receptor is always more sophisticated than the corresponding enzyme and is absolutely specific in recognizing a chemical entity. For instance, 5HTO, as depicted in the scheme (Fig. 1), is thought to be originally built for the recognition of 5HT but probably because of a somewhat looser architecture than the 5HT receptor it may confuse for example 5HT and tyramine. The

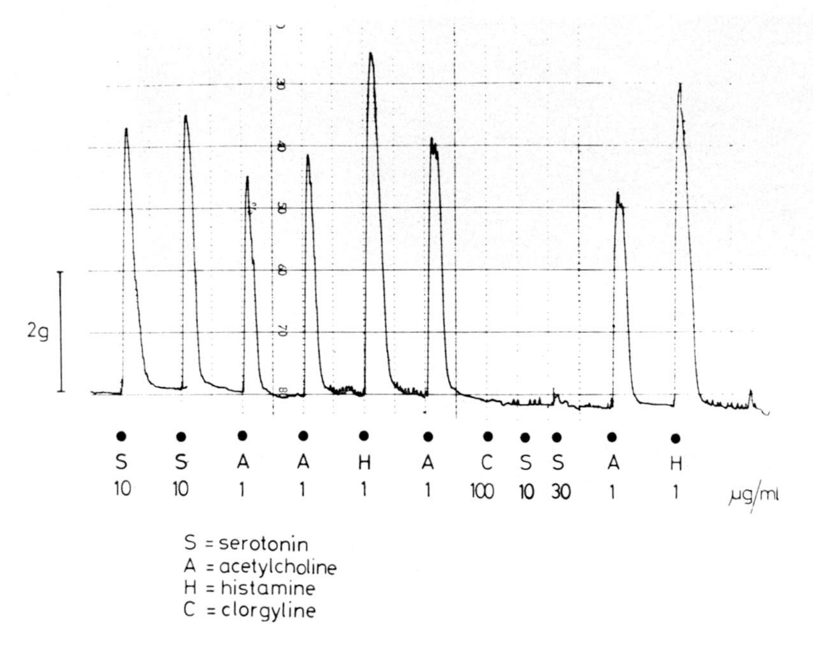

FIG. 5A. The selective inhibition of the effect of 5-hydroxytryptamine (serotonin) on the guinea pig ileum by clorgyline.

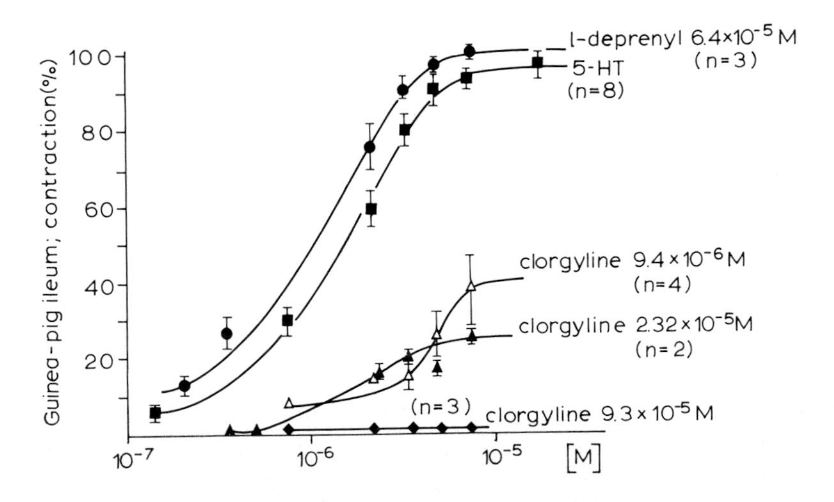

FIG. 5B. The non-competitive character of the clorgyline-induced inhibition of the effect of 5HT on the guinea pig ileum. Note the ineffectiveness of (—)-deprenyl in the test.

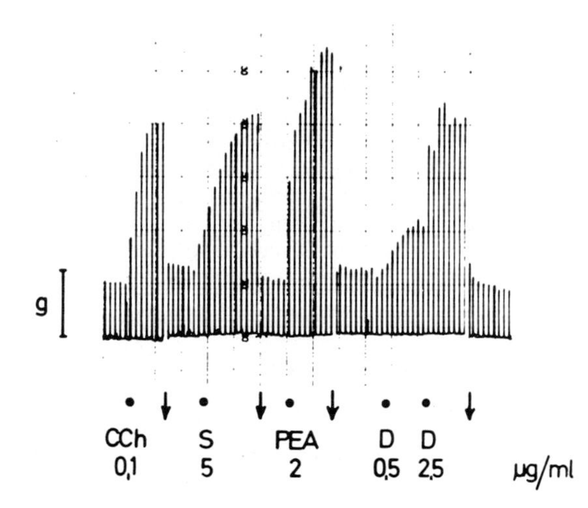

FIG. 6A. Facilitation of motor transmission in the isolated guinea pig vas deferens by carbamylcholine (CCh), 5-hydroxytryptamine (serotonin) (S), phenylethylamine (PEA) and (−)-deprenyl (D). Contractions of the vas deferens evoked by supramaximal field stimulation. Train of stimuli (25 pulses of 0.3 ms duration, 5Hz frequency) delivered every minute.

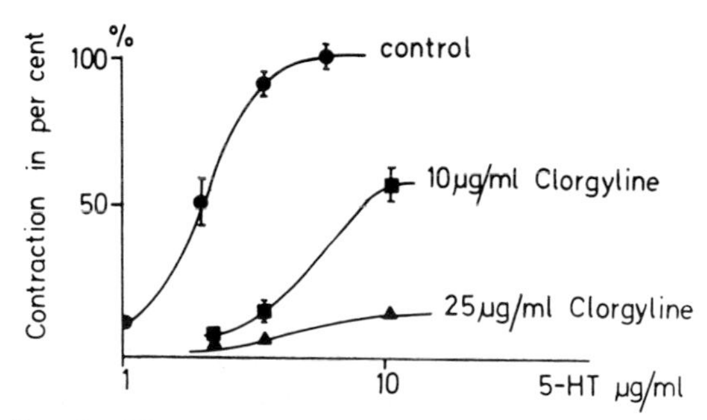

FIG. 6B. The non-competitive inhibition of the effect of 5HT on the isolated guinea pig vas deferens by clorgyline. Each curve represents an average of four experiments.

5HT receptor of the ileum, on the other hand, is adjusted to the 5HT molecule to a higher level of precision (even tryptamine is 50–100 times less potent than 5HT) and provides an opportunity for demonstrating the selective affinity of clorgyline for the indole-recognizing structures.

(2) Guinea pig vas deferens

Carbamylcholine, 5HT and PEA facilitate motor transmission in the field-stimulated isolated guinea pig vas deferens (Fig. 6). These effects are of nervous

TABLE 7

Characteristic differences in the responsiveness of the field-stimulated, isolated guinea pig and rat vas deferens to different compounds

Treatment		*Effect on motor transmission*		*Described by:*
		Guinea pig vas	*Rat vas*	
6-OH-dopamine (pretreatment:	2 × 70 mg/kg, i.v.)	Complete blockade	Complete blockade	Furness *et al.* (1970)
Tetrodotoxin	(5×10^{-1} g/ml)	Complete blockade	Complete blockade	Knoll *et al.* (1972*a*)
Phentolamine	(5×10^{-5} g/ml)	Facilitation	Facilitation	Vizi *et al.* (1973)
(—)-Deprenyl	(5×10^{-5} g/ml)	Facilitation	Facilitation	Knoll (this study)
Carbamylcholine	(5×10^{-1} g/ml)	Facilitation	Facilitation	Bentley & Sabine (1963)
Phenylethylamine	(5×10^{-5} g/ml)	Facilitation	Nil	Knoll (this study)
5-Hydroxytryptamine	(5×10^{-5} g/ml)	Facilitation	Nil	Knoll (this study)
Prostaglandins (PGE_1, PGE_2)	(10^{-5} g/ml)	Inhibition	Nil	Illés *et al.* (1973)
Single stimulation (0.05-0.1 Hz)		Ineffective	Effective	

origin because tetrodotoxin (0.1 μg/ml) blocks them. The effect of carbamylcholine is blocked by atropine, that of 5HT is competitively antagonized by methysergid, and the PEA effect is prevented by cocaine.

(—)-Deprenyl stimulates (see Fig. 6A) and clorgyline usually inhibits activity in this test. Because 5HT is facilitatory, the inhibitory effect of clorgyline is probably due to its anti-5HT effect. Fig. 6B demonstrates that clorgyline inhibits the effects of 5HT in a non-competitive manner, in this test too.

There are many contradictions in the physiological and pharmacological literature on the nature of transmission in the vas deferens of the rat and guinea pig (e.g. Furness *et al.* 1970; Ambache & Zar 1971; Ambache *et al.* 1972). I have now found that PEA is completely ineffective on the rat organ but surprisingly effective on the guinea pig vas (see Fig. 6). In some cases even 0.5–1 μg/ml enormously facilitates transmission. Cocaine (5 μg/ml) completely blocks the effects of PEA.

Table 7 shows the differences in the behaviour of the rat and guinea pig vas. To explain these discrepancies I propose the working hypothesis illustrated in Fig. 7. According to this hypothesis both systems operate with noradrenaline transmission. The important difference lies in the transmitter machinery of the

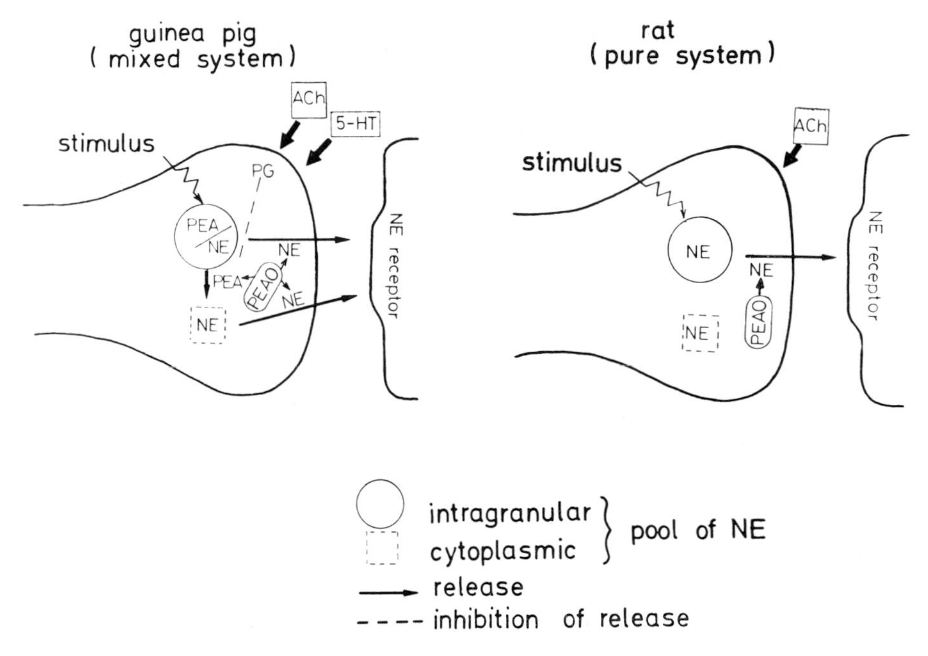

FIG. 7. The hypothetical noradrenergic neuromuscular transmission machinery in the vas deferens of the guinea pig and the rat. One of the possible sites of the modulatory effects of acetylcholine (ACh) and 5-hydroxytryptamine (5HT) is shown only. For the sake of clarity the uptake mechanisms are omitted. NE, noradrenaline; PEA, phenylethylamine; PEAO, phenylethylamine oxidase.

nerve endings. It appears that there is a 'pure' system in the rat organ and a much more complicated 'mixed' system in the guinea pig vas. In the pure system an unrestricted release of noradrenaline from the granules follows the stimulation of the nerve endings and in the mixed system the final concentrations of noradrenaline on the smooth muscle receptors depend on a dual effect. The stimulus releases noradrenaline and PEA (and/or related phenylalkylamines) from the granules and PEA releases noradrenaline from cytoplasmic mobile pools. The release of transmitter from the granules is controlled by prostaglandins in the mixed system only. The pure system is modulated by acetylcholine, the mixed system also by 5HT. Regarding the mechanism of facilitation of motor transmission by acetylcholine and 5HT in the guinea pig organ and by acetylcholine in the rat vas, the possibilities of presynaptic, postsynaptic or combined effects should be considered in planning future experiments. In the scheme in Fig. 7 only a presynaptic version is indicated.

The proposed structure of the 'mixed' noradrenergic nerve ending gives us a temporarily sufficient explanation why the guinea pig vas does not respond to a single shock and why PEA, 5HT and prostaglandins are effective on the guinea pig organ only. Each link, however, of the suggested machinery needs further detailed analysis.

(3) Cat papillary muscle

Field stimulation of the isolated papillary muscle releases noradrenaline and the increase in the height of the contractions depends on the number of pulses. In this organ small amounts of 5HT and PEA inhibit the effect of field stimulation, and the same was observed with clorgyline (Fig. 8). There was no significant change in the field-stimulated contractions after the administration of (—)-deprenyl (10 μg/ml). Thus this test again demonstrates the striking difference between the two selective MAO inhibitors and furnishes further evidence that clorgyline has a selective affinity for the 5HT system.

The inhibitory effects of small amounts of PEA might be interpreted as the uptake of this amine by the noradrenaline stores with the consequence that field stimulation releases PEA in part, instead of noradrenaline. The observations that pretreatment of the cat with reserpine (1 mg/kg, intraperitoneally, 16 hours before the experiment), as well as the addition of cocaine (1 μg/ml) to the organ bath, completely inhibit the effects of PEA are in good agreement with this assumption.

THERAPEUTIC ASPECTS OF SELECTIVE MAO INHIBITORS

Comparison of the pharmacological effects of (—)-deprenyl and clorgyline indicate that PEAO and 5HTO blockers differ in their modes of action and careful analysis is needed for delineating their appropriate use in therapy. Progress in this direction might lead to their replacing the currently used non-specific MAO inhibitors.

At present, (—)-*deprenyl seems to be the most remarkable selective MAO inhibitor from the therapeutic point of view.* The reasons for this statement are the following. There is rapidly increasing evidence that phenylethylamine is a physiological constituent of the brain which might play a stimulatory role in the CNS (Sabelli & Giardina 1973). The observation that the urinary excretion of free PEA is reduced in depressed patients supports the hypothesis that a PEA deficit may be one of the biochemical lesions in depression (Fischer *et al.* 1968). Because of the selective affinity of (—)-deprenyl for PEAO the combined administration of phenylalanine (a precursor of PEA) and (—)-deprenyl might be highly efficient in certain types of depression.

On the other hand, (—)-deprenyl combined with an L-dopa decarboxylase

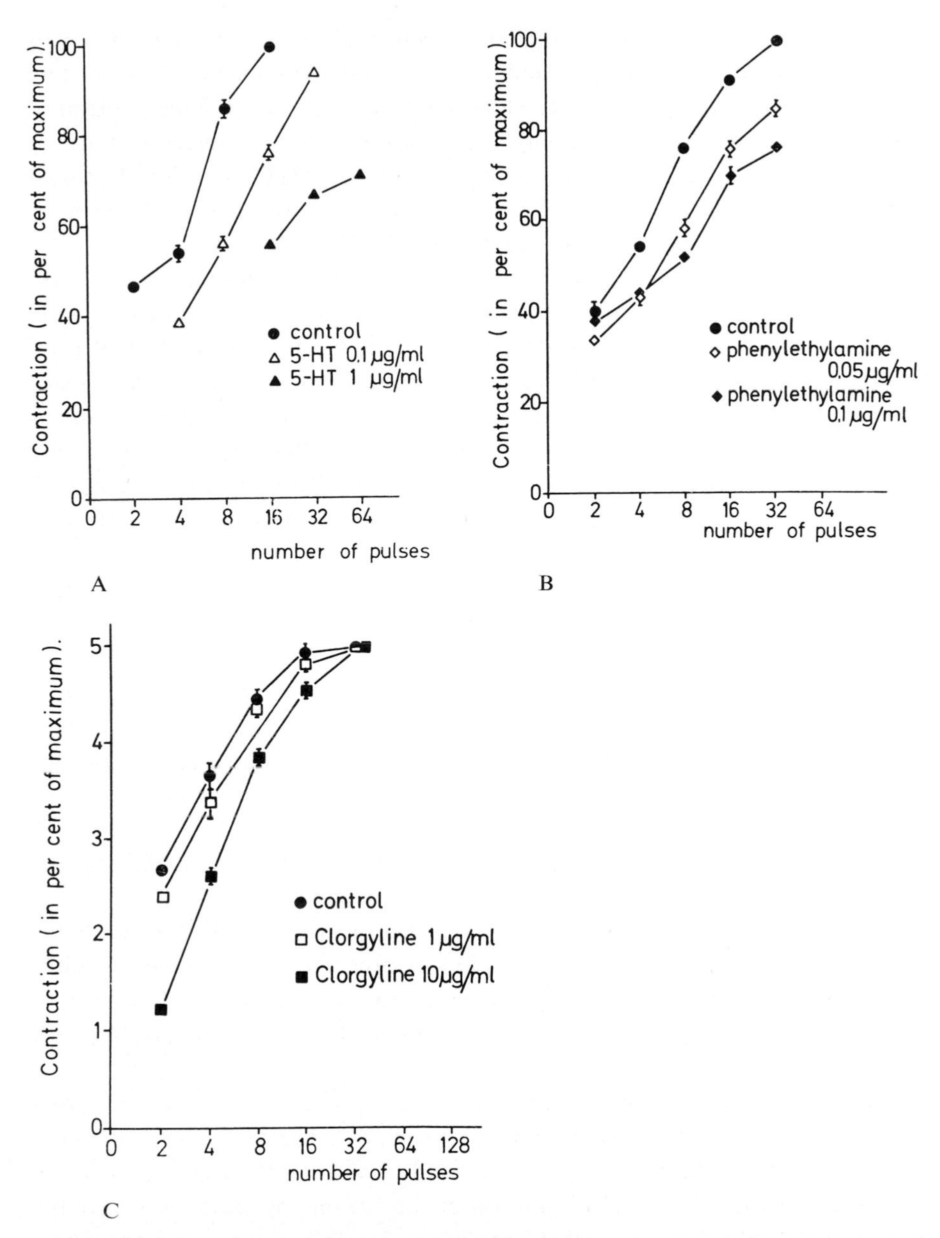

FIG. 8. The inhibitory effect of 5-hydroxytryptamine (5HT) (A), phenylethylamine (B) and clorgyline (C) on the positive inotropic responses to field stimulation of atropinized (1 µg/ml) cat papillary muscle. Basic electric driving (30% above threshold voltage; 2 Hz; 3 ms pulse duration) was applied throughout. Positive inotropic responses were produced by field stimulation (supramaximal voltage; 1 ms pulse duration; 2–32 pulses, respectively, at 2 Hz every minute).

inhibitor was found to be of high therapeutic efficacy in 44 patients with Parkinson's disease (M. B. H. Youdim, personal communication). According to Neff *et al.* (1974), in contrast to clorgyline, which curtails the metabolism of 5HT and dopamine, (—)-deprenyl increases the brain concentrations of dopamine only and leaves the concentrations of 5HT and noradrenaline unchanged. Neff concluded that dopamine is a substrate of both types of MAO (see Neff & Fuentes, this volume, pp. 163-173).

By contrast, the data presented in this symposium by Sharman (pp. 203–216) and by Maître (p. 127) support the assumption that dopamine is oxidized mainly by the A type enzyme. It may therefore be that the beneficial effect of (—)-deprenyl, the selective B type blocker, in Parkinson's disease is a consequence of the involvement of phenylethylamine in the physiological activity of dopaminergic neurons. The mechanism proposed, and illustrated in Fig. 7, for noradrenergic transmission in the guinea pig vas deferens might operate in certain types of catecholaminergic neurons of the central nervous system as well.

The most important side-effect of the MAO inhibitors is the 'cheese' reaction. There are a number of important differences between clorgyline and (—)-deprenyl in this regard:

(1) (—)-Deprenyl only slightly inhibits the MAO activity of the intestine in different species (Squires 1972), while clorgyline is highly effective. This means that the first barrier, which by splitting tyramine in foodstuffs controls the access of this monoamine to the circulation, is practically unaffected in (—)-deprenyl-treated animals and is neutralized in clorgyline-treated ones.

(2) The effects of tyramine are potentiated by clorgyline and inhibited by (—)-deprenyl (Knoll *et al.* 1968, 1972*b*).

(3) In contrast to the MAO inhibitors in therapeutic use, which increase the outflow of [^{3}H]noradrenaline from brain synaptosomes, (—)-deprenyl inhibits it (Knoll & Magyar 1972). Clorgyline was found to act similarly to the non-specific MAO inhibitors. Fig. 9 shows the striking difference between clorgyline and (—)-deprenyl in this test.

Collating all the data we might expect at least as high an incidence of hypertonic crisis with selective 5HTO inhibitors such as clorgyline as with the non-selective MAO inhibitors in clinical use, and because of their high affinity for the 5HT system, not only an inhibition of the enzyme but also, as I have demonstrated, a blockade of the 5HT receptors may follow their administration. Selective PEAO inhibitors such as (—)-deprenyl are devoid of this type of side-effect and seem to be the selective MAO inhibitors with greater potential from a practical point of view.

Varga & Tringer (1967) and Tringer *et al.* (1971) demonstrated the anti-

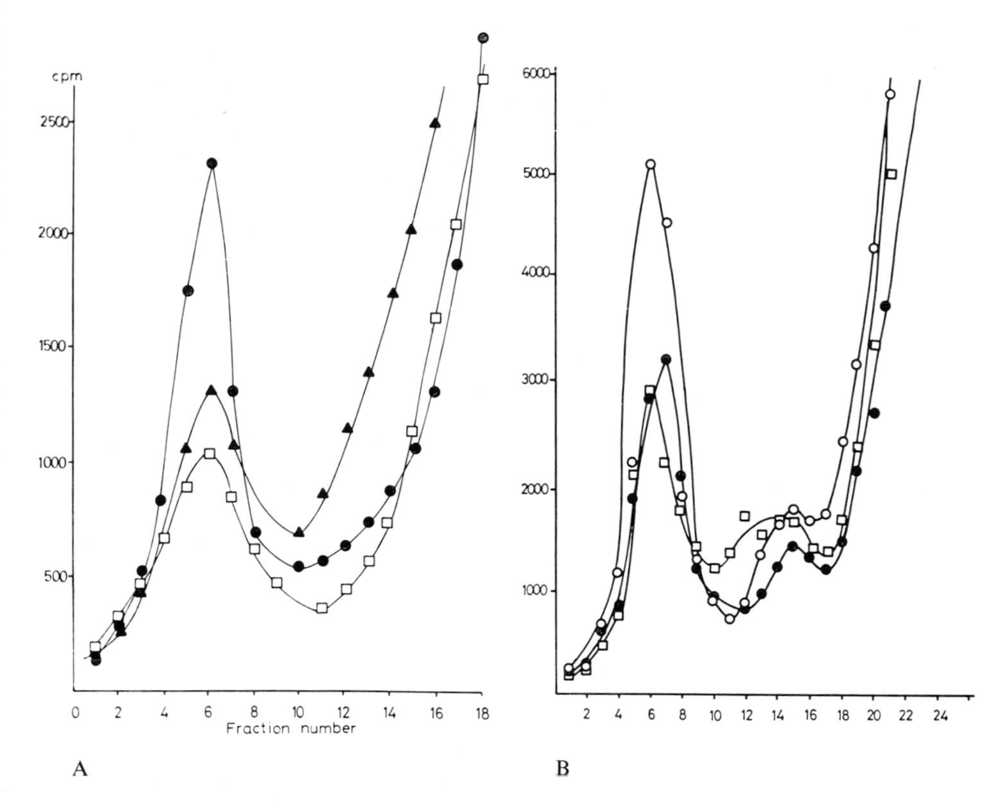

FIG. 9A. The effect of (—)-deprenyl on the distribution of (±)-[^{3}H]noradrenaline (sp. activity: 10.1 Ci/mmol) in the microsomal fraction of rat heart homogenate, isolated on a sucrose gradient (Michaelson *et al.* 1964). Rats, 100 g body weight, were injected with [^{3}H]noradrenaline (100 μCi/100 g) intravenously, and divided into three groups. One group of rats was used as control; the second and third groups were injected intraperitoneally with 12.5 and 25.0 mg/kg of deprenyl, respectively; 30 min after the injection of [^{3}H]noradrenaline, animals were decapitated, their hearts were homogenized and the subcellular distribution of [^{3}H]noradrenaline was carried out on a sucrose gradient. For each experiment, three hearts were pooled. □—□, control; ▲—▲, (—)-deprenyl (12.5 mg/kg; ●—●, (—)-deprenyl (25 mg/kg).

FIG. 9B. The effect of clorgyline on the distribution of (±)-[^{3}H]noradrenaline in the microsomal fraction of rat heart homogenate, isolated on a sucrose gradient. (Experimental conditions same as in Fig. 9A). ○—○, control; ●—●, clorgyline (7.5 mg/kg); □—□, clorgyline (15.0 mg/kg).

depressant effect of racemic deprenyl (denoted E-250) and (—)-deprenyl and also the complete lack of the cheese effect, but in the light of the accumulated data on the high selectivity of action of (—)-deprenyl on PEAO a re-evaluation of these data and a trend towards clinical investigations focused on the brain metabolism of PEA and dopamine seem to be justified.

References

AMBACHE, N. & ZAR, M. A. (1971) Evidence against adrenergic motor transmission in the guinea-pig vas deferens. *J. Physiol. (Lond.) 216*, 359-389

AMBACHE, N., DUNK, L. P., VERNEY, J. & ZAR, M. A. (1972) Inhibition of postganglionic motor transmission in vas deferens by indirectly acting sympathomimetic drugs. *J. Physiol. (Lond.) 227*, 433-456

BELLEAU, B. & MORAN, J. (1963) Deuterium isotope effects in relation to the chemical mechanism of monoamine oxidase. *Ann. N.Y. Acad. Sci. 107*, 822-839

BENTLEY, G. A. & SABINE, J. R. (1963) The effects of ganglion-blocking and postganglionic sympatholytic drugs on preparations of the guinea-pig vas deferens. *Br. J. Pharmacol. 21*, 190-201

BLACKWELL, B., MARLEY, E., PRICE, J. & TAYLOR, D. (1967). Hypertensive interactions between monoamine oxidase inhibitors and foodstuffs. *Br. J. Psychiatr. 113*, 349-365

COSTA, E. & SANDLER, M. (eds.) (1972) *Monoamine Oxidases—New Vistas (Adv. Biochem. Psychopharmacol. 5)*, Raven Press, New York and North-Holland, Amsterdam

FISCHER, E., HELLER, B. & MIRÓ, A. H. (1968) β-Phenylethylamine in human urine. *Arzneimittelforsch. 18*, 1486

FURNESS, J. B., CAMPBELL, G. R., GILLAND, S. M., MALMYOSS, T., COBB, J. L. S. & BURNSTOCK, G. (1970) Cellular studies of sympathetic denervation produced by 6-hydroxydopamine in the vas deferens. *J. Pharmacol. Exp. Ther. 174*, 111-132

GOLDBERG, L. I. (1964) Monoamine oxidase inhibitors: adverse reactions and possible mechanisms. *J. Am. Med. Assoc. 190*, 456-462

ILLÉS, P., HADHÁZY, P., TORMA, Z., VIZI, E. S. & KNOLL, J. (1973) The effect of number of stimuli and rate of stimulation on the inhibition by PGE_1 of adrenergic transmission. *Eur. J. Pharmacol. 24*, 29-36

JOHNSTON, J. P. (1968) Some observations upon a new inhibitor of monoamine oxidase in brain tissue. *Biochem. Pharmacol. 17*, 1285-1297

KIER, L. B. (1971) *Molecular Orbital Theory in Drug Research*, Academic Press, New York & London

KLINE, M. S. (1958) Clinical experience with iproniazid (Marsilid). *(Symp. Marsilid, N.Y., 1957) J. Clin. Exp. Psychopathol. 19*, 72-78 (Suppl. spec.)

KNOLL, J. & MAGYAR, K. (1972) Some puzzling pharmacological effects of monoamine oxidase inhibitors. In *Monoamine Oxidases—New Vistas* (Costa, E. & Sandler, M., eds.) *(Adv. Biochem. Psychopharmacol. 5*), pp. 393-408, Raven Press, New York and North-Holland, Amsterdam

KNOLL, J., ECSERY, Z., NIEVEL, J. & KNOLL, B. (1964) Phenylisopropil-methylpropinylamin HCl, E-250 egy uj hatásspektrumu pszichoenergetikum. *MTA V. Oszt. Közl. 15*, 231-239 [*Proc. 5th class Hung. Acad. Sci.*]

KNOLL, J., ECSERI, Z., KELEMEN, K., NIEVEL, J. G. & KNOLL, B. (1965) Phenylisopropylmethylpropinylamine (E-250): a new spectrum psychic energizer. *Arch. Int. Pharmacodyn. Ther. 155*, 154-164

KNOLL, J., VIZI, E. S. & SOMOGYI, G. (1968) Phenylisopropyl-methylpropinylamine (E-250): a monoaminooxidase inhibitor antagonizing the effects of tyramine. *Arzneimittelforsch. 18*, 109-112

KNOLL, J., SOMOGYI, G. T., ILLÉS, P. & VIZI, E. S. (1972*a*) Acetylcholine release from isolated vas deferens of the rat. *Naunyn Schmiedebergs Arch. Pharmakol. 274*, 198-202

KNOLL, J., VIZI, E. S. & MAGYAR, K. (1972*b*) in *Recent Developments of Neurobiology in Hungary. III. Results in Neuroanatomy, Neurophysiology, Neuropathophysiology and Neuropharmacology* (Lissák, K., ed.), pp. 167-217, Publishing House of the Hungarian Academy of Sciences, Budapest

MAGYAR, K., VIZI, E. S., ECSERI, Z. & KNOLL, J. (1967) Comparative pharmacological analysis of the optical isomers of phenyl-isopropylmethyl-propinylamine (E-250). *Acta Physiol. Acad. Sci. Hung. 32*, 377-387

MICHAELSON, I. A., RICHARDSON, K. C., SNYDER, S. H. & TITUS, E. O. (1964) The separation of catecholamine storage vesicles from rat heart. *Life Sci. 3*, 971-978

NEFF, N. H., YANG, H.-Y. T. & FUENTES, J. A. (1974) in *Neuropsychopharmacology of Monoamines and Their Regulatory Enzymes* (Usdin, E., ed.), pp. 49-57, Raven Press, New York

SABELLI, A. C., & GIARDINA, W. J. (1973) Amine modulation of affective behavior. In *Chemical Modulation of Brain Function* (Sabelli, A. C., ed.), pp. 225-259, Raven Press, New York

SQUIRES, R. F. (1972) Multiple forms of monoamine oxidase in intact mitochondria as characterized by selective inhibitors and thermal stability: A comparison of eight mammalian species. In *Monoamine Oxidases—New Vistas* (Costa, E. & Sandler, M., eds.) *(Adv. Biochem. Psychopharmacol. 5)*, pp. 355-370, Raven Press, New York and North-Holland, Amsterdam

TRINGER, L., HAITS, G. & Varga, E. (1971) The effect of (—)phenyl-isopropylmethyl-propinylamine-HCl in depressions. *Societas Pharmacologica Hungarica. Vth Conferentia Hungarica Pro Therapia et Investigatione in Pharmacologia* (Leszkovszky, G. P., ed.), pp. 111-114, Publishing House of the Hungarian Academy of Sciences, Budapest

VARGA, E. & TRINGER, L. (1967) Clinical trial of a new type of promptly acting psychoenergetic agent (phenyl-isopropylmethyl-propinylamine-HCl, E-250). *Acta Med. Acad. Sci. Hung. 23*, 289-295

VIZI, E. S., SOMOGYI, G. T., HADHÁZY, P. & KNOLL, J. (1973) Effect on duration and frequency of stimulation on the presynaptic inhibition by α-adrenoreceptor stimulation of the adrenergic transmission. *Naunyn Schmiedeberg's Arch. Pharmacol. 280*, 79-91

WILLIAMS, C. H. (1974) Monoamine oxidase. I. Specificity of some substrates and inhibitors. *Biochem. Pharmacol. 23*, 615-628

YOUDIM, M. B. H. (1972) Multiple forms of monoamine oxidase and their properties. In *Monoamide Oxidases—New Vistas* (Costa, E. & Sandler, M., eds.) *(Adv. Biochem. Psychopharmacol. 5)* pp. 67-77, Raven Press, New York and North-Holland, Amsterdam

YOUDIM, M. B. H. (1974) Heterogeneity of rat brain mitochondrial monoamine oxidase. *Adv. Biochem. Psychopharmacol. 11*, 59-63

Discussion

Neff: You say that clorgyline blocks the receptor for 5-hydroxytryptamine (5HT), and it is also a monoamine oxidase inhibitor that prevents the deamination of 5HT. Perhaps clorgyline does not block the 5HT receptor but blocks the formation of 5-hydroxyindoleacetaldehyde, which might be responsible for the biological activity you observed in your preparation.

Knoll: Clorgyline acts primarily as an MAO inhibitor and I consider the 5HT receptor blocking activity as one of its side-effects. The fact, on the other hand, that not only the MAO A enzyme, but also the most selective 5HT-recognizing macromolecule, the 5HT receptor, can be blocked more effectively by clorgyline than by (—)-deprenyl, supports my view that two of the changes—the introduction of the oxygen bridge between the aromatic ring and the isopropyl group, and the chlorine substitution in the *para* position, made by Johnston in the previously described MAO inhibitor, E-250 (later: deprenyl)—have endowed the substance with higher selectivity towards the 5HT system.

Fuller: Perhaps your data do not supply the answer to this question, because the concentrations of clorgyline that inhibit 5HT oxidation are too low to inhibit the receptor.

Knoll: It may well be that low doses of clorgyline will have negligible effects *in vivo* on the receptor. But, firstly, the experiments have been done with high doses many times and, secondly, we cannot exclude the possibility that a higher sensitivity towards this receptor-blocking effect exists in the CNS. Anyway, as I said, clorgyline is first of all an MAO inhibitor and its effect on the 5HT receptor is of greater theoretical than practical interest.

Trendelenburg: Since agents like clorgyline are irreversible inhibitors of MAO, could you tell us whether the receptor antagonism you mentioned is also of the irreversible type?

Knoll: It can be washed out.

Maître: Usually, inhibition of the pressor response to tyramine reflects the inhibition of noradrenaline uptake. How do you explain the inhibition of the pressor response to tyramine by the (—)-form of deprenyl, if uptake-inhibiting properties are a feature of the (+)-form, as you mentioned?

Knoll: The (+)-form of deprenyl is more potent as an uptake inhibitor than the (—)-form, but the latter is not completely devoid of this effect.

Maître: What about the central nervous stimulating properties of deprenyl—are they related to (—)- or to (+)-deprenyl?

Knoll: The amphetamine-like CNS-stimulating effect of (+)-deprenyl is more pronounced.

Iversen: There is some evidence for amphetamine-like properties in the sample of (+)-deprenyl that Dr Knoll provided for a study done by Dr. P. H. Kelly in the Department of Experimental Psychology at Cambridge. He examined the effect of the drug on intracranial self-stimulation behaviour in the rat. Deprenyl transiently enhanced this behaviour, but the time-course of that effect was short by comparison with the long-lasting effects of deprenyl on dopamine concentration and MAO activity in the brain.

Knoll: In a self-stimulating experiment one cannot rule out the possibility that the blockade of the oxidation of phenylethylamine enhanced this behaviour.

Sharman: In our laboratory, using mice, we find an unusual behavioural effect with some (—)-deprenyl that Dr Youdim gave us. It closely resembles the effect we see with tranylcypromine. I describe it as a 'military' response because the animals line up in rows!

Knoll: There are many papers describing different CNS stimulatory effects of phenylethylamine, and Sabelli & Giardina (1973) claim that in contrast to noradrenaline, which has mixed effects, it is a pure stimulator. Its possible involvement in this 'military' response could easily be checked.

van Praag: The phenylethylamine story in man is intriguing but rather obscure, so far. Decreased excretion in urine has been reported by two groups, in 100% cases of so-called endogenous depression (Sabelli & Mosnaim 1974; Fischer *et al.* 1972*b*). As far as I know, nobody has tried to measure it in cerebrospinal fluid.

Sandler: We have measured phenylacetic acid, the major metabolite of phenylethylamine in c.s.f. (L. Fellows & M. Sandler, unpublished). There is a very low concentration, about 5 ng/ml. But there is a problem here. Large amounts of phenylacetic acid are excreted every day in human urine, but most of it (about 150 mg) is as phenylacetylglutamine and only about 1 mg is in the free form (Goodwin *et al.* 1975). We haven't yet developed a method for measuring this conjugate in the c.s.f.

van Praag: What is the function of phenylethylamine? Professor Knoll says that it is a 'modulator', but what is really known about its activity in brain?

Knoll: I demonstrated in my paper that in spite of the fact that rat and guinea pig vas deferens are both noradrenergic organs, phenylethylamine facilitates neuromuscular transmission only in the guinea pig vas. To explain these results I proposed a mechanism of phenylethylamine modulation, shown in Fig. 7 (p. 149). I think, now, that 'pure' noradrenergic neurons exist in the rat vas and 'mixed' ones in the guinea pig vas. The possibility of similar functional differences in the catecholaminergic neurons of the brain seems to me a reasonable subject for further research.

Sourkes: Do both vas deferens types have the intramural ganglion?

Knoll: There are no apparent anatomical and histological differences; the noradrenergic system of the vas deferens differs functionally in the two species.

Neff: The phenylethylamine model is intriguing, but you quoted the studies of Sabelli and of Fischer *et al.* (1972*a*) and I think the model needs to be re-evaluated in the light of observations by Saavedra (1974) and by Willner *et al.* (1974), who are finding quantities of phenylethylamine in brain that are about 1/200th of that reported by the other two groups.

Knoll: The important point is that phenylethylamine is an endogenous substance and has clear-cut CNS stimulatory effects. Methods are improving, and there are also differences in the accuracy of measurements in different laboratories. Knowledge of the real amounts of phenylethylamine in the brain is of course a basic requirement for further research but the physiological significance of a substance does not depend on the magnitude of the absolute quantities. Phenylethylamine might be a very important modulator in the brain even if it is only present as 1/200th of the amount described earlier as present in brain tissue.

Fuller: I want to make some comments which extend the pharmacological effects of selective MAO inhibitors and also relate to some of Dr Neff's comments. I showed some data in the Sardinia meeting (Fuller 1972) comparing Lilly 51641 and other inhibitors. If one compares inhibitors in terms of the potentiation of various amines or amine precursors, there are some remarkable differences in potency. We compared 51641 with tranylcypromine. The latter potentiated the behavioural effects of L-dopa and phenylethylamine in mice, and did so roughly equally, in terms of dose–response curves. 51641 was about 10 times better than tranylcypromine in potentiating the effects of L-dopa but only about one-tenth as good in potentiating the effects of phenylethylamine. Since then we have seen some dopa-potentiating effects in rats, and these results are consistent with dopamine being primarily oxidized by a type A enzyme. In mice, dopa plus an MAO inhibitor produces a characteristic syndrome of increased behaviour, piloerection, exophthalmos, irritability and aggressive behaviour; the mice attack each other. We don't normally see this response in rats given dopa with an MAO inhibitor, but we can produce this type of response in rats by treating them in addition with an inhibitor of dopamine β-hydroxylase (disulfiram or diethyldithiocarbamate). One explanation might be that the effect primarily relates to dopamine and that in the rat much of the dopamine is metabolized further to noradrenaline. With the combination of the three drugs, the rats become excited, alert and extremely aggressive. This is much the same kind of activity as we see in mice, produced by L-dopa and the MAO inhibitor alone. This sort of activity is potentiated far more effectively by type A inhibitors than by type B inhibitors.

Knoll: We obtained similar results by combining (−)-deprenyl and higher doses of L-dopa.

Fuller: We see effects with as little as 0.25mg/kg of 51641.

Sharman: The effect of L-dopa and monoamine oxidase inhibition is nothing like the response to (−)-deprenyl.

Green: We only obtain the amphetamine-like effect of (+)-deprenyl at high concentrations (10 mg/kg). It is a locomotor effect, very much like that of metamphetamine, and we know the enzyme has been inhibited well below this dose, oxidation of a B substrate being inhibited by 2.5 mg/kg.

With regard to Dr Fuller's work, we suspect that in the L-dopa potentiation test, some of the effects are due to serotoninergic activity in the brain, because it seems likely that dopa is being decarboxylated in 5HT neurons. You can decrease the locomotor response to tranylcypromine and L-dopa by about 30–40% if you give *p*-chlorophenylalanine (PCPA). In work done in collaboration with Dr Peter Kelly we found that PCPA did not decrease the circling produced by metamphetamine in rats with unilateral nigro-striatal lesions, which is presumably a pure response due to dopamine release only. We also find that in intact rats PCPA does not decrease metamphetamine stereotypy (A. R. Green & P. H. Kelly, unpublished observations). We feel that part of the tranylcypromine/L-dopa response is due to the dopa releasing 5HT, which is one component of the behavioural response, and one is seeing a dopamine and 5HT response together.

Fuller: You are suggesting that the effects of L-dopa are related partly to depletion of 5HT by dopamine, so that there is less 5HT to be released by L-dopa: consistent with that is our finding that L-dopa is far more effective in depleting 5HT levels in mice than in rats and, as I mentioned, also in producing this behavioural syndrome.

van Praag: In human beings, the combination of MAO inhibitors and dopa gives rise to an increase in blood pressure which is sometimes very great. Do you know anything about the blood pressure of your animals after this particular combination, and what the influence could have been on their behaviour?

Fuller: We have no information in our animals, but we have some data in man. Dr W. Gillen has studied Lilly 51641 in patients with Parkinson's disease, since this is the most potent MAO inhibitor we found in potentiating L-dopa effects in animals. These are not anti-Parkinsonian effects and may be unrelated to the nigro-striatal dopamine receptors, but that fact caused us to wonder whether this MAO inhibitor would potentiate the anti-Parkinsonian effects of L-dopa in man, and therefore allow the normally very high doses of L-dopa to be reduced. Before these studies were begun, Dr Gillen gave 51641 to seven patients with Parkinsonism. All seven showed a definite improvement in their Parkinson symptoms. The degree of improvement, however, was not as great as one would expect to see with L-dopa, and that possibly relates to the fact that with impaired dopaminergic neurons one can only enhance function to a certain degree by means of MAO inhibition. He then gave low doses of L-dopa to patients being treated with 51641 and gradually increased the dose of L-dopa. He saw an effect at doses smaller than are normally used to treat Parkinson symptoms. The effect, however, included an elevation of blood pressure in two patients given exactly the same dose (400 mg). This made us suspect that any further study might be dangerous and we stopped the investigation. If we were to extrapolate back to animals, I would not be surprised if the same rise in blood pressure occurred.

Sandler: It is known that the combination of L-dopa plus MAO inhibitor is a dangerous one and sends the blood pressure up (see Sandler 1972). The combination of these two drugs with a peripheral decarboxylase inhibitor, however, is of interest to a number of us.

Knoll: Regarding the blood pressure raising effects, I would like to stress the greater safety of an MAO B blocker which, in contrast to the A blockers, inhibits intestinal MAO poorly.

Youdim: We have results on deprenyl as an anti-Parkinsonian agent (Birkmeyer *et al.* 1975) in patients developing an 'on-off' phase after some years on L-dopa. Patients given a combination of peripheral decarboxylase inhibitor, dopa and 10 mg (—)-deprenyl come out of these 'off' periods within 30 minutes. We have observed excellent effects in 44 cases.

Even with 5 mg of deprenyl, given intravenously or intramuscularly, we have had a good improvement. Deprenyl was also given orally but it was not as effective as when given by intravenous injection. However, the effect of oral administration was longer-lasting. We have not observed side-effects such as raised blood pressure. In three cases, patients developed psychosis; when the dose of deprenyl was reduced this disappeared. We are now studying the long-term effect of deprenyl.

Pletscher: One wouldn't expect a rise in blood pressure, because extra-cerebral decarboxylation is inhibited. Did you try the same with clorgyline?

Youdim: No, we did not.

Neff: Did you give deprenyl with L-dopa but without the decarboxylase inhibitor?

Youdim: Yes, but the results were not as good. There was no improvement.

van Praag: What does deprenyl alone do?

Youdim: We haven't tried this yet.

Pletscher: What happens if one combines clorgyline with 5-hydroxytryptophan? Stimulation of the 5HT receptor should result in some behavioural modification.

Green: We find that if tryptophan is given after a large dose of clorgyline, there are no overt behavioural changes in rats, in contrast to the effect of giving a non-specific inhibitor such as pargyline or tranylcypromine.

Sandler: Like Dr Youdim, we have done a few experiments, in collaboration with Dr D. B. Calne (unpublished), using tranylcypromine in combination with L-dopa and a peripheral decarboxylase inhibitor, but the results were somewhat equivocal and there was no real benefit for the Parkinsonian patients.

References

BIRKMAYER, W., RIEDERER, P., YOUDIM, M. B. H. & LINAUER, W. (1975) The potentiation of the anti akinetic effect after L-dopa treatment by an inhibitor of Mao-B, deprenil. *J. Neural Transm. 36*, 303-326

FISCHER, E., SPATZ, H., HELLER, B. & REGGIANI, H. (1972*a*) Phenethylamine content of human urine and rat brain, its alterations in pathological conditions and after drug administration. *Experientia 15*, 307-308

FISCHER, E., SPATZ, H., SAAVEDRA, J. M., REGGIANI, H., MIRO, A. H. & HELLER, B. (1972*b*) Urinary elimination of phenylethylamine. *Biol. Psychiatr. 5*, 139-147

FULLER, R. W. (1972) Selective inhibition of monoamine oxidase, in *Monoamine Oxidases—New Vistas* (Costa, E. & Sandler, M., eds.) *(Adv. Biochem. Psychopharmacol. 5)*, pp. 339-354, Raven Press, New York and North-Holland, Amsterdam

GOODWIN, B. L., RUTHVEN, C. J. R. & SANDLER, M. (1975) Gas chromatographic assay of phenylacetic acid in biological fluids. *Clin. Chim. Acta 62*, 443-446

SAAVEDRA, J. M. (1974) Enzymatic isotopic assay for and presence of *p*-phenylethylamine in brain. *J. Neurochem. 22*, 211-218

SABELLI, H. C. & GIARDINA, W. J. (1973) Amine modulation of affective behavior, in *Chemical Modulation of Brain Function* (Sabelli, H. C., ed.), pp. 225-259, Raven Press, New York

SABELLI, H. C. & MOSNAIM, A. D. (1974) Phenylethylamine hypothesis of affective behavior. *Am. J. Psychiatry 131*, 695-699

SANDLER, M. (1972) Catecholamine synthesis and metabolism in man (with special reference to parkinsonism), in *Catecholamines* (Blaschko, H. & Muscholl, E., eds.), vol. 33, *Handbook of Experimental Pharmacology*, pp. 845-899, Springer, Berlin

WILLNER, J., LEFEVRE, H. F. & COSTA, E. (1974) Assay by multiple ion detection of phenylethylamine and phenylethanolamine in rat brain. *J. Neurochem. 23*, 857-859

The use of selective monoamine oxidase inhibitor drugs for evaluating pharmacological and physiological mechanisms

NORTON H. NEFF and JÓSE A. FUENTES

Laboratory of Preclinical Pharmacology, National Institute of Mental Health, Saint Elizabeths Hospital, Washington, D.C.

Abstract Noradrenaline is a specific substrate for type A monoamine oxidase (MAO) and this enzyme is the predominant MAO associated with sympathetic neurons. The enzyme associated with the biogenic amine-containing neurons of brain is not clear. However, the amines that are deaminated by type A and B MAO can be evaluated *in vivo* by administering specific inhibitory drugs. Apparently type A enzyme deaminates noradrenaline, 5-hydroxytryptamine and dopamine. Type B enzyme deaminates dopamine, but not the other two amines. Drugs that block type A MAO were capable of preventing depression of motor activity, palpebral ptosis and the initial hyperthermia induced by treatment with reserpine. Drugs that block type B MAO were ineffective. By combining treatments with neurotoxic drugs and specific MAO inhibitors it was possible to evaluate the role of 5-hydroxytryptamine in modulating body temperature after reserpine treatment.

Many investigators have postulated that multiple forms of monoamine oxidase (amine: oxygen oxidoreductase [deaminating] [flavin-containing]; EC 1.4.3.4) (MAO) exist *in vivo*, on the basis of substrate and inhibitor studies, and that not all forms of the enzyme are found in all tissues (see Fuller 1972). Johnston (1968), however, provided the first systematic approach for studying the various MAO enzymes when he described his studies with the inhibitory drug clorgyline. He showed that there was a stepwise inhibition of MAO activity when homogenates were preincubated with increasing concentrations of clorgyline prior to adding the substrate tyramine. To account for this, he postulated that there were two forms of enzyme: an enzyme sensitive to clorgyline and an enzyme resistant to clorgyline. He designated the enzyme sensitive to clorgyline as enzyme A and the resistant form as enzyme B. We will call them type A and type B enzyme, as there is evidence that they may represent classes of enzymes with similar characteristics rather than single enzymes (Neff & Goridis 1972).

TABLE 1

A partial list of substrates and inhibitory drugs of type A and B monoamine oxidase

	Type A MAO	*Type B MAO*
Preferred substrates	Noradrenaline 5-Hydroxytryptamine	Benzylamine 2-Phenylethylamine
Relatively specific inhibitory drugs	Clorgyline Lilly 56141 Harmaline	Deprenyl Pargyline
Common substrates	Dopamine Tyramine Tryptamine	

Compiled from Johnston (1968); Hall *et al.* (1969); Fuller (1972); Squires (1972); Yang *et al.* (1972); and Yang & Neff (1974).

Johnston (1968) found that 5-hydroxytryptamine (5HT) was a preferred substrate for type A MAO while tyramine was a substrate common for both types of enzyme. Since Johnston's original observation other specific drugs and substrates have been recognized (Table 1). In this paper we review the types of MAO found in neuronal tissues and demonstrate how selective inhibitors can be used to evaluate functions that are modulated by the various MAO enzymes.

TYPES OF MONOAMINE OXIDASE ASSOCIATED WITH NEURONAL TISSUES

Both types of MAO are found in most tissues; however, there is a considerable difference in the ratio of these activities when they are estimated with tyramine (common substrate) and 5HT (type A MAO substrate) as substrates. For example, there is almost a four-fold difference of the ratio when pineal, superior cervical ganglion and cerebral hemispheres are compared (Table 2). Apparently type A enzyme is the predominant enzyme associated with sympathetic neurons, as destruction of the sympathetic neurons that innervate the pineal gland decreases 5HT deamination by 70% without significantly reducing the deamination of tyramine (Goridis & Neff 1971). Moreover, the ratio of 5HT to tyramine deamination is highest in the superior cervical ganglion. A loss of type A enzyme following destruction of sympathetic neurons has also been demonstrated in mesenteric and femoral artery (Coquil *et al.* 1973) and the vas deferens (Jarrott 1971).

In brain and spinal cord there is no clear pattern of enzyme distribution.

TABLE 2

Comparison of monoamine oxidase activity in rat tissues using tyramine or 5-hydroxytryptamine (5HT) as substrates

Tissues	*Tyramine*	*5HT*	*Ratio of activity 5HT/tyramine*
	(nmol/mg tissue/h ± S.E.M.) (N)		
Superior cervical ganglion[a]	8.5 ± 0.5 (4)	13 ± 3 (4)	1.5
Pineal gland	9.4 ± 0.5 (4)	3.4 ± 0.8 (7)	0.36
Cerebral hemispheres	11 ± 0.9 (4)	6.7 ± 1 (6)	0.61

Mitochondria were isolated from tissues and incubated with either 2.1 mM-tyramine or 1.2 mM-5HT.
[a] Activity expressed as nmol/ganglion/h ± S.E.M. (*N*). (From Goridis & Neff 1971.)

The cell bodies of the 5HT- and noradrenaline-containing neurons of the spinal cord are located in the brain stem (Dahlstrom & Fuxe 1965). Transection of the cord results in the loss of the transmitter amines and the enzymes associated with these neurons caudal to a transection. We studied MAO of rat spinal cord after transection and found no change of enzyme activity above or below a transection of the cord (Goridis *et al.* 1972). These results suggest that most of the MAO activity in the caudal portion of the spinal cord is not associated with neurons that have their cell bodies above the transection, and that type A MAO of spinal cord is found in structures other than biogenic amine-containing neurons.

Human brain contains type A and B enzyme activities. These activities were identified with the drugs clorgyline and deprenyl and with the substrates 5HT and 2-phenylethylamine. Enzyme activity was rather uniformly distributed when evaluated in 15 regions of the human brain (Schwartz *et al.* 1974). These studies are consistent with the notion that type A enzyme is not limited to biogenic amine-containing neurons of the brain. In peripheral sympathetic neurons type A MAO is the predominant enzyme. Perhaps the predominant enzyme in biogenic amine-containing neurons of brain is also type A enzyme, but we have not as yet devised a simple procedure to identify it.

AMINE METABOLISM IN BRAIN AFTER BLOCKADE OF TYPE A OR TYPE B MONOAMINE OXIDASE

Although it is not possible to determine the type of MAO found within the biogenic amine-containing neurons of brain, it is possible to demonstrate that different enzymes are responsible for the metabolism of the various transmitter amines. 5HT and noradrenaline are deaminated by type A MAO, while

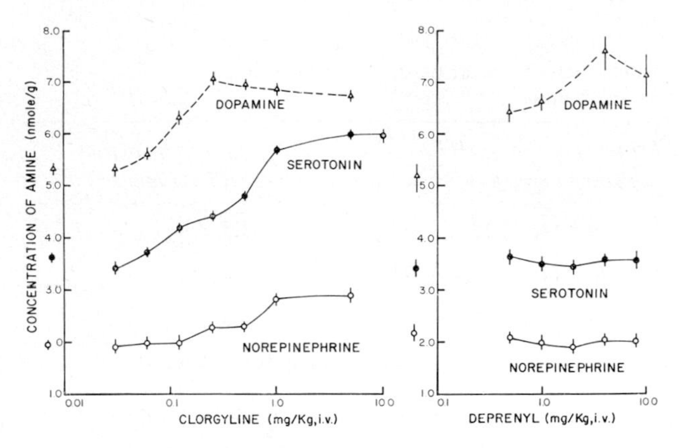

FIG. 1. The concentrations of dopamine, 5-hydroxytryptamine (serotonin) and noradrenaline (norepinephrine) in rat brain 2 h after the intravenous injection of increasing doses of clorgyline or deprenyl.

dopamine is deaminated by type A or type B MAO (Table 1). After the administration of clorgyline, the amines noradrenaline, 5HT and dopamine increase in brain (Fig. 1). After the administration of deprenyl, only dopamine increases in brain. Dopamine apparently has access to both enzymes, because administering clorgyline in a dose that almost completely inhibits type A enzyme, together with deprenyl, increases the level of dopamine even further than after either drug alone (Yang & Neff 1974). Differential inhibition of amine metabolism can also be evaluated from the concentrations of the acidic metabolites of 5HT and dopamine in brain. 5-Hydroxyindoleacetic acid decreases after clorgyline treatment, but not after deprenyl treatment. In contrast, 3,4-dihydroxyphenylacetic acid concentrations decline after deprenyl or clorgyline treatment (Yang & Neff 1974).

EVALUATING PHARMACOLOGICAL MECHANISMS WITH THE AID OF SPECIFIC MONOAMINE OXIDASE INHIBITORS

Regulation of body temperature

In vivo clorgyline, harmaline and Lilly 51641 block type A MAO at lower doses than are required to block type B MAO, while pargyline and (—)-deprenyl are more active against type B MAO than type A MAO (Table 3). These drugs have diverse chemical structures and therefore any common pharmacological properties they exhibit are most likely due to their ability to block MAO. All the drugs mentioned above induced hypothermia (Table 4).

TABLE 3

Apparent *in vivo* inhibition of type A and B monoamine oxidase as evaluated from the blockade of the oxidative deamination of 5-hydroxytryptamine and 2-phenylethylamine

Drug	*Dose (mg/kg)*	*% Inhibition of MAO activity*	
		5HT	*2-Phenylethylamine*
Clorgyline	1.0	99 ± 1	10 ± 4
Lilly 51641	0.1	91 ± 3	27 ± 5
Harmaline	1.0	64	20
(−)-Deprenyl	1.0	10 ± 6	83 ± 2
Pargyline	0.5	21 ± 13	82 ± 10

Rats were killed 2 h after the intravenous administration of the drugs (harmaline, 15 min) and MAO activity was measured.
Values for harmaline were obtained by extrapolation from serial dilution of homogenate.
Data presented as mean ± S.E.M. for 4–6 homogenates.

TABLE 4

Hypothermic effect of monoamine oxidase inhibitory drugs

Treatment	*Dose (mg/kg)*	*Δ Tr °C*
Saline	–	−0.1 ± 0.2
Clorgyline	1.0	−0.8 ± 0.1[a]
	30.0	−1.1 ± 0.1[a]
Harmaline	0.1	−0.7 ± 0.1[a]
Lilly 51641	1.0	−0.5 ± 0.1[a]
(−)-Deprenyl	1.0	−0.8 ± 0.2[a]
Pargyline	0.5	−0.5 ± 0.2[a]

Values represent the mean differences ± S.E.M. (5–8) between the rectal temperature before and 2 h (harmaline, 15 min) after the intravenous administration of the drugs. Rectal temperature was 37.6 ± 0.2 °C (10) before treatment.
[a] $P < 0.05$ when compared with saline-treated animals.

Perhaps the hypothermic activity of all the drugs is the consequence of their preventing the metabolism of an amine that is a common substrate, such as dopamine, tyramine or tryptamine. Dopamine produces hypothermia when injected into rat brain (Kruk 1972), a finding that is consistent with this hypothesis.

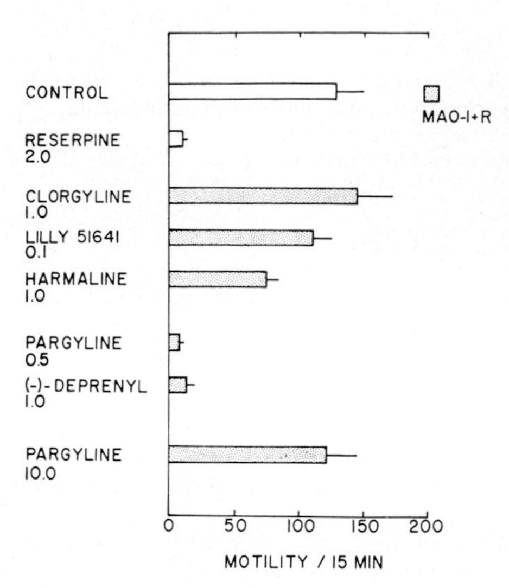

FIG. 2. Spontaneous motor activity after reserpine alone or after MAO inhibitory drugs plus reserpine. Rats were injected with MAO inhibitors 2 h (harmaline, 15 min) before reserpine. Motor activity was recorded 30 min later for 15 min. Each column represents a mean value ± S.E.M. for 10–15 animals. Doses in mg/kg, given intravenously, are shown to the left.

Prevention of motor depression induced by reserpine treatment

Reserpine releases endogenous amines from their storage sites (Brodie *et al.* 1955; Holzbauer & Vogt 1956) and they are rapidly deaminated by MAO. By administering selective MAO inhibitory drugs to reserpine-treated animals, thus sparing some amines from metabolism and not others, we have attempted to evaluate the functions that are modulated by the different enzymes. By identifying the enzymes, we may in turn narrow the search for the amines that might be associated with a specific physiological function.

Reserpine treatment depresses motor activity. We found that only type A MAO inhibitory drugs (clorgyline, Lilly 51641, and harmaline) prevent motor depression induced by reserpine (Fig. 2). However, high doses of pargyline, doses that block A as well as B enzyme, prevented motor depression induced by reserpine. Apparently amines that are metabolized by type A MAO (noradrenaline and 5HT) are responsible for preventing the depression induced by reserpine. Moreover, endogenous amines that are common substrates (dopamine, tyramine and tryptamine) and amines that are specific substrates for type B MAO (2-phenylethylamine) are probably not responsible for preventing the reserpine syndrome. Our results are consistent with reports from other laboratories (Johnston 1968; Christmas *et al.* 1972; Fuller 1972).

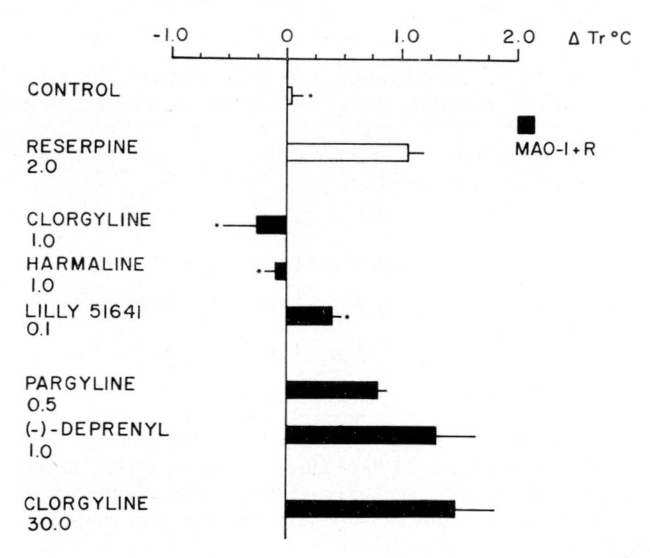

FIG. 3. Rectal temperature after reserpine alone or after MAO inhibitors plus reserpine. Data are reported as the difference between the initial rectal temperature and the temperature 2.5 h (45 min harmaline) after the administration of the inhibitory drugs. Reserpine was injected 30 min before the second temperature reading. Data are presented as the mean ± S.E.M. for 10–15 rats. Doses in mg/kg, given intravenously, are shown to the left.
* $P < 0.05$ when compared to reserpine-treated animals.

Prevention of hyperthermia induced by reserpine administration

In the rat, the initial response to an injection of reserpine is hyperthermia followed within hours by hypothermia. Pretreating rats with MAO inhibitors modifies the response to reserpine, depending on the type of drug administered. Type A inhibitors (clorgyline and harmaline) prevent the initial hyperthermia (Fig. 3). Hyperthermia was evident after pretreatment with type B inhibitors (pargyline and deprenyl) and also after high doses of clorgyline. Lilly 51641, in the dose administered, produced about a 30% blockade of type B MAO *in vivo* (Table 3) and rectal temperature was elevated after reserpine treatment although not to the same extent as after treatment with pargyline or deprenyl (Table 4). The rise of rectal temperature after reserpine was prevented by prior blockade of type A enzyme and not by blockade of type B enzyme. After a high dose of clorgyline (30 mg/kg) hyperthermia was not blocked. Perhaps endogenous 2-phenylethylamine or an unstudied monoamine that is a specific substrate for type B MAO is responsible for this hyperthermia.

The neurotoxic compound 6-hydroxydopamine (Tranzer & Thoenen 1968) destroys catecholamine-containing nerve terminals while 5,6-dihydroxytrypt-

TABLE 5

The influence of 6-hydroxydopamine (HDA) and 5,6-dihydroxytryptamine (DHT) treatment on body temperature changes induced by clorgyline alone or given together with reserpine

Treatment (intravenous)	*Rectal temperature (°C ± S.E.M.) (5)*		
	Pretreatment (intraventricular)		
	Saline	*HDA*	*DHT*
Saline	37.1 ± 0.1	37.0 ± 0.1	37.3 ± 0.2
Reserpine	38.3 ± 0.1[a] (↑)	38.9 ± 0.3[a] (↑)	37.9 ± 0.2[a] (↑)
Clorgyline	36.2 ± 0.1[a] (↓)	35.9 ± 0.2[a] (↓)	37.5 ± 0.2
Clorgyline + reserpine	36.7 ± 0.2[a] (↓)	36.6 ± 0.1[a] (↓)	39.2 ± 0.2[a] (↑)

Rats were treated with either 2 × 1.63 μmol/rat HDA or 0.39 μmol/rat DHT. Animals were killed 7 days after the second dose of HDA and 14 days after the dose of DHT. Rectal temperature was measured 2.5 h after clorgyline (1 mg/kg) or 0.5 h after reserpine (2 mg/kg). Reserpine was administered 0.5 h before temperature measurement when given with clorgyline.
[a] $P < 0.05$ when compared with saline-treated animals.

amine (Baumgarten *et al.* 1971) destroys 5HT-containing nerve terminals. By selectively destroying these neuronal systems with these neurotoxins we have been able to demonstrate that the accumulation of 5HT in brain is probably responsible for the ability of clorgyline to prevent the hyperthermia induced by reserpine (Table 5). Treatment with 5,6-dihydroxytryptamine, but not 6-hydroxydopamine, reversed the hypothermia normally seen after combined treatment with clorgyline and reserpine. The ability of 5HT to induce hypothermia when injected into rat brain is consistent with this finding (Feldberg & Lotti 1967).

Prevention of palpebral ptosis induced by reserpine administration

Palpebral ptosis after reserpine is probably the result of depletion of noradrenaline from sympathetic neurons (Tedeschi *et al.* 1967). Sympathetic neurons contain primarily type A MAO and noradrenaline is a substrate for type A MAO (Table 1). In support of these observations we found that blockade of type A enzyme by clorgyline, Lilly 51641 and harmaline prevented the palpebral ptosis induced by reserpine, whereas pargyline and (—)-deprenyl did not (Table 6). Pargyline could prevent ptosis if given in large doses (10 mg/kg).

TABLE 6

Prevention by monoamine oxidase inhibitors of palpebral ptosis induced by reserpine administration

Treatment	*Dose (mg/kg)*	*Ptosis rating*
Saline	–	4.0
Reserpine	2.0	1.2
MAO inhibitor + reserpine:		
Clorgyline	1.0	4.0
Lilly 51641	0.1	4.0
Harmaline	1.0	4.0
Pargyline	0.5	1.1
(–)-Deprenyl	1.0	1.0
Pargyline	10.0	4.0

The MAO inhibitors were administered intravenously 2 h before reserpine (harmaline, 15 min). By themselves the drugs had no apparent effect on the eyes. Ptosis was evaluated 60 min after the injection of reserpine. Scoring was based on a scale from 0 to 4. Zero corresponded to closed eyes and 4 to normally opened eyes. Each value represents a mean for groups of 10–15 animals.

CONCLUSIONS

Type A MAO appears to be the predominant enzyme associated with sympathetic neurons and it is the enzyme that deaminates noradrenaline. In brain, type A and type B enzyme appear to be rather uniformly distributed. It is possible, however, to identify their endogenous substrates from the accumulation of the transmitter amines after administering inhibitory drugs. Apparently type A enzyme deaminates noradrenaline, 5HT and dopamine. Type B enzyme deaminates dopamine, but not the other two amines. Prior treatment of rats with type A drugs prevents the depression of motor activity, palpebral ptosis and the initial hyperthermia induced by treatment with reserpine. Type B drugs are ineffective in these tests, suggesting that dopamine and the other common substrates as well as the specific substrates for type B MAO are not responsible for much of the pharmacology of reserpine. By using neurotoxic drugs, such as 6-hydroxydopamine and 5,6-dihydroxytryptamine, in combination with selective MAO inhibitors, it may now be possible to evaluate many of the physiological roles suggested for the neurotransmitters.

Our observations also suggest that type A inhibitors might be more useful as antidepressants than the non-specific inhibitor drugs that are currently used in the clinic. Furthermore, type B inhibitors might be useful in treating Parkinson's disease, especially if combined with L-dopa therapy, where the desired effect is to prolong the action of dopamine.

ACKNOWLEDGEMENTS

We thank Mr J. Rubenstein for his expert technical assistance. Clorgyline was generously supplied by Dr K. Gaimster of May and Baker, Ltd, Dagenham, England; deprenyl was generously supplied by Professor J. Knoll of the Semmelweis University of Medicine, Budapest, Hungary; Lilly 51641 was generously supplied by Dr R. Fuller, Lilly Research Laboratories, Indianapolis, Ind.

References

BAUMGARTEN, H. G., BJORKLUND, A., LACHENMAYER, L., NOBIN, A. & STENEVI, V. (1971) Long-lasting selective depletion of brain serotonin by 5, 6-dihydroxytryptamine. *Acta Physiol. Scand.*, Suppl. 373

BRODIE, B. B., PLETSCHER, A. & SHORE, P. A. (1955) Evidence that serotonin has a role in brain function. *Science (Wash. D.C.) 122*, 968

CHRISTMAS, A. J., COULSON, C. J., MAXWELL, D. R. & RIDDELL, D. (1972) A comparison of the pharmacological and biochemical properties of substrate-selective monoamine oxidase inhibitors. *Br. J. Pharmacol. 45*, 490-503

COQUIL, J. F., GORIDIS, C., MACK, G. & NEFF, N. H. (1973) Monoamine oxidase in rat arteries: evidence for different forms and selective localization. *Br. J. Pharmacol. 48*, 590-599

DAHLSTROM, A. & FUXE, K. (1965) Evidence for the existence of monoamine neurons in the central nervous system. *Acta Physiol. Scand. 64*, Suppl. 247

FELDBERG, W. & LOTTI, V. J. (1967) Temperature responses to monoamines and an inhibitor of MAO injected into the cerebral ventricles of rats. *Br. J. Pharmacol. Chemother. 31*, 152-161

FULLER, R. W. (1972) Selective inhibition of monoamine oxidase. In *Monoamine Oxidases—New Vistas* (Costa, E. & Sandler, M., eds.) *(Adv. Biochem. Psychopharmacol. 5)*, pp. 339-354, Raven Press, New York

GORIDIS, C. & NEFF, N. H. (1971) Evidence for a specific monoamine oxidase associated with sympathetic nerves. *Neuropharmacology 10*, 557-564

GORIDIS, C., MEEK, J. L. & NEFF, N. H. (1972) Monoamine oxidase activity of rat spinal cord after transection. *Life Sci. 11*, 861-866

HALL, D. W. R., LOGAN, B. W. & PARSONS, G. H. (1969) Further studies on the inhibition of monoamine oxidase by M & B 9302 (Clorgyline). I. Substrate specificity in various mammalian species. *Biochem. Pharmacol. 18*, 1447-1454

HOLZBAUER, M. & VOGT, M. (1956) Depression by reserpine of the noradrenaline concentration in the hypothalamus of the cat. *J. Neurochem. 1*, 8-11

JARROTT, B. (1971) Occurrence and properties of monoamine oxidase in adrenergic neurons. *J. Neurochem. 18*, 7-16

JOHNSTON, J. P. (1968) Some observations upon a new inhibitor of monoamine oxidase in brain tissue. *Biochem. Pharmacol. 17*, 1285-1297

KRUK, Z. L. (1972) The effect of drugs acting on dopamine receptors on the body temperature of the rat. *Life Sci. 11*, 845-850

NEFF, N. H. & GORIDIS, C. (1972) Neuronal monoamine oxidases: specific enzyme types and their rates of formation. In *Monoamine Oxidases—New Vistas* (Costa, E. & Sandler, M., eds.) *(Adv. Biochem. Psychopharmacol. 5)*, pp. 307-323, Raven Press, New York

NEFF, N. H. & YANG, H.-Y. T. (1974) Another look at the monoamine oxidases and the monoamine oxidase inhibitor drugs. *Life Sci. 14*, 2061-2074

SCHWARTZ, M. A., WYATT, R. J., YANG, H.-Y. T. & NEFF, N. H. (1974) Multiple forms of brain monoamine oxidase in schizophrenic and normal individuals. *Arch. Gen. Psychiatr. 31*, 557-560

SQUIRES, R. F. (1972) Multiple forms of monoamine oxidase of intact mitochondria as characterized by selective inhibitors and thermal stability: a comparison of eight mammalian species. In *Monoamine Oxidases—New Vistas* (Costa, E. & Sandler, M., eds.) *(Adv. Biochem. Psychopharmacol. 5)*, pp. 355-370, Raven Press, New York

TEDESCHI, D. H., FOWLER, P. J., FUJITA, T. & MILLER, R. B. (1967) Mechanisms underlying reserpine-induced ptosis and blepharospasm: evidence that reserpine decreases central sympathetic outflow in rats. *Life Sci. 6*, 515-523

TRANZER, J. P. & THOENEN, H. (1968) An electron microscopic study of selective, acute degeneration of sympathetic nerve terminals after administration of 6-hydroxydopamine. *Experientia 24*, 155-158

YANG, H.-Y. T. & NEFF, N. H. (1974) The monoamine oxidases of brain: selective inhibition with drugs and the consequence for the metabolism of the biogenic amines. *J. Pharmacol. Exp. Ther. 189*, 733-740

YANG, H.-Y. T., GORIDIS, C. & NEFF, N. H. (1972) Properties of monoamine oxidases in sympathetic nerve and pineal gland. *J. Neurochem. 19*, 1241-1250

Discussion

Maître: You have shown a decrease in the endogenous levels of DOPAC after deprenyl—a decrease of about 50% after 5mg/kg, given intravenously (Yang & Neff 1974). Would you exclude the possibility that this decrease is due to the inhibition of the MAO A enzyme, by this dose?

Neff: I can't exclude the possibility that this effect is related to inhibition of type A MAO. However, it seems unlikely, as deprenyl treatment did not change the concentration of 5-hydroxyindoleacetic acid.

Green: We agree with your classification of A and B substrates, from our experiments on injecting clorgyline and deprenyl and then taking out the brain. However, using the drug quipazine (Fig. 1) we found a different pattern, with brain tissue. Dopamine and 5-hydroxytryptamine (5HT) were inhibited equally effectively by quipazine. Tryptamine and kynuramine were inhibited less (Fig. 2). This doesn't agree with the inhibitory effects seen with clorgyline and deprenyl. (Figs. 1 and 2, see p. 174.)

Pletscher: Dr Neff, what is the best evidence that the A and B forms of MAO are really two entities and that it is not, for instance, one enzyme with two sites? I am becoming increasingly puzzled. Earlier it was said that the more or less clear-cut effects seen *in vitro* could be due to artifacts. *In vivo*, the differences are much less clear. You mentioned problems of distribution and it has also

2-(1-Piperazinyl) quinoline maleate
QUIPAZINE

FIG. 1 (Green). Structure of quipazine (2-(1-piperazinyl) quinoline maleate).

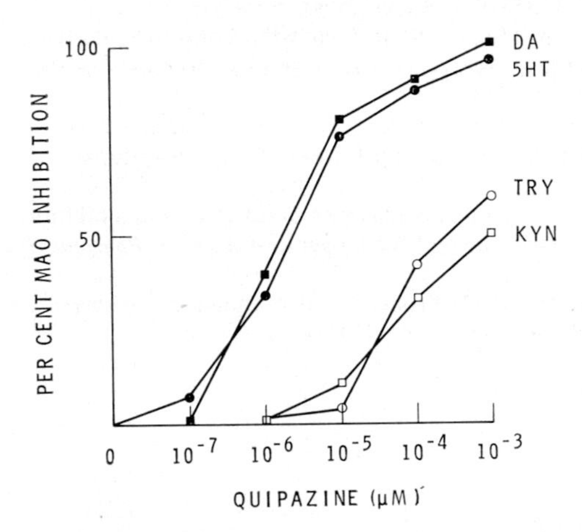

FIG. 2 (Green). Effect of quipazine on monoamine oxidase activity. The crude brain mitochondrial enzyme preparation in 0.05 M-Tris–HCl buffer pH 7.4 was preincubated for 15 minutes with various concentrations of quipazine. The enzyme was then assayed with various substrates: 5-hydroxytryptamine (5HT) (●); dopamine (DA) (■); tryptamine (TRY) (○); and kynuramine (KYN) (□). The final concentration of each amine was 1 mM during the assay.

been suggested that differences in uptake might be involved *in vivo*. What is the hard evidence that there are two enzymes?

Neff: The only good evidence that there is more than one enzyme is a study by McCauley & Racker (1973), where they isolated B type enzyme and made antibodies to it.

Pletscher: The process of solubilization itself could be producing an artifact.

Neff: I agree. One must separate A and B enzyme activities, make antibodies to each, and demonstrate that they react with different enzyme species.

Pletscher: This will be valid only if you are sure that you are not dealing with artifacts or configurational changes due to the experimental procedure.

Neff: It depends how you define types of enzyme activity. There may be only one enzyme but because of its location it may not be able to deaminate all substrates. Or there may be two separate enzyme proteins. The differences are purely semantic for the pharmacologist. What is clear is that you can give drugs and inhibit one form of the enzyme preferentially. In both situations we are dealing with two functionally different forms of enzyme activity.

Tipton: From the physiological standpoint the important factor is that the substrates appear to encounter two species of enzyme. However, the observations that these two species result from membrane binding raises the possibility that the properties of monoamine oxidase *in vivo* may alter in response to changes in its lipid environment. From the biochemical point of view it is, of course, important to know if there are several enzymes or a single enzyme that is modified by its environment.

Dr Neff, your results (Fig. 1 below) indicate that the inhibition of phenethylamine oxidation by clorgyline gives a 'dose–response' curve that extends over several orders of magnitude in clorgyline concentration. Such a response would not be expected for a single enzyme. Do you think there may be more than one species of enzyme involved in the oxidation of this substrate?

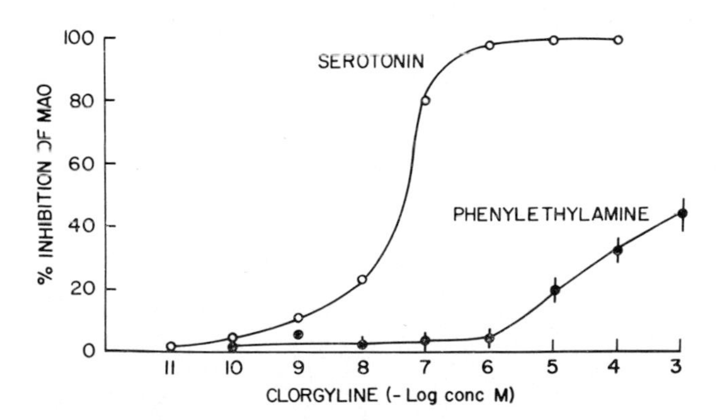

FIG. 1 (Neff). Percentage inhibition of rat brain monoamine oxidase activity plotted against the concentration of clorgyline with 5-hydroxytryptamine (serotonin) (0.2 mM) or phenylethylamine (0.2 mM) as substrates.

Neff: Yes. We originally discussed whether to call them 'species' A and B or 'type' A and B, 'type' being a more general term which would include 'species' that were similar. As far as I am concerned, type A and type B MAO are multiple enzymes. Within type A there may be A1, A2, and so on. At least two types of enzyme can be distinguished with the selective inhibitor drugs; but the picture may be more complicated.

Youdim: We have similar results to the work of McCauley & Racker (1973). Antibody induced to purified rat liver monoamine oxidase in rabbit shows two immunoprecipitants with the liver MAO which are still active. Cross-reaction of anti-liver MAO antibody with the rat brain MAO preparations also gives two immunoprecipitants, which appears to be very similar to the results observed for the liver enzyme (Youdim & Collins 1975).

Our studies (Bahkle & Youdim 1975) with the isolated perfused rat lung suggest that the lung MAO can differentiate between A and B type substrates and, when amines are perfused at equal molarity through the lung, MAO behaves like two different enzymes and the amines are metabolized at the same rates as when they are perfused individually. Furthermore, MAO types A and B can be selectively inhibited. I think we should now be more concerned about the physiology of monoamine deamination rather than with the question of whether there is more than one protein.

Neff: As a pharmacologist, I want to be capable of blocking one form of the enzyme and not the other with drugs. With such drugs we may learn the role of the enzymes as well as understand how to develop new drugs.

Youdim: Could I ask you about the subcellular distribution on the sucrose density gradient (Fig. 2 below)? You have plotted the ratio of phenylethylamine to 5HT. I wonder how many peaks you obtain? For tyramine you had two mitochondrial populations. What is the distribution for 5HT and phenylethylamine?

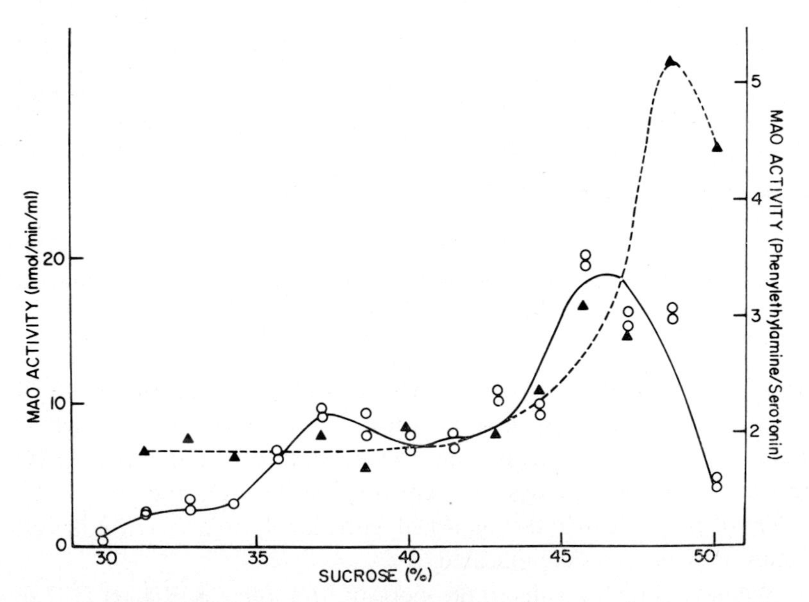

Fig. 2 (Neff). The distribution of monoamine oxidase activity in a continuous sucrose gradient. ○—○, activity using tyramine (2.1 mM) as substrate; ▲ - - - ▲, the ratio of the activity of phenylethylamine (0.2 mM) to 5-hydroxytryptamine (serotonin).

Neff: They follow essentially the same pattern as tyramine.

Oreland: Dr Ekstedt and I have also done some immunological studies on MAO (B. Ekstedt & L. Oreland, unpublished work). We found that antibodies against our purified pig liver enzyme (Oreland 1971) also precipitated rat liver MAO. This made us believe that it might be difficult to get clear-cut results about different forms of the enzyme using immunological techniques.

Singer: Much has been made of the work in Racker's laboratory on the immunochemical differentiation of A and B as perhaps the only real quasi-separation of the two forms of the enzyme. I understand from Dr Youdim that the antibody was made using Yasunobu's liver preparation of MAO as antigen. Is this correct?

Neff: No. McCauley & Racker have their own purification procedure, which is similar to Yasunobu's method.

Singer: The method used is important, because if it is Yasunobu's method this is, by the definition commonly used, a B type preparation, in the sense that it doesn't oxidize A type substrates. But how sure is one that the A type, or A conformation, of the enzyme is in reality absent from it, and not just inactivated in it? The procedure is such that it is apt to inactivate rather than completely separate different forms of enzymes. If you have an inactive form of the enzyme, it is not so certain that the antibody formed is exclusively to the B form. Then if we add the further uncertainty created by the possibility that A and B forms may differ because of their lipid composition, the value of the immunochemical method becomes very dubious in differentiating them. I wonder in fact if anyone has ever claimed to have purified an A form enzyme free from B form?

Youdim: Dr d'Iorio has claimed electrophoretic separation of A and B type enzyme activity (Diaz Borges & D'Iorio 1973).

Neff: The question is whether anyone has purified A over B. As I understand their paper, McCauley & Racker purified B enzyme, made an antibody to B enzyme, and then determined whether it would precipitate A activity or B activity, or both. They could only precipitate B activity, so it is unlikely that they purified a B enzyme containing inactive A enzyme.

Trendelenburg: Dr Youdim, you suggested that your experiments with the perfused lung provide conclusive proof of the existence of MAO of both the A and B type. In the intact tissue the enzyme is located within cells that may have different uptake mechanisms. Can you distinguish between discrimination by the uptake mechanisms and discrimination by the enzyme types?

Youdim: The only thing I can say is that our studies on homogenates of lung also gave the A and B type enzyme activity. Therefore no uptake mechanism is involved. We were able to selectively inhibit the A or B type enzyme in rats treated with deprenyl or clorgyline and then perfuse the lungs with various monoamines.

Trendelenburg: In perfusion experiments there is inevitably the possibility that 5HT is taken up into one type of cell and the other substrate is taken up into a different type of cell. Thus, you may be delivering your substrates to two differently located monoamine oxidases.

Youdim: This is possible. In the homogenates, as well as in the solubilized form of the enzyme, both populations of MAO exist. I don't know how we can overcome this problem.

Iversen: One of the most interesting things you have described, Dr Neff, is the possible heterogeneity of brain mitochondria with possible enrichment of the A and B forms in the two types, but I was puzzled by your experiments because one peak appeared to be not really mitochondrial but in a position on the density gradient that was synaptosomal.

Neff: To evaluate whether MAO was associated with free mitochondria or mitochondria found in synaptosomes we took the two peaks that we found in the gradient, froze and thawed them three times, and looked to see if the enzyme activity had changed. The idea was that if the mitochondria were in synaptosomes there would be a barrier that would prevent substrates from being deaminated. Our treatment disrupted the synaptosomes. We found that freezing and thawing did not change the enzyme activity. Apparently enzyme activity was not related to the presence of synaptosomes.

Murphy: Our brain localization studies and data from J. Saavedra & J. Axelrod (personal communication 1975), who have sampled 160 brain areas, show the same dismal picture that you report, namely that there is very little suggestion of the segregation of A type MAO in any location with the exception that Saavedra & Axelrod found that the locus coeruleus had a 5:1 ratio of A:B enzyme.

Neff: That is interesting, because the locus coeruleus would be an area where one would expect to find a high ratio of A:B enzyme.

Fuller: You described the failure of type B inhibitors to prevent the hyperthermia induced by reserpine in rats. Have you given a type B inhibitor in combination with a type A inhibitor, to see whether the effect of the type A would still be evident?

Neff: No, we did not do that study. We did give a large dose of clorgyline (30 mg/kg) which would presumably inhibit A and B enzyme. After this treatment, clorgyline did not prevent hyperthermia after reserpine.

Oreland: Dr Neff, you suggested that A form inhibitors might be of particular value in the treatment of depressive states, a treatment which would certainly be a long-term treatment. In your experiments you gave single injections of the inhibitors to the rats. Do you think it is possible in the long run to distinguish between A and B form inhibition? We have found that the selectivity of deprenyl and clorgyline is partly a matter of the time of incubation (T. Egashira & L. Oreland, in preparation). This result seems logical since we are dealing with irreversible inhibitors with the selectivities probably depending on differences in reaction rates towards the different forms of the enzyme. Thus, if you wait long enough, the selectivity should diminish.

Neff: We have never given multiple doses of clorgyline or deprenyl.

References

BAHKLE, Y. & YOUDIM, M. B. H. (1975) Inactivation of phenylethylamine and 5-hydroxytryptamine in rat isolated lungs: evidence for monoamine oxidase A and B in lung. *J. Physiol. (Lond.) 248*, 23-25*P*

DIAZ BORGES, J. M. & D'IORIO, A. (1973) Polyacrylamide gel electrophoresis of rat liver mitochondrial monoamine oxidase. *Can. J. Biochem. 51*, 1089-1095

ORELAND, L. (1971) Purification and properties of pig liver mitochondrial monoamine oxidase. *Arch. Biochem. Biophys. 146*, 410-421

MCCAULEY, R. & RACKER, E. (1973) Separation of two monoamine oxidases from bovine brain. *Mol. Cell. Biochem. 1*, 73-81

YANG, H.-Y. T. & NEFF, N. H. (1974) The monoamine oxidase of brain: selective inhibition with drugs and the consequences for the metabolism of the biogenic amines. *J. Pharmacol. Exp. Ther. 189*, 733-740

YOUDIM, M. B. H. & COLLINS, G. G. S. (1975) Properties and physiological significance of multiple forms of mitochondrial monoamine oxidase (MAO), in *Isozymes. II: Physiological Function* (Markert, C. L., ed.), pp. 619-636, Academic Press, New York

The part played by monoamine oxidase in the inactivation of catecholamines in intact tissues

U. TRENDELENBURG, K. H. GRAEFE and M. HENSELING

Institut für Pharmakologie und Toxikologie der Universität Würzburg, Federal Republic of Germany

Abstract 1. *Experiments with isolated hearts perfused with* [^3H]*(—)-noradrenaline*. In the cat and in the rabbit heart, the deaminated metabolites of noradrenaline are predominantly of neuronal origin, although there is a well-developed extraneuronal uptake system in the cat (but not in the rabbit) heart. In the rat heart, on the other hand, considerable extraneuronal deamination of noradrenaline takes place in addition to the intraneuronal deamination. Thus, there is considerable species variability in the role played by the extraneuronal monoamine oxidase (MAO), partly because the enzyme is in series with an uptake mechanism the effectiveness of which is subject to pronounced species differences, partly because of additional reasons (access of substrate to the enzyme within the cell? competition between MAO and catechol *O*-methyltransferase, COMT?).

2. *Experiments with rabbit aortic strips*. Strips were first exposed to labelled noradrenaline and then washed out with amine-free solution for 250 minutes. In spite of inhibition of both MAO and COMT the late efflux (i.e., the efflux of radioactivity collected between the 200th and the 250th minute of wash-out; this efflux being due to loss of activity from adrenergic nerve endings) contained a high percentage of metabolites. The reasons for this phenomenon are discussed. In addition, analysis of this late efflux showed that the pattern of deaminated metabolites obtained for (—)-noradrenaline differed from that obtained for the (+)-isomer.

Determinations of the activity of monoamine oxidase (amine: oxygen oxidoreductase [deaminating] [flavin-containing]; EC 1.4.3.4) (MAO) in homogenates obtained from normal organs, on the one hand, and from denervated organs or from organs of immunosympathectomized animals, on the other hand, show that MAO is located both intra- and extraneuronally (Lowe & Horita 1970; Jarrott 1971; Jarrott & Iversen 1971; Jarrott & Langer 1971), the ratio 'intra-/extraneuronal MAO activity' showing considerable species and organ variability. While these observations provide valuable information on the distribution of MAO, they are not necessarily directly relevant to the role played by MAO in the degradation of exogenous noradrenaline in the intact tissue. After all, MAO is an intracellular enzyme, and the amine has to be taken up into cells before it can be deaminated. Thus, irrespective of its intra- or extraneuronal localization, MAO is in series with uptake processes for noradrenaline. In the case of the adrenergic nerve ending, the uptake mechanism obeys Michaelis–Menten kinetics, has a high affinity for noradrenaline, is cocaine-sensitive and is highly effective in all species. The uptake mechanism associated with smooth and cardiac muscle, on the other hand, also obeys Michaelis–Menten kinetics, but is sensitive to three different groups of inhibitors: *O*-methylated catecholamines, corticosteroids and β-halogenoalkylamines. The affinity of this mechanism for noradrenaline is lower than that of the neuronal uptake mechanism (Iversen 1965; Graefe & Trendelenburg 1974). Given this arrangement, it is not only possible but likely that the degree of deamination of noradrenaline is determined not so much by the total activity of MAO (as determined in homogenates) but by the rate of inward transport of the amine (which determines the concentration of the substrate at the enzyme).

A second factor to be considered is access to the enzyme after uptake into the cell. While there is good evidence that noradrenaline has easy access to MAO after its uptake into adrenergic nerve endings, this may not be so for extraneuronally located MAO. However, since the extraneuronal system has been studied much less than the neuronal one, the evidence is still fragmentary.

In the following, results are described which were obtained in studies designed to determine the metabolic fate of exogenous noradrenaline in intact tissues.

THE PERFUSED HEART

The cat heart

Hearts were obtained from cats pretreated with reserpine (in order to reduce vesicular uptake and retention of the exogenous amine) and were perfused with a constant concentration of $[^3H](-)$-noradrenaline (0.3 μM). In the venous effluent we determined the concentration of the unchanged ^3H-labelled amine as well as of the various ^3H-labelled metabolites (NMN, normetanephrine; DOPEG, dihydroxyphenylglycol; DOMA, dihydroxymandelic acid; OMDA, a fraction containing methoxyhydroxyphenylglycol (MOPEG) + vanillylmandelic acid (VMA)). From these values, the rate of removal of $[^3H]$noradrenaline by the heart was calculated as well as the rate of appearance

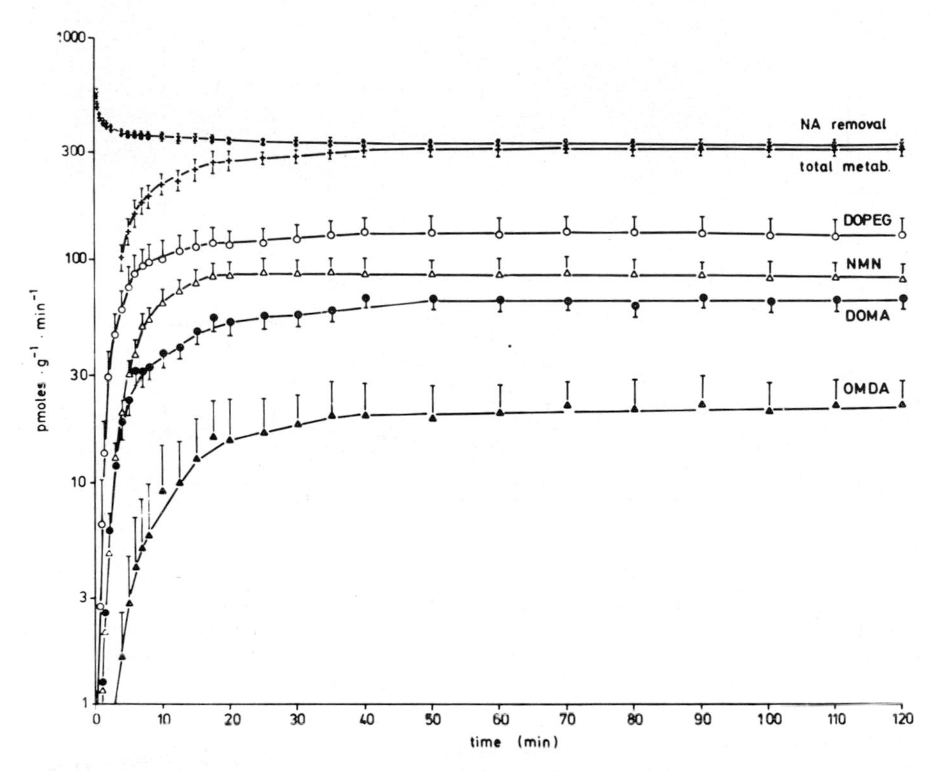

FIG. 1. Removal of noradrenaline by the perfused cat heart and appearance of metabolites in venous effluent. Cat hearts (from reserpine-pretreated animals) were perfused with 0.3 μM-[^{3}H](−)-noradrenaline. *Ordinate:* rates (in pmol g^{-1}min^{-1}; log scale). *Abscissa:* time (in min) after onset of perfusion. Means (± S.E.) for five experiments are shown. From above down: rate of removal of [^{3}H]noradrenaline from the perfusion fluid (as calculated from arterio-venous difference, weight of heart and rate of perfusion; ×—×), rate of appearance (in the venous effluent) of sum of all metabolites (+—+), of DOPEG (○—○), of NMN (△—△), of DOMA (●—●) and of OMDA (▲—▲).

of the various metabolites in the venous effluent. The rate of removal of the unchanged amine declined first quickly and then slowly (Fig. 1), to approach a steady-state value. At steady state, the rate of removal was virtually identical with the rate of appearance of the sum of all metabolites; or in other words, during steady state the removal of the unchanged amine was virtually entirely accounted for by the metabolism of the amine.

Fig. 1 also shows that the most important metabolites appearing in the venous effluent were DOPEG and NMN, the other metabolites appearing at lower rates (during steady state). Of interest is also the speed with which the various metabolites approach a steady state. It is evident from Fig. 1 that NMN

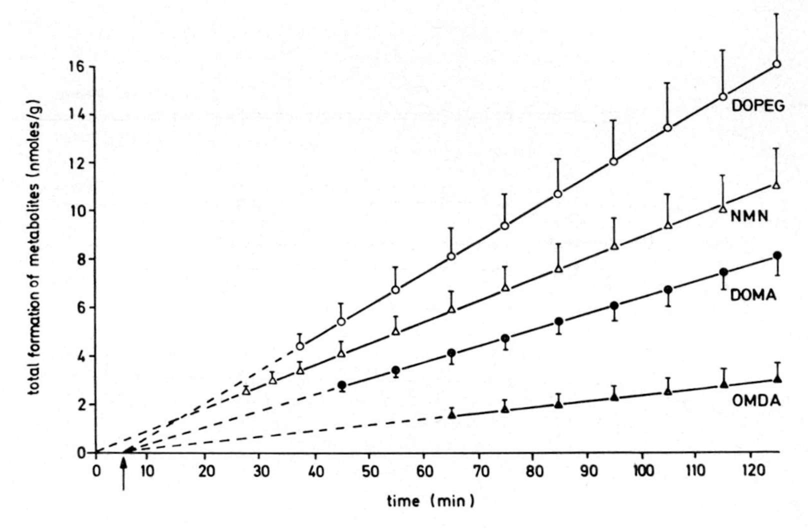

FIG. 2. Total formation of the metabolites of [^{3}H](−)-noradrenaline in the cat heart during perfusion with 0.3 μM-[^{3}H](−)-noradrenaline. *Ordinate:* total formation of metabolite (in nmol g^{-1}). *Abscissa:* time (in min) after onset of perfusion. Same experiments as in Fig. 1. Arrow shows intersection of the regression lines for DOPEG, DOMA and OMDA with abscissa. The regression line for NMN intersects at zero time.

approached a steady state more quickly than the other metabolites did. More detailed analysis of this approach to the steady-state rate of appearance of the metabolite in the venous effluent confirmed that the half-time for this approach was shorter for NMN (4.4 min) than for the other metabolites (of which DOPEG and DOMA approached steady state biphasically).

From the results presented in Fig. 1 one can calculate the 'total formation of the various metabolites', if one adds the metabolites recovered from the heart and from the fluid of the heart cavities (determined at the end of the experiment) to the cumulative efflux of the metabolites into the venous effluent. These values were calculated for that period of the perfusion during which the metabolite under study appeared in the venous effluent at a steady rate. For NMN the line for 'total formation of the metabolite during the perfusion' intersected the abscissa at zero time (Fig. 2). In other words, the formation of NMN by the perfused heart began virtually without any delay and remained constant throughout the perfusion. For the other metabolites the corresponding curves intersected the abscissa at 7.0 ± 1.2 min; that is, at a time that was significantly different from zero. For the reasons given below these other metabolites (DOPEG, DOMA and OMDA) are known to be of neuronal origin, while NMN is of extraneuronal origin. Hence, there is an initial delay in the production of neuronal metabolites, before they are formed at a constant rate. It is likely that this initial delay is due to an initial build-up of the intraneuronal concentration of noradrenaline (i.e., of the concentration of the substrate at the enzyme).

Separate experiments were carried out under similar conditions (reserpine-pretreated cats, 0.3 μM of [^{3}H]noradrenaline perfused continuously) but well after a steady state had been reached (i.e., after 90 min of perfusion). 30μM-cocaine was added to the perfusion fluid which continued to contain noradrenaline. Fig. 3 shows that the addition of this inhibitor of neuronal

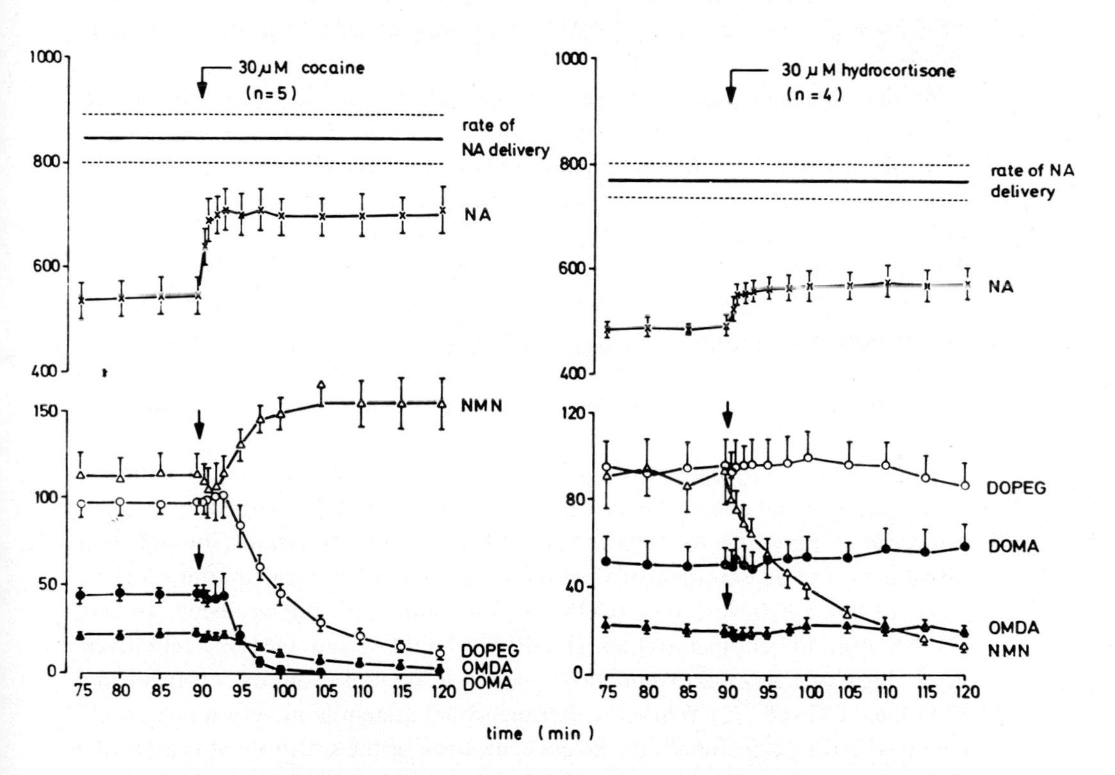

FIG. 3. The effect of cocaine (*left panel*) and hydrocortisone (*right panel*) on the rate of appearance (in the venous effluent) of noradrenaline and its metabolites in cat hearts perfused with 0.3 μM-[^{3}H](−)-noradrenaline. *Ordinate:* rates of appearance in venous effluent (in pmol g^{-1} min^{-1}; linear scale). *Abscissa:* time (in min) after onset of perfusion with noradrenaline. Means (± S.E.) are shown. Cats were pretreated with reserpine. After steady-state conditions had been reached for some time, either cocaine or hydrocortisone was added to the perfusion fluid (in the continued presence of [^{3}H]noradrenaline). Symbols (from above down): solid line, rate of delivery of [^{3}H]noradrenaline to the heart; ×—×, rate of appearance of [^{3}H]noradrenaline in the venous effluent; other symbols as in Fig. 1.

uptake resulted in pronounced changes in the removal of the amine (i.e., it reduced the arterio-venous difference for [^{3}H]noradrenaline) as well as in the rate of appearance of the various metabolites in the venous effluent. Since the rates of appearance of DOPEG, DOMA and OMDA declined (and either reached zero or came very close to zero), one is justified in concluding that these metabolites are of predominantly neuronal origin. For NMN, on the other hand, the rate of appearance in the venous effluent increased and reached a new steady state level. Hence, NMN is of extraneuronal origin. Moreover, it is evident that—in the absence of cocaine—extraneuronal metabolism (as an index of extraneuronal uptake) was decreased by the simultaneous neuronal uptake, since this is the only reasonable explanation for the increased steady-state level of the appearance of NMN in the venous effluent in the presence of cocaine.

Analogous experiments were carried out with the inhibitor of extraneuronal uptake, hydrocortisone (Graefe & Trendelenburg 1974). In the presence of 30 μM-hydrocortisone, only the metabolite of extraneuronal origin (i.e., NMN) disappeared from the venous effluent, while there was very little or no change in the rate of the appearance of the neuronal metabolites in the venous effluent. This observation confirms the conclusions regarding the neuronal and extraneuronal origin, respectively, of the various metabolites. It is of interest to note that hydrocortisone failed to affect significantly the steady-state level of the appearance of the neuronal metabolites. Thus, while neuronal uptake appears to be able to restrict extraneuronal uptake (see above), the reverse does not apply.

The results lead to the following conclusions (for cat hearts): (1) Only NMN is of extraneuronal origin; hence, the extraneuronal MAO of the cat heart does not seem to play any role under these experimental conditions, though it is possible that some extraneuronal deamination might be revealed after block of catechol *O*-methyltransferase (COMT). This point has not yet been studied. (2) All deaminated metabolites (DOPEG, DOMA and OMDA) are overwhelmingly of neuronal origin; thus, the cat heart must contain intraneuronal MAO *and* COMT. (3) While the extraneuronal system is in very quick equilibrium with the perfusion fluid (the concentration of the substrate at the enzyme appears to equilibrate instantaneously with the concentration in the perfusion fluid), equilibration for nerve endings requires time.

In experiments with normal hearts (not pretreated with reserpine), very similar observations were made, but the rate of appearance of the deaminated metabolites (of neuronal origin) approached steady-state conditions considerably more slowly than in reserpine-pretreated hearts. This quantitative difference is obviously due to the storage of unchanged amine in the intraneuronal vesicles, since a steady state can be reached only after the net flux of [^{3}H]noradrenaline into the vesicles has ceased.

Rabbit hearts

Under experimental conditions identical with those described above for experiments with cat hearts (i.e., reserpine-pretreated rabbits, constant perfusion with 0.3 μM-[^{3}H](—)-noradrenaline) the metabolite pattern differed from that observed for cat hearts (Fig. 4). The most striking difference is represented by NMN, which appeared at very low rates. As already pointed out by Bönisch & Trendelenburg (1974) and Lindmar & Löffelholz (1974), rabbit hearts seem to have a very poorly developed extraneuronal uptake mechanism, since very small arterio-venous differences are obtained for isoprenaline (a substrate for extraneuronal but not for neuronal uptake; Hertting 1964). Thus, the rabbit heart appears to represent a tissue with an avid neuronal uptake (note that the steady-state rate of removal is greater for rabbit than for cat hearts; Figs. 4 and 1) and very poor extraneuronal uptake. In this case, it is virtually impossible to determine the role of extraneuronal MAO, since it

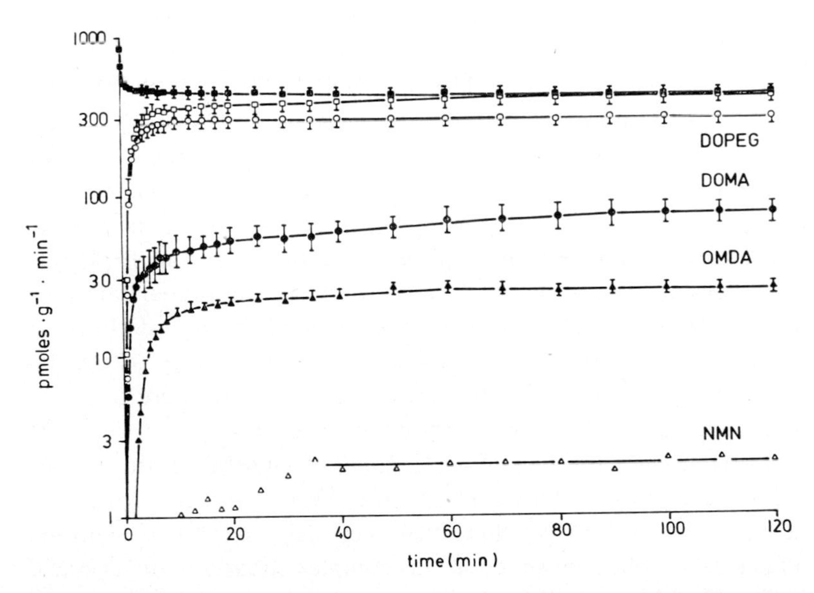

FIG. 4. Removal of noradrenaline by the perfused rabbit heart and appearance of metabolites in venous effluent. Means ± S.E. are shown from four experiments. Experimental conditions, ordinate and abscissa as in Fig. 1. ■—■, rate of removal of [^{3}H]noradrenaline from perfusion fluid; □—□, rate of appearance of sum of all metabolites in the venous effluent. Other symbols as in Fig. 1.

cannot play an important part when the extraneuronal uptake mechanism is poorly developed.

Comparison of Figs. 1 and 4 shows other interesting species differences. In the rabbit heart the rate of appearance of DOPEG in the venous effluent approaches a steady state very quickly and monophasically (half-time: 1.4 min). For DOMA and OMDA, on the other hand, the approach to a steady state takes longer and is biphasic. These differences between cat and rabbit hearts indicate (*a*) that there is more than one pool in the adrenergic nerve ending into which the deaminated metabolites distribute after their formation, and (*b*) that there are species differences between these pools. Moreover, the plot of total formation of metabolite against time (see above) yielded regression lines which intersected the abscissa at zero time. Hence, in contrast to the results obtained for DOPEG, DOMA and OMDA in cat hearts (see Fig. 2), the equilibration of the concentration of the substrate at the intraneuronal enzyme with the perfusion concentration seemed to be very quick in rabbit hearts.

Rat hearts

We were interested in the extraneuronal fate of noradrenaline in this tissue. Hearts were obtained from reserpine-pretreated rats and perfused with 0.95 μM of [^{3}H](−)-noradrenaline in the presence of a concentration of cocaine (30 μM) which blocked neuronal uptake by 96%. Fig. 5 shows the rate of the appearance of the various metabolites in the venous effluent. It should be noted that, although cocaine was present, there was a considerable efflux of deaminated metabolites. It is unlikely that *all* deaminated metabolites were of neuronal origin. In order to test this point, 87 μM-corticosterone (an inhibitor of extraneuronal uptake in the rat heart, see Bönisch *et al.* 1974) was added to the perfusion fluid which continued to contain noradrenaline plus cocaine (Fig. 6). *All* metabolites (including the deaminated ones) declined in the venous effluent. This observation indicates that the majority of the deaminated metabolites formed in the absence of corticosterone were of extraneuronal origin. It cannot be stated with confidence that they were *all* of extraneuronal origin, since the rate of appearance of DOPEG in the venous effluent did not decline to zero during 60 min of perfusion with corticosterone. However, it would be false to conclude that they were (partly) of neuronal origin; two facts have to be considered in this complex situation. (*a*) The acid metabolites (i.e., DOMA and VMA, the latter part of our OMDA fraction) do not easily leave the tissue (Henseling *et al.* 1973; Levin 1974); hence an ongoing efflux of these two metabolites may be representing efflux from a store of the metabolites rather than ongoing formation of these two metabolites.

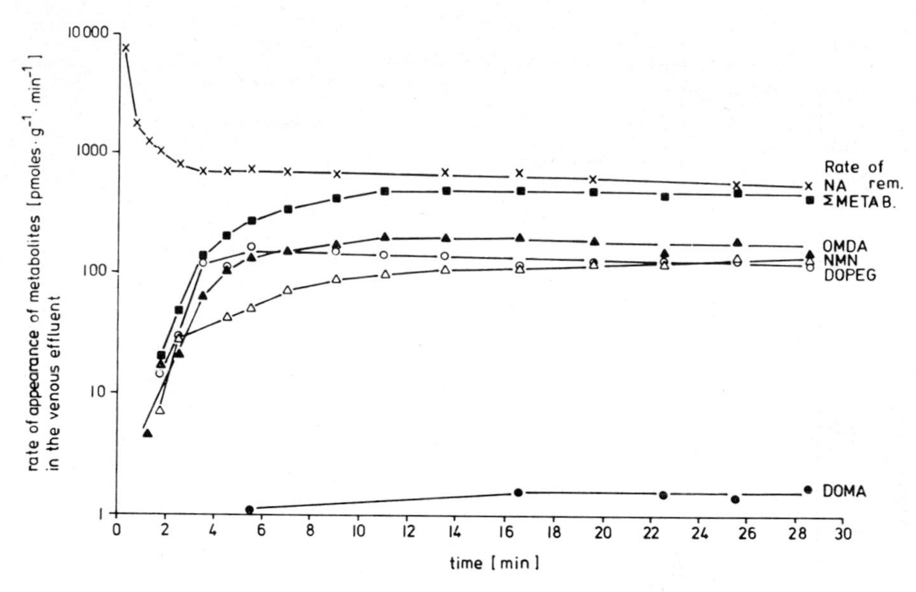

FIG. 5. Rate of removal of [^{3}H](−)-noradrenaline by the perfused rat heart and rate of appearance of metabolites in the venous effluent. The animals were pretreated with reserpine, and hearts were perfused with 0.95 μM-[^{3}H](−)-noradrenaline plus 30 μM-cocaine. *Ordinate:* rates (in pmol g^{-1} min^{-1}; log scale). *Abscissa:* time (in min) after onset of perfusion. Means of three experiments are shown. From above down: rate of removal of [^{3}H]noradrenaline (×—×); rate of appearance (in the venous effluent) of the sum of all metabolites (■—■), of OMDA (▲—▲), of NMN (△—△), of DOPEG (○—○) and of DOMA (●—●).

However, this argument does not apply to DOPEG, which seems to be able to penetrate membranes easily (Henseling *et al.* 1973; Levin 1974). (*b*) At the time when corticosterone was added to the perfusion fluid, the heart had accumulated a substantial amount of unchanged [^{3}H]noradrenaline. Since corticosterone impairs the efflux of the unchanged amine (Bönisch *et al.* 1974; Uhlig *et al.* 1974; Eckert *et al.* 1976), this considerable amount of noradrenaline is trapped in extraneuronal stores which contain the metabolizing enzymes. Hence, the efflux of metabolites observed after the addition of corticosterone to the perfusion fluid represents in all probability the metabolism of 'trapped parent amine that was taken up some time earlier' rather than metabolism of 'noradrenaline immediately after its extraneuronal uptake'. Or, in other words, while steady-state conditions existed just before the addition of corticosterone to the perfusion fluid, corticosterone interrupted the steady state. For other examples of the ability of corticosterone to increase the metabolism of 'amine trapped in the extraneuronal stores', see Uhlig *et al.* (1974) and Eckert *et al.* (1976).

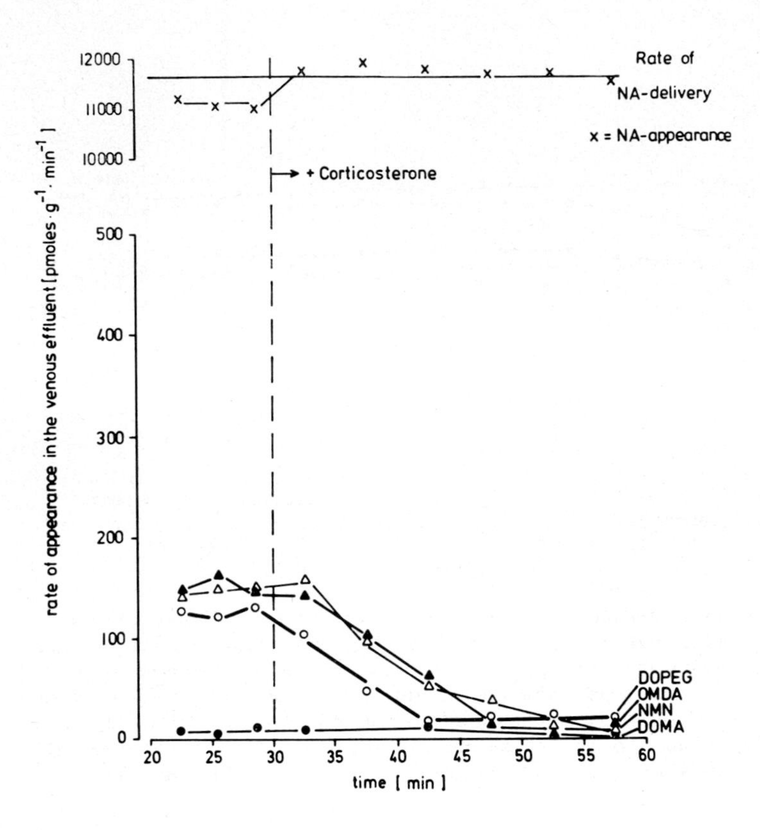

FIG. 6. The effect of corticosterone on the appearance of metabolites in the venous effluent of rat hearts perfused with noradrenaline plus cocaine. Conditions as in Fig. 5. Hearts were perfused with 0.95 μM-[^{3}H](−)-noradrenaline plus 30 μM-cocaine throughout the experiment. After the rate of appearance of the metabolites in the venous effluent had reached a steady state (i.e., after 30 min), 87 μM-corticosterone was added to the perfusion fluid. *Ordinate:* rates (in pmol g^{-1} min^{-1}; linear scale). *Abscissa:* time (in min) after onset of perfusion with amine. From above down: rate of delivery of [^{3}H]noradrenaline to the heart (horizontal line); rate of appearance of [^{3}H]noradrenaline in venous effluent (×—×); other symbols as in Fig. 5. Means from two experiments are shown. Note that corticosterone abolished the arterio-venous difference and suppressed the appearance of all metabolites in the venous effluent.

These results confirm that the extraneuronal uptake system of the rat heart is highly effective (Iversen 1965; Lightman & Iversen 1969; Bönisch & Trendelenburg 1974). Moreover, they show that extraneuronal MAO plays an important role in the metabolism of extraneuronally located noradrenaline even when COMT remains intact.

Conclusions

While there is good evidence that intraneuronal MAO plays an important role in the metabolism of exogenous noradrenaline in all mammalian species so far studied, there is very pronounced species variability with regard to the role played by extraneuronal MAO. The results presented here show that there are at least two factors which determine the role of extraneuronal MAO: (*a*) since the enzyme is in series with uptake$_2$, the efficiency of uptake$_2$ is of decisive importance. Extraneuronal uptake is very effective in the hearts of some species (e.g. cat and rat) but poorly developed in others (e.g. rabbit and also guinea pig; see Bönisch & Trendelenburg 1974). (*b*) Even for species with highly developed uptake$_2$ systems, the role played by extraneuronal MAO differs greatly; while the rat heart produces substantial amounts of extraneuronal deaminated metabolites, the cat heart does not. At the moment, it is not clear whether this difference is due to differences in extraneuronal MAO activity, to differences in intracellular access of the substrate to the extraneuronal enzyme, or to a more successful competition by the COMT of the cat heart.

RABBIT AORTIC STRIPS

Isolated aortic strips were obtained from normal or from reserpine-pretreated animals (1 mg/kg, 18 h before the experiment). They were incubated with 1.18 μM of labelled (−)-, (±)- or (+)-noradrenaline for 30 min and then washed out for 250 minutes with amine-free solution. The metabolizing enzymes were inhibited throughout; MAO was inhibited by exposure of the strips *in vitro* to 0.5 mM-pargyline for 30 minutes (followed by wash-out), COMT by the presence of 0.1 mM-U-0521 throughout the experiment.

Some experiments were terminated at the end of the initial incubation with noradrenaline. The total formation of metabolites (recovered from tissue *and* bath) by reserpine-pretreated strips was reduced by inhibition of both metabolizing enzymes by 95% (as compared with corresponding controls with intact enzymes).

The efflux of radioactivity from the strips after 200 to 250 minutes of wash-out (i.e., the 'late efflux') is of virtually exclusively neuronal origin. This is evident from the results presented in Fig. 7 (top line) which show that the rate of efflux of total radioactivity from 'nerve-free strips' (obtained by removal of the adventitia and by exposure to 30 μM-cocaine to prevent uptake into possibly surviving nerve endings) was only a very small fraction of the rate of efflux of total radioactivity from either normal or reserpine-pretreated strips. Compartmental analysis of the efflux curves confirmed that efflux from extraneuronal

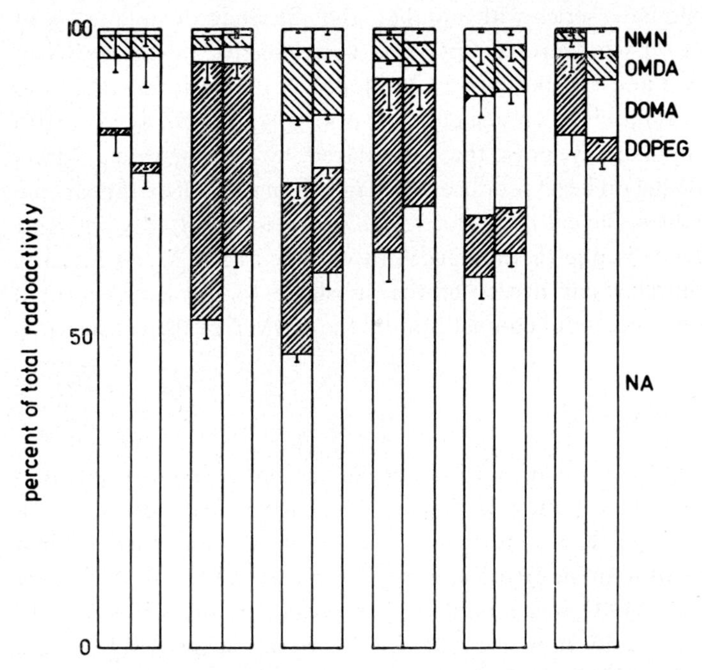

FIG. 7. Metabolite pattern of the 'late' efflux of total radioactivity from rabbit aortic strips first incubated with 1 18 μM of labelled noradrenaline for 30 min and then washed out with amine-free solution for 250 minutes. *Ordinate:* Contribution of amine or metabolite (in % of total radioactivity) to the efflux determined between the 200th and 250th minute of wash-out. Abbreviations: D, nerve-free strips; P, *in vitro* pretreatment with pargyline; U, in the presence of U-0521; COC, in the presence of 30 μM-cocaine; R, after pretreatment with reserpine; PP, after *in vivo* and *in vitro* pretreatment with pargyline; W, cocaine was present during wash-out only. For concentrations of inhibitors and further details, see text. From above down: type of preparation; isomer of noradrenaline; number of observations; mean of rate of efflux (in pmol g^{-1} min^{-1}); S.E. Means ± S.E. are shown for noradrenaline (NA, □), DOPEG (▨), DOMA (□), OMDA (▧) and NMN (□).

stores has half-times of 3 and 10 minutes (for two different extraneuronal compartments), which ensure that there is hardly any radioactivity left in these stores after 250 minutes of wash-out.

This 'late neuronal efflux' (Fig. 7) contained a surprisingly high percentage of deaminated metabolites (in spite of inhibition of MAO and COMT). There are several possible explanations for this unexpected finding.

(*a*) The adrenergic nerve ending might contain a pool of MAO which is resistant to pargyline. However, pretreatment of the animals with 100 mg/kg of pargyline (3 h before the experiment, in addition to the usual *in vitro* treatment with pargyline, see above) greatly reduced the contribution by the deaminated metabolites to the efflux of total radioactivity (see group PPU in Fig. 7). Hence, the MAO responsible for the appearance of deaminated metabolites in the late efflux seemed to be pargyline-sensitive.

(*b*) If the efflux constant for a metabolite is lower than that for the parent amine, prolonged wash-out should result in an increase in the *percentage* contribution of the metabolite to the total radioactivity remaining in the strip. In this case, relatively high rates of the 'late' efflux of the metabolite can appear. While this argument may be partly applicable to DOMA and OMDA (whose *percentage* contribution to the total radioactivity remaining in the strip did increase significantly), it does not apply to DOPEG. Irrespective of whether strips were analysed at the beginning or at the end of the wash-out period, DOPEG always represented about 2 to 3% of the total radioactivity recovered from the strip.

(*c*) The phenomenon discussed here might be generated by neuronal re-uptake. The unidirectional movement of noradrenaline out of the nerve ending might be much greater than is evident from the results of Fig. 7, since a considerable proportion of this efflux of unchanged amine might be transported back into the nerve ending by the neuronal uptake mechanism. Or, in other words, the existence of highly effective re-uptake for the parent amine might result in a selective loss of the metabolites (which are not subject to re-uptake). If this were so, cocaine (which blocks this re-uptake) should greatly reduce the percentage contribution by the deaminated metabolites to the efflux of total radioactivity. Fig. 7 shows that this was not the case (although a minor effect of cocaine is apparent).

(*d*) The same argument can be applied to the loss of the parent amine from, and its re-uptake into the storage vesicles. However, in this case there should be a very pronounced difference in the metabolite pattern of the late efflux from normal (intact vesicles) and from reserpine-pretreated strips (reduced vesicular storage capacity). Again, Fig. 7 shows that this was not the case.

Since the known uptake mechanisms (neuronal and vesicular) cannot account for the phenomenon, we have to entertain the possibility that axoplasmic binding sites for noradrenaline exist which protect the extravesicularly located amine from degradation by intraneuronal MAO.

These hypothetical axoplasmic binding sites cannot explain a second phenomenon presented in Fig. 7: the difference in metabolite pattern of the late efflux after initial incubation with either (−)- or (+)-noradrenaline. These differences agree with the finding of Levin (1974) that (in rabbit aortic strips with intact enzymes) DOPEG is the main neuronal metabolite for (−)-noradrenaline, while (+)-noradrenaline is also metabolized to DOMA and VMA (which appears in our experiments in the OMDA fraction). While our results show no evidence for any stereoselectivity of MAO, the (−)-aldehyde seems to be metabolized further predominantly by the aldehyde reductase, while the aldehyde dehydrogenase appears to represent an important metabolic pathway for the (+)-aldehyde. However, it is by no means certain that these findings indicate stereoselectivity of the aldehyde dehydrogenase and/or aldehyde reductase. The difference in the metabolite pattern may well be the consequence of differences in the distribution of the two isomers into the various neuronal compartments (perhaps containing different populations of these two enzymes).

Although a conclusive explanation of the phenomena presented in Fig. 7 has not been found, the late neuronal efflux from rabbit aortic strips provides evidence for a highly complex compartmentation of the axoplasmic pool of noradrenaline.

ACKNOWLEDGEMENT

The studies described here were supported by the Deutsche Forschungsgemeinschaft.

References

BÖNISCH, H. & TRENDELENBURG, U. (1974) Extraneuronal removal, accumulation and *O*-methylation of isoprenaline in the perfused heart. *Naunyn-Schmiedeberg's Arch. Pharmacol. 283*, 191-218

BÖNISCH, H., UHLIG, W. & TRENDELENBURG, U. (1974) Analysis of the compartments involved in the extraneuronal storage and metabolism of isoprenaline in the perfused heart. *Naunyn-Schmiedeberg's Arch. Pharmacol. 283*, 223-244

ECKERT, E., HENSELING, M. & TRENDELENBURG, U. (1976) The effect of inhibitors of extraneuronal uptake on the distribution of ^{3}H-(±)-noradrenaline in nerve-free rabbit aortic strips. *Naunyn-Schmiedeberg's Arch. Pharmacol.* in press

GRAEFE, K.-H. & TRENDELENBURG, U. (1974) The effect of hydrocortisone on the sensitivity of the isolated nictitating membrane to catecholamines. Relationship to extraneuronal uptake and metabolism. *Naunyn-Schmiedeberg's Arch. Pharmacol. 286*, 1-48

HENSELING, M., ECKERT, E., GRAEFE, K. H. & TRENDELENBURG, U. (1973) Differences in retention and efflux of ^{3}H-DOMA (dihydroxymandelic acid) and ^{3}H-DOPEG (dihydroxyphenylglycol) during wash out of rabbit aortic strips preloaded with ^{3}H-(±)-noradrenaline. *Naunyn-Schmiedeberg's Arch. Pharmacol. 279*, R16

HERTTING, G. (1964) The fate of ^{3}H-isoproterenol in the rat. *Biochem. Pharmacol. 13*, 1119-1128

IVERSEN, L. L. (1965) The uptake of catechol amines at high perfusion concentrations in the rat isolated heart: A novel catechol amine uptake process. *Br. J. Pharmacol. 25*, 18-33
JARROTT, B. (1971) Occurrence and properties of monoamine oxidase in adrenergic neurons. *J. Neurochem. 18*, 7-16
JARROTT, B. & IVERSEN, L. L. (1971) Noradrenaline metabolizing enzymes in normal and sympathetically denervated vas deferens. *J. Neurochem. 18*, 1-6
JARROTT, B. & LANGER, S. Z. (1971) Changes in monoamine oxidase and catechol-*O*-methyl transferase activities after denervation of the nictitating membrane of the cat. *J. Physiol. (Lond.) 212*, 549-559
LEVIN, J. A. (1974) The uptake and metabolism of ^{3}H-1- and ^{3}H-dl-norepinephrine by intact rabbit aorta and by isolated adventitia and media. *J. Pharmacol. Exp. Ther. 190*, 210-226
LIGHTMAN, S. L. & IVERSEN, L. L. (1969) The role of uptake$_2$ in the extraneuronal metabolism of catecholamines in the isolated rat heart. *Br. J. Pharmacol. 37*, 638-649
LINDMAR, R. & LÖFFELHOLZ, K. (1974) Neuronal and extraneuronal uptake and efflux of catecholamines in the isolated rabbit heart. *Naunyn-Schmiedeberg's Arch. Pharmacol. 284*, 63-92
LOWE, M. C. & HORITA, A. (1970) Stability of cardiac monoamine oxidase. Activity after chemical sympathectomy with 6-hydroxydopamine. *Nature (Lond.) 228*, 175-176
UHLIG, W., BÖNISCH, H. & TRENDELENBURG, U. (1974) The *O*-methylation of extraneuronally stored isoprenaline in the perfused heart. *Naunyn-Schmiedeberg's Arch. Pharmacol. 283*, 245-261

Discussion

Neff: In the arteries and small blood vessels of the rat, at least, there are more than just type A and type B monoamine oxidases. There is a connective tissue MAO that is not inhibited by pargyline or by the other classical inhibitors, but is inhibited by carbonyl reagents (Coquil *et al.* 1973). Perhaps this should be taken into account in your studies.

Trendelenburg: As you will have noticed, I did not mention A or B forms of MAO, because we have not done any experiments along these lines. The pretreatment of the animals with pargyline (in addition to the *in vitro* treatment) clearly reduced the appearance of deaminated metabolites in the medium; hence, it does not seem to be a 'pargyline-resistant enzyme', although it may be *relatively* resistant. We were worried that the efflux constants of these metabolites might be lower than for the amine, so that one would obtain a *relative* accumulation of the metabolites in the tissue. However, this is not the case, at least not for DOPEG. This never accumulates in the tissue; it always represents a very small percentage of the total radioactivity of the tissue. Nevertheless, about 50% of the efflux consists of DOPEG.

Tipton: The pattern of amine metabolism that would be expected from a consideration of the properties of the enzymes involved has been evaluated by Duncan & Sourkes (1974) and by ourselves (Turner *et al.* 1974), and the former group have listed some data on the stereospecificities of the aldehyde-metabolizing enzymes.

We found that the K_m values for the aldehyde reductases were decreased by the presence of a β-hydroxyl group in the substrate, and since aldehyde dehydrogenase was adversely affected by this substitution it was suggested that this could account for the greater formation of alcoholic metabolites that is seen with substrates of this type (Turner & Tipton 1972). With the β-hydroxylated amines, the lower K_m value exhibited by the reductases would result in their becoming saturated at lower aldehyde levels than those necessary to saturate the dehydrogenase (Turner *et al.* 1974). Your results indicated that DOPEG reached a steady state more rapidly than the acid metabolites; do you think that this could be explained in terms of the kinetics of the enzymes involved?

Trendelenburg: I don't think so. In perfusion experiments with reserpine-pretreated rabbit hearts we determined the total formation of the acid metabolite (DOMA) and of the glycol (DOPEG) with time. The formation of these two metabolites proceeded throughout the perfusion at a constant rate, and, more importantly, it reached this constant rate virtually at the beginning of the perfusion. Thus, the biphasic shape of the curve depicting the *appearance* of these metabolites in the venous effluent is not related to the formation of the metabolites but rather to their distribution. At the end of the experiment, a small amount of DOPEG is found in the heart at a time when the rate of efflux is very high. On the other hand, the heart then contains large amounts of DOMA and OMDA, although their rates of efflux are then low. Therefore, the heart needs to accumulate a lot of DOMA and OMDA to sustain a low rate of efflux, while the reverse is true for DOPEG.

Fuller: Have you any data on the relative amounts of catechol *O*-methyltransferase (COMT) in cat and rabbit heart? And if so, does the amount of enzyme account for the differences in the amount of normetanephrine formed, or is it something else?

Trendelenburg: I am very hesitant to homogenize a tissue. All the evidence from experiments of this and similar kinds indicates that there are various neuronal and extraneuronal stores. When we measure normetanephrine formation, I should like to know the tissue responsible for this formation. It might be the smooth muscle of the coronary vascular bed; it might be the entire myocardium. Homogenization would not provide the answer. I would prefer to know the cell type responsible, and it is very unlikely that all cells of the heart are equally involved in the formation of this metabolite.

Jarrott (1970) measured COMT in homogenates of rat and guinea pig hearts and found about equal activity. For the *O*-methylation of isoprenaline we find a high *O*-methylating capacity in the perfused rat heart and a poor one in perfused guinea pig heart (Bönisch & Trendelenburg 1974). For the *O*-methylation of noradrenaline, the perfused cat heart has a high capacity, while the capacity of the perfused rabbit heart is poor. Thus, while it is important to study homogenates, the results can be quite different from those obtained with perfused hearts.

Blaschko: The beautiful analysis in your paper makes me ask one question: how do you translate these experiments into terms of physiological significance? In these experiments the heart is perfused; that is to say, it is more like the condition with which the heart will be confronted when it receives catecholamine—for example, adrenaline—as hormone, than when it receives released noradrenaline after stimulation of the sympathetic nerves. Do you think noradrenaline released from nerve endings would suffer a fate similar or analogous to that in these experiments?

Trendelenburg: The only evidence for a similarity of the fate of exogenous and (released) endogenous noradrenaline comes from studies of the cat's nictitating membrane. There is an extraneuronal system for *O*-methylation with high affinity for catecholamines; this system is sensitive to hydrocortisone (Graefe & Trendelenburg 1974). After prelabelling of the neuronal stores with [^{3}H]noradrenaline, Luchelli-Fortis & Langer (1975) found that [^{3}H]normetanephrine accounted for 15% of the overflow of radioactivity (elicited by nerve stimulation) in the absence of hydrocortisone, and for only 4% in the presence of hydrocortisone. Thus, there is a hydrocortisone-sensitive *O*-methylation of the transmitter after its release. While there are qualitative similarities in the fate of exogenous and endogenous noradrenaline, one would expect quantitative differences.

Iversen: Professor Trendelenburg has given us a very powerful demonstration of the advantages of using the intact tissue. In the heart, for example, complete destruction of the sympathetic innervation by chemical or surgical sympathectomy leads to no detectable difference in total MAO activity. Nevertheless, by your technique you are able to look at the very small neuronal compartment of MAO.

I wonder if you have thought of using this system to look at the properties of the neuronal compartment of MAO? For example, you might pretreat animals with different MAO inhibitors *in vivo* and see what effect different drugs have; or one could give pargyline *in vivo* and measure the rate of *de novo* synthesis and turnover of MAO in the adrenergic nerve terminal, which would be a very interesting question that, as far as I can see, cannot be answered in any other way.

Trendelenburg: You saw the many points on the graphs. They are rather horrible to look at, but they are even more horrible to determine. There are many samples, and each sample has to go over three columns; this is very tedious, and we do not lightly volunteer to employ this method. Nevertheless, we should like to get some idea of the relative importance of these enzymes, and if the discussion of the A and B forms becomes exciting enough, we may well be tempted to use specific inhibitors. At the moment, the rat heart is the best candidate for measurable extraneuronally deaminated metabolites; the cat heart is a failure in this respect. It is excellent for studies of the extraneuronal formation of normetanephrine, but there are virtually no extraneuronally deaminated metabolites.

Iversen: The cat, however, might have a great advantage for looking at the neuronal compartment, which is perhaps the most difficult one to study.

Knoll: Perhaps one could approach this problem better by using the papillary muscle of the heart. If you feed the tissue with [^{3}H]noradrenaline and then give field stimulation of different pulses, you can measure what happens to the metabolites.

Trendelenburg: There are several methods available for the study of the fate of released transmitter. However, Langer and his co-workers are doing this type of experiment, and I see no reason to compete with them. Secondly, when you get involved in problems of released transmitter, the experimental complications are multiplied. Ours are very simple experiments; we simply perfuse with noradrenaline. When, on the other hand, you release the transmitter you have to worry much more about the timing of the collection of the samples, about whether the activity comes from neuronal or extraneuronal sources, and so on.

Murphy: The point I want to raise also concerns complexity, and it fits in with Dr Iversen's suggestion. We all have many things we would like to see you do, Dr Trendelenburg! This is a natural experiment. Instead of using drugs one might make use of the recent findings of Callingham & Lyles (1975) reporting a marked increase in the percentage of MAO A activity in the rat heart with age (as measured in terms of body weight). This remarkable change goes from 0% A activity in the immature (36 g) rat to a much higher percentage of MAO A activity (70%) in the older (414 g) rat. These findings, although they have only been reported in abstract form, raise the possibility that not only the species but the age of the animal might contribute to measured differences between A and B enzyme activity. I presume yours were middle-aged animals?

Trendelenburg: Yes; our rats weighed 200–250 g.

Gorkin: Professor R. Imaizumi and his co-workers claimed that they could demonstrate stimulation of MAO activity in heart by treatment with reserpine

(Izumi *et al.* 1967). Do you think that in the conditions of your experiment MAO activity may be stimulated?

Trendelenburg: While the phenomenon is well known, the question is unresolved. Palm *et al.* (1970) showed electron micrographs of cardiac mitochondria obtained from rats pretreated with very high doses of reserpine. The mitochondria were swollen and looked damaged. When such an alteration of mitochondria is caused by very high doses of reserpine, access to monoamine oxidase might be improved. There is some evidence for the view that the rate of deamination is increased after fairly high doses of reserpine—that is, increased beyond what one would expect from any block of vesicular uptake (Lindmar & Löffelholz 1974). However, in our studies, we tried to stay in the range of low doses of reserpine (1.5 mg/kg).

Gorkin: So in your case this effect was excluded?

Trendelenburg: I can't swear to it that it was excluded. However, the pretreatment with reserpine did not affect the ability of the heart to remove noradrenaline from the perfusion fluid, although the retention of noradrenaline in storage vesicles fell by 95–98%. If one were to give higher doses of reserpine, an increase in monoamine oxidase activity might well be observed.

Sourkes: Dallman and others have shown that you can get very sick-looking mitochondria in the liver of rats with iron deficiency, riboflavin deficiency, or copper deficiency (Dallman & Goodman 1970, 1971; Tandler *et al.* 1969). The mitochondria are bloated and distorted. Yet MAO alters in activity in riboflavin and in iron deficiencies but not in copper deficiency. So distortion of the mitochondrion and the 'stretching' of the outer membrane does not necessarily cause changes in MAO activity.

Trendelenburg: I did not want to imply that I know the explanation. I simply wanted to draw a parallel to what others have been seeing after pretreatment with high doses of reserpine.

Sourkes: I am entirely sympathetic! I looked up these papers wondering whether the distortion of the membrane in this way might affect the MAO activity, but it did not.

Youdim: Treatment of rats with 5 mg/kg reserpine causes a significant increase in the MAO activity of heart (Youdim & Sandler 1968; Izumi *et al.* 1967). Furthermore, reserpine affects the subcellular distribution of MAO activity. Centrifugation of heart homogenates (160 000 g for two hours) prepared from reserpinized rats shows the presence of 'solubilized' MAO in the supernatant. It seemed to us that reserpine was increasing MAO activity by affecting membrane permeability.

Sandler: Incidentally, in certain circumstances, such as those that Dr

Blaschko invoked, anaesthesia (halothane) seems to facilitate MAO activity (Schneider *et al.* 1974).

Fuller: In all your experiments reserpine was given *in vivo*, Professor Trendelenburg. Have you ever added reserpine *in vitro*, say 20 minutes after perfusion? At that time, according to your theory, noradrenaline would have accumulated inside the neuron, but it would not be accessible to MAO. Then you would suddenly make it accessible to MAO and you should get a burst of deaminated metabolites, which might support that interpretation.

Trendelenburg: Very few people have used reserpine *in vitro*, partly because of problems of solubility, partly because one would need high concentrations in order to see an effect. Adler-Graschinsky *et al.* (1972) have used the reserpine-like compound Ro 4-1284 on guinea pig atria *in vitro*; there was a very pronounced increase in the appearance of deaminated metabolites.

Fuller: I am particularly interested in the timing. If the lag, as you suggested, is due to the fact that you are from the beginning accumulating intraneuronal noradrenaline, but in storage granules, you should see a burst of activity, depending on how quickly the reserpine activity develops.

References

ADLER-GRASCHINSKY, E., LANGER, S. Z. & RUBIO, M. C. (1972) Metabolism of norepinephrine released by phenoxybenzamine in isolated guinea-pig atria. *J. Pharmacol. Exp. Ther. 180*, 286-301

BÖNISCH, H. & TRENDELENBURG, U. (1974) Extraneuronal removal, accumulation and *O*-methylation of isoprenaline in the perfused heart. *Naunyn-Schmiedeberg's Arch. Pharmacol. 283*, 191-218

CALLINGHAM, B. A. & LYLES, G. A. (1975) Some effects of age upon irreversible inhibition of cardiac MAO. *Br. J. Pharmacol. 53*, 458-459*P*

COQUIL, J. F., GORIDIS, C., MACK, G. & NEFF, N. H. (1973) Monoamine oxidase in rat arteries: evidence for different forms and selective localization. *Br. J. Pharmacol. 48*, 590-599

DALLMAN, P. R. & GOODMAN, J. R. (1970) Enlargement of mitochondrial compartment in iron and copper deficiency. *Blood 35*, 496-505

DALLMAN, P. R. & GOODMAN, J. R. (1971) The effects of iron deficiency on the hepatocyte: a biochemical and ultrastructural study. *J. Cell. Biol. 48*, 79-90

DUNCAN, R. J. S. & SOURKES, T. L. (1974) Some enzymic aspects of the production of oxidised or reduced metabolites of catecholamines and 5-hydroxytryptamine by brain tissues. *J. Neurochem. 22*, 663-669

GRAEFE, K.-H. & TRENDELENBURG, U. (1974) The effect of hydrocortisone on the sensitivity of the isolated nictitating membrane to catecholamines. Relationship to extraneuronal uptake and metabolism. *Naunyn-Schmiedeberg's Arch. Pharmacol. 286*, 1-48

IZUMI, F., OKA, M., YOSHIDA, H. & IMAIZUMI, R. (1967) Effect of reserpine on monoamine oxidase activity in guinea pig heart. *Life Sci. 6*, 2333-2343

JARROTT, B. (1970) Uptake and metabolism of catecholamines in the perfused hearts of different species. *Br. J. Pharmacol. 38*, 810-821

LINDMAR, R. & LÖFFELHOLZ, K. (1974) Neuronal and extraneuronal uptake and efflux of catecholamines in the isolated rabbit heart. *Naunyn-Schmiedeberg's Arch. Pharmacol. 284*, 63-92

LUCHELLI-FORTIS, M. A. & LANGER, S. Z. (1975) Selective inhibition by hydrocortisone of ^{3}H-normetanephrine formation during ^{3}H-transmitter release elicited by nerve stimulation in the isolated nerve-muscle preparation of the cat nictitating membrane. *Naunyn-Schmiedeberg's Arch. Pharmacol. 287*, 261-275

PALM, D., GROBECKER, H. & BAK, I. J. (1970) Membrane effects of catecholamine releasing drugs, in *Bayer Symposium II*, pp. 188-198, Springer-Verlag, Berlin, Heidelberg & New York

SCHNEIDER, D. R., HARRIS, S. G., GARDIER, R. W., O'NEILL, J. J. & DELAUNOIS, A. L. (1974) Increased monoamine oxidase activity produced by general inhalation anaesthetic agents. *Arch. Int. Pharmacodyn. 211*, 64-73

TANDLER, B., ERLANDSON, R. A., SMITH, A. L. & WYNDER, E. L. (1969) Riboflavin and mouse hepatic cell structure and function. II. Division of mitochondria during recovery from simple deficiency. *J. Cell Biol. 41*, 477-493

TURNER, A. J. & TIPTON, K. F. (1972) The purification and properties of an NADPH-linked aldehyde reductase from pig brain. *Eur. J. Biochem. 30*, 361-368

TURNER, A. J., ILLINGWORTH, J. A. & TIPTON, K. F. (1974) Simulation of biogenic amine metabolism in the brain. *Biochem. J. 144*, 353-360

YOUDIM, M. B. H. & SANDLER, M. (1968) Activation of monoamine oxidase and inhibition of aldehyde dehydrogenase by reserpine. *Eur. J. Pharmacol. 4*, 105-108

Can the intra- and extra-homoneuronal metabolism of catecholamines be distinguished in the mammalian central nervous system?

D. F. SHARMAN

Agricultural Research Council Institute of Animal Physiology, Babraham, Cambridge

Abstract The catecholamines, dopamine and noradrenaline, are metabolized in mammalian brain to *O*-methylated, deaminated metabolites and to metabolites which are only deaminated. In the rodent brain there is evidence which shows that there is a separation of the sites at which the two types of metabolites are formed. It has been suggested that changes in the concentrations of the deaminated metabolites might reflect the metabolism of the catecholamines within those neurons which form the catecholamines for use as transmitter substances (intra homoneuronal metabolism) and that changes in the concentrations of the *O*-methylated deaminated metabolites reflect, at least in part, the metabolism of catecholamines at other sites. The development of monoamine oxidase inhibiting drugs which can distinguish between different types of monoamine oxidase and the demonstration of multiple forms of monoamine oxidase showing different substrate specificities suggested that the monoamine oxidases involved in the formation of the two types of catecholamine metabolites might have different properties. The results of studies on the effect of (—)-deprenyl and clorgyline on the formation of the two acidic metabolites of dopamine—3,4-dihydroxyphenylacetic acid (DOPAC) and 4-hydroxy-3-methoxyphenylacetic acid (homovanillic acid, HVA)—in the brains of rats and mice have shown that the formation of both metabolites appears to involve the same type or types of enzyme but that the enzymes involved in their formation in the rat differ from those in the mouse.

From their investigations into the effects of reserpine and tyramine on the urinary excretion of noradrenaline and its metabolites, Kopin & Gordon (1962, 1963) concluded that, in general, the overflow of noradrenaline released from sympathetic nerve endings was metabolized mainly by *O*-methylation, and that the noradrenaline metabolized within the nerve endings is predominantly deaminated. This paper describes some experiments which have been directed towards an explanation of the changes in the cerebral metabolism of catecholamines which can be induced by drug treatment or occur in association with defined patterns of behaviour, in terms of the sites in the brain at which the catecholamines are metabolized. In particular, the experiments test the possibility of distinguishing between the metabolism of a catecholamine within the neurons which form that catecholamine for use as a transmitter substance (intra-homoneuronal metabolism) and the metabolism which might occur in other types of neurons or in glial cells (extra-homoneuronal metabolism). The demonstration that the cerebral monoamine oxidase appears to exist in multiple forms, and that monoamine oxidase activity could be classified according to its activity towards different substrates or by its sensitivity to different inhibitory drugs (Johnston 1968; Neff & Goridis 1972; Sandler & Youdim 1972; Yang & Neff 1974; Youdim 1974; Youdim *et al.* 1972), suggested that the enzyme involved in the deamination of catecholamines in the brain might possess different properties at the different sites at which the catecholamines are metabolized and that such differences might be of physiological significance.

THE LOCUS OF THE METABOLISM OF 3,4-DIHYDROXYPHENYLETHYLAMINE (DOPAMINE) IN THE BRAIN

The major metabolic products of dopamine in the mammalian brain are 3,4-dihydroxyphenylacetic acid (DOPAC) and 4-hydroxy-3-methoxyphenylacetic acid (homovanillic acid; HVA). DOPAC is formed by the sequential actions of the two enzymes monoamine oxidase (amine: oxygen oxidoreductase [deaminating] [flavin-containing]; EC 1.4.3.4) and aldehyde dehydrogenase (EC 1.2.1.3) on the parent catecholamine and HVA is formed similarly from the 3-*O*-methyl derivative of dopamine, 3-methoxytyramine, or could be formed by the action of catechol *O*-methyltransferase (EC 2.1.1.6) on DOPAC. These metabolic changes are illustrated in Fig. 1.

In studies on the effects of drugs, which act on the central nervous system, on the metabolism of catecholamines in the brain it has been found that the concentrations of these two acidic metabolites in the brain can be altered independently by drug treatment or by environmental changes (Andén *et al.* 1964; Ceasar *et al.* 1974; Hutchins *et al.* 1975; Laverty & Sharman 1965; Murphy *et al.* 1969; Roffler-Tarlov *et al.* 1971; Sharman 1966, 1967). Some treatments which cause changes in the cerebral concentrations of HVA and DOPAC and the directions of such changes are given in Table 1.

Neff *et al.* (1967) demonstrated, in the rat, the presence of an active transport system transferring the acidic metabolite of 5-hydroxytryptamine (5HT, serotonin), 5-hydroxyindol-3-ylacetic acid (5HIAA), from brain to blood.

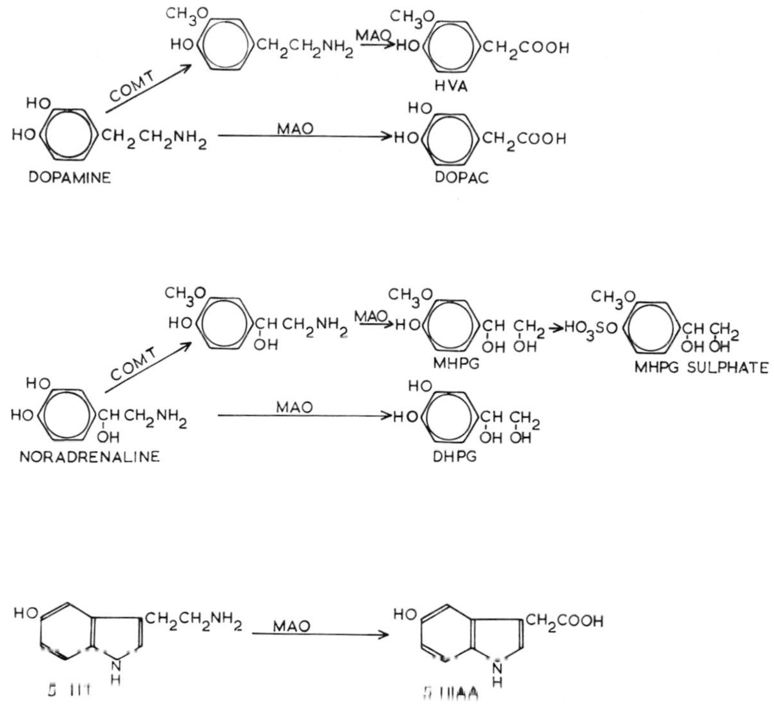

FIG. 1. The metabolism of dopamine, noradrenaline and 5-hydroxytryptamine in mammalian brain.

Abbreviations:		
	HVA	4-hydroxy-3-methoxyphenylacetic acid, homovanillic acid.
	DOPAC	3,4-dihydroxyphenylacetic acid.
	MHPG	1-(4-hydroxy-3-methoxyphenyl)ethane-1,2-diol.
	DHPG	1-(3,4-dihydroxyphenyl)ethane-1,2-diol.
	5-HT	5-hydroxytryptamine.
	5-HIAA	5-hydroxyindol-3-ylacetic acid.
	COMT	Catechol *O*-methyltransferase (E.C. 2.1.1.6).
	MAO	Monoamine oxidase (E.C. 1.4.3.4).

(The conversion of MHPG to its conjugate with sulphuric acid, $MHPGSO_3H$, does not appear to take place to any great extent in the brain of the mouse.)

Such an active transport system is also present in the mouse. This active transport system can be inhibited by probenecid, a drug which inhibits the active transport of acidic substances across the wall of the renal tubules. When rats or mice are treated with probenecid, there is an increase in the concentration of HVA in the brain, but the concentration of DOPAC remains unchanged. The striatal concentration of HVA can also be increased in the striatum of the mouse independently of that of DOPAC by environmental changes which are

TABLE 1

Some drug treatments and environmental conditions which change the concentrations of 3,4-dihydroxyphenylacetic acid (DOPAC) and homovanillic acid (HVA) in the brain of rats or mice

Treatment	*Change in concentration*		
	DOPAC		*HVA*
Probenecid	→		↑
Amphetamine-like drugs	↓		↑
Reserpine γ-OH-butyric acid (late phase)	↑	followed by	↑
Tropolone	↑		↓
Phenothiazine and butyrophenone neuroleptic drugs	↑		↑
Monoamine oxidase inhibiting drugs	↓		↓
Restraint at 20–22 °C	→		↑
Exposure to −10 °C			↑
Exposure of male mice made aggressive by isolation to a fresh environment	↑		→

↑ = increase in concentration
↓ = decrease in concentration
→ = no change in concentration

associated with a fall in body temperature (Hutchins *et al.* 1975) and such changes may originate in an effect of temperature on the active transport system. In male mice, made aggressive by isolation, only the concentration of DOPAC is increased in the striatum when they are exposed to a fresh environment. Amphetamine and 2-aminotetralin cause an increase in the striatal concentration of HVA accompanied by a fall in the concentration of DOPAC and the administration of neuroleptic drugs such as chlorpromazine or haloperidol results in an increase in the concentration of HVA which is accompanied by an increase in the concentration of DOPAC. Tropolone, which includes the inhibition of catechol *O*-methyltransferase among its actions, brings about a small fall in the concentration of HVA and a small increase in the concentration of DOPAC in the striatal tissue of normal mice. However, in mice which have been treated with a neuroleptic drug to increase the rate of formation of the acidic metabolites, tropolone reduces the striatal concentration of HVA almost to zero, but the concentration of DOPAC is increased only by an amount similar in size to that seen in normal animals (Roffler-Tarlov *et al.* 1971). The administration of the monoamine oxidase inhibiting drug, pargyline, reduces the concentration of both HVA and DOPAC. The administration of reserpine causes an increase in the concentration of DOPAC which precedes an increase in the concentration of HVA. A similar response is seen in the later stages of treatment with γ-hydroxybutyric acid (Hutchins *et al.* 1972).

These results are interpreted as follows.

1. The effect of probenecid in increasing the striatal concentration of HVA without altering that of DOPAC suggests that the two acids occur at different sites in the brain. As there appears to be a similar active transport system for the removal of 5HIAA from the rat and mouse brain that can also be inhibited by probenecid it seems unlikely that such a mechanism would be able to distinguish between HVA and DOPAC if both of these substances existed at the same site. Also it has been demonstrated (Sharman 1967) that some HVA can be formed at a site where it is unaffected by the action of probenecid.

2. Reserpine is thought to act by preventing the ATP-dependent storage of biogenic amines in intraneuronal granules. After the administration of reserpine to animals, there is a fall in the concentration of dopamine in the striatum which is accompanied by an increase in the concentration of DOPAC. An increase in the concentration of HVA occurs some 30 minutes later. This observation indicates that the metabolism of dopamine which occurs closest to the intraneuronal storage granule results in the formation of DOPAC. The administration of γ-hydroxybutyric acid has been shown to inhibit the firing of dopamine-containing neurons in the brain (Roth *et al.* 1973), thus reducing the release of dopamine at the nerve terminals. In the rat, there is an initial reduction in the formation of the acidic metabolites of dopamine but the synthesis of dopamine is accelerated and the cerebral concentration of this catecholamine is increased. In the rat and the mouse (Walters & Roth 1972; Hutchins *et al.* 1972) the striatal concentration of DOPAC then increases while the concentration of dopamine is increasing, whereas the concentration of HVA does not increase until the concentration of dopamine begins to fall towards the control value. The increase in DOPAC could represent the metabolism of excess dopamine within the dopamine-containing neuron.

3. The effect of tropolone on the concentrations of the two acidic metabolites suggests that some of the HVA present in the brain is formed by the methylation of DOPAC but that this amount remains constant under conditions when the rate of metabolism of dopamine is greatly increased.

It was concluded, for the mouse, that that HVA which was removed from the brain by the probenecid-sensitive active transport system might represent the extra-homoneuronal metabolism of dopamine. In those species which do not possess an active transport system for the removal of acidic metabolites from the brain, in contrast to the removal from the cerebrospinal fluid (see Sharman 1974), it is much more difficult to obtain evidence for a separation of the location of HVA and DOPAC.

Experiments to try to localize the site or sites of formation of the two deaminated metabolites of noradrenaline, 1-(3,4-dihydroxyphenyl)-ethan-1,2-diol (DHPG) and 1-(4-hydroxy-3-methoxyphenyl)-ethan-1,2-diol (MHPG), similar to those described above for the metabolism of dopamine, have not produced such clear answers (Ceasar *et al.* 1974). However, Braestrup & Nielsen (1975) have shown that, in the rat, the formation of DHPG and MHPG occurs with some preference within and external to, respectively, the noradrenaline-containing neurons.

Because of the difficulty in determining the relative contributions of the possible alternative metabolic pathways it was decided to examine the rates of formation of deaminated metabolites from endogenous catecholamines by estimating the rates at which the metabolites disappeared from the brain after inhibition of monoamine oxidase.

Tozer *et al.* (1966) estimated the turnover rate of 5HIAA in rat brain from the rate at which the cerebral concentration of this metabolite declined after monoamine oxidase inhibition. These authors found that pargyline (75 mg/kg i.p.) or tranylcypromine (10 mg/kg i.p.) caused the cerebral 5HIAA concentration to decline exponentially with almost identical rate constants. A similar result was obtained with phenelzine (10 mg/kg i.p.) except that the onset of the decline was delayed for about 15 min. In order to study the turnover rate of the metabolites of noradrenaline in the hypothalamus, Ceasar *et al.* (1974) injected mice with pargyline (100 mg/kg i.p.) or tranylcypromine (10 mg/kg i.p.). The results are illustrated in Fig. 2 which shows that the turnover of MHPG, estimated from the rate of decline in its concentration after tranylcypromine and assuming first-order kinetics, was 2.8 times as large as that estimated from the rate of decline measured after treatment with pargyline. When the two acidic metabolites of dopamine are estimated in the mouse striatum after treatment with the same doses of pargyline and tranylcypromine, the difference in the rates of decline is not as great as that seen with the metabolites of noradrenaline. However, the concentrations of HVA and DOPAC decrease at a slightly faster rate after tranylcypromine (10 mg/kg i.p.) than after pargyline (100 mg/kg i.p.).

These results suggest that monoamine oxidase responsible for the metabolism of the catecholamines in the mouse brain differs in some respect from that involved in the metabolism of 5HT in the rat brain, or that in the mouse the metabolism of pargyline is different from the rat, or that pargyline does not penetrate the brain in the mouse as effectively as it does in the rat. Furthermore, it is apparent that more hypothalamic noradrenaline is deaminated directly than is metabolized by catechol*O*-methyltransferase in the mouse, whereas approximately the same amounts of dopamine are metabolized to the

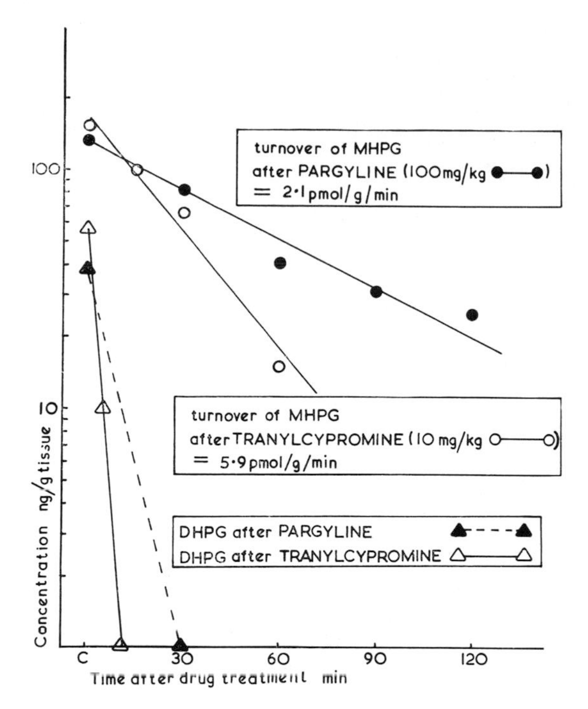

FIG. 2. The effect of pargyline (100 mg/kg, i.p.) and tranylcypromine (10 mg/kg, i.p.) on the concentrations of the glycol metabolites of noradrenaline (1-(4-hydroxy-3-methoxyphenyl)-ethane-1,2-diol; MHPG and 1-(3,4-dihydroxyphenyl)ethane-1,2-diol; DHPG) in the hypothalamus of the mouse.

two acidic metabolites. If pargyline is used to inhibit monoamine oxidase, the turnover of cerebral catecholamines, calculated from the rates of decline of the concentrations of the deaminated metabolites, may be underestimated.

The development of monoamine oxidase inhibiting drugs which will differentiate between two types of monoamine oxidase activity, type A and type B (Johnston 1968; Yang & Neff 1974), encouraged the re-examination of the metabolism of catecholamines in the brain of the mouse. Yang & Neff (1974) reported that (—)-deprenyl, an inhibitor of type B enzyme activity, would bring about a reduction in the striatal concentration of DOPAC in the rat without altering the concentration of 5HIAA, a result which might be predicted from the substrate selectivity of type B enzyme activity determined *in vitro*. In addition, in rats which had been treated with clorgyline, a specific inhibitor of the type A enzyme activity, there was a reduction in the concentrations of both DOPAC and 5HIAA in the brain, showing that dopamine is apparently

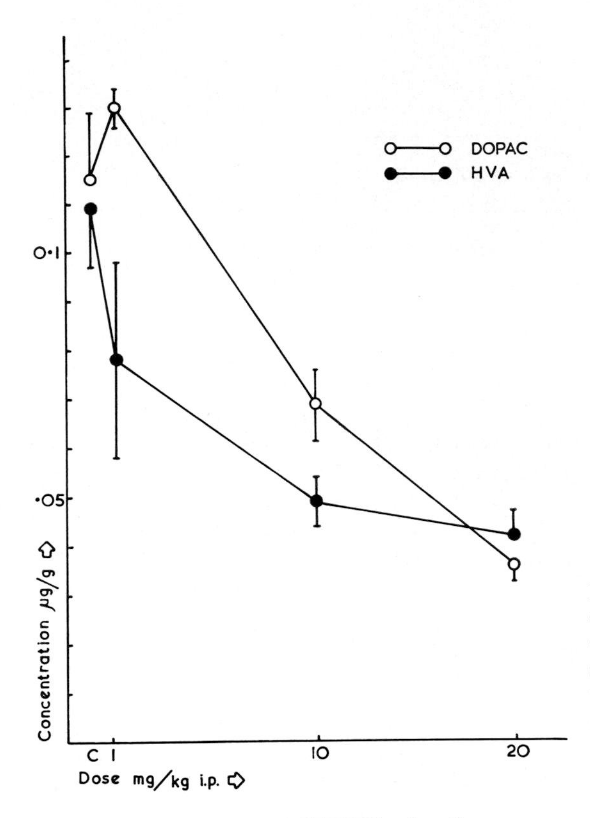

FIG. 3. The effect of (−)-deprenyl on the concentrations of 3,4-dihydroxyphenylacetic acid (DOPAC) and homovanillic acid (HVA) in the striatum of the rat. (Results are means ± S.E.M. from three observations. Duration of treatment, 2 h.)

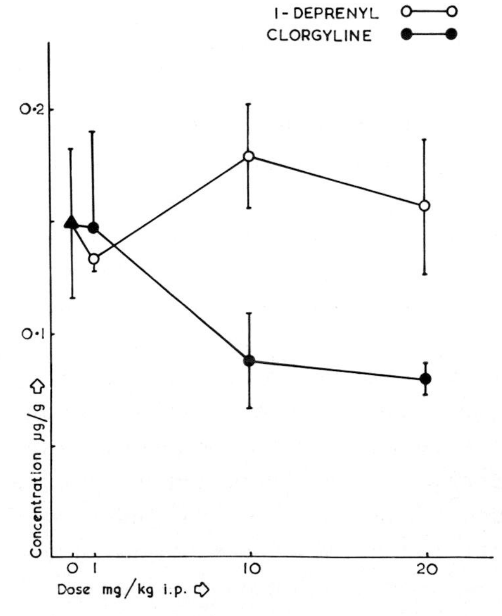

FIG. 4. The effect of (−)-deprenyl and clorgyline on the concentration of 5-hydroxyindol-3-ylacetic acid in the brain of the mouse. (Results are means ± S.E.M. from three observations. Duration of treatment, 2 h.)

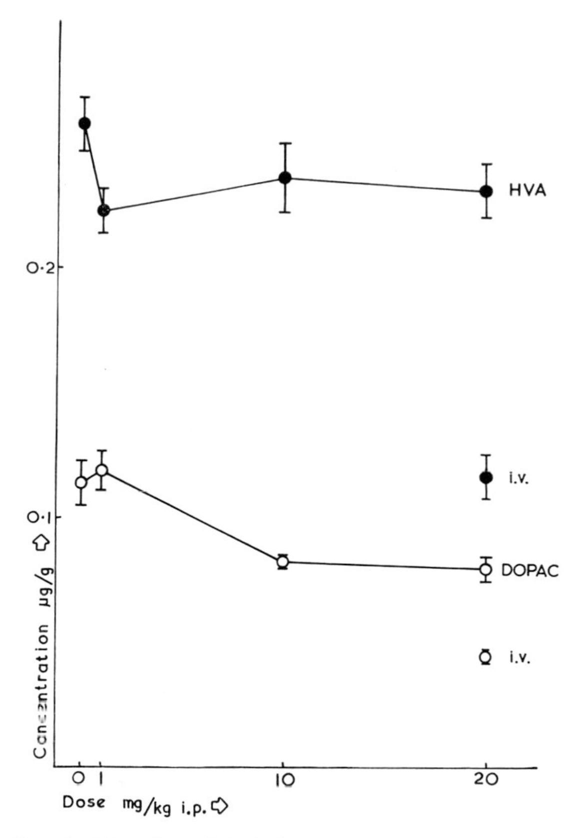

FIG. 5. The effect of (—)-deprenyl on the concentrations of 3,4-dihydroxyphenylacetic acid (DOPAC) and homovanillic acid (HVA) in the striatum of the mouse. (Results are means ± S.E.M. from 5-11 observations on pooled tissue from six mice. Duration of treatment, 2 h.)

metabolized by both type A and type B enzyme and that 5-hydroxytryptamine is metabolized preferentially by type A enzyme. Fig. 3 shows the effect of increasing doses of (—)-deprenyl administered intraperitoneally on the concentrations of HVA and DOPAC in the striatum of the rat. There appears to be little or no separation between the effects of (—)-deprenyl on the two acidic metabolites of dopamine, indicating that the type B activity is involved in the formation of both acidic metabolites. The effect of (—)-deprenyl and clorgyline on the concentration of 5HIAA in mouse brain is illustrated in Fig. 4. As observed in the rat (Yang & Neff 1974), clorgyline reduced the cerebral concentration of 5HIAA whereas (—)-deprenyl was ineffective. When the effect of intraperitoneal injections of (—)-deprenyl on the striatal concentrations of DOPAC and HVA was examined in the mouse, an unexpected result was obtained. This is illustrated in Fig. 5 which shows that (—)-deprenyl in doses

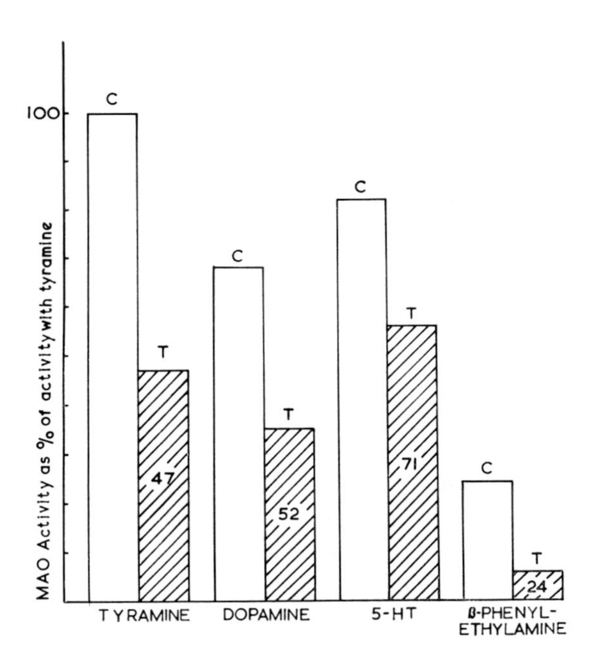

FIG. 6. Monoamine oxidase activity towards different substrates in the striatal tissue of mice treated with (—)-deprenyl (20 mg/kg i.p.). C, control mice. T, treated mice. The monoamine oxidase activity was estimated 2 h after the administration of (—)-deprenyl. The activity is expressed as a % of the activity towards tyramine. The figures in the hatched columns give the monoamine oxidase activity in the tissue from treated mice as a % of the activity observed in the corresponding controls.

up to 20 mg/kg i.p. had no effect on the striatal concentration of HVA but that at doses of 10 mg/kg and 20 mg/kg (—)-deprenyl caused a small ($P < 0.05$) fall in the concentration of DOPAC. In order to test whether (—)-deprenyl penetrated the blood–brain barrier, mice were injected intraperitoneally with a dose of 20 mg/kg and after two hours the monoamine oxidase activity in the striatum towards different substrates was estimated by a method based on that described by Robinson *et al.* (1968). The results are given in Fig. 6 and show that the activity towards β-phenylethylamine, a preferred substrate of type B enzyme, was inhibited by 75%. The dose of 20 mg/kg i.p. of (—)-deprenyl was sufficiently large to cause some inhibition of the monoamine oxidase activity towards 5-hydroxytryptamine.

When the same dose of (—)-deprenyl (20 mg/kg) was injected intravenously into mice there was a reduction in the striatal concentrations of both DOPAC and HVA. The effect of clorgyline on the striatal concentrations of DOPAC and HVA was also examined. Fig. 7 shows that the intraperitoneal admini-

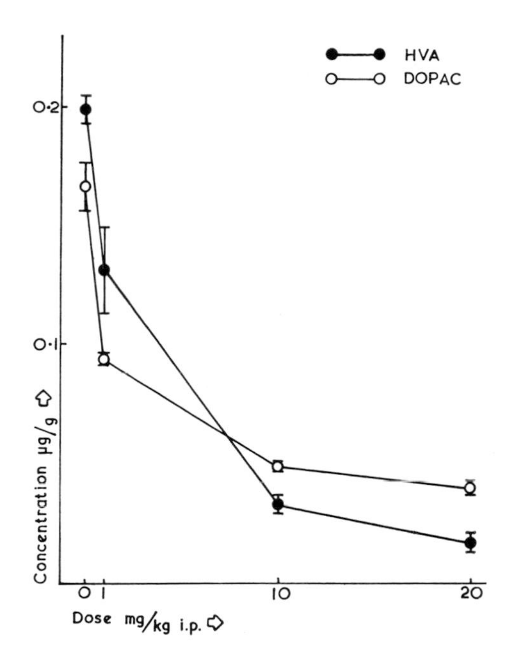

FIG. 7. The effect of clorgyline on the concentrations of 3,4-dihydroxyphenylacetic acid (DOPAC) and homovanillic acid (HVA) in the striatum of the mouse. (Results are means + S.E.M. from three observations on pooled tissue from six mice. Duration of treatment, 2 h.)

stration of this substance brings about very large reductions in the striatal concentrations of both acidic metabolites.

Among other drugs which have been shown to reduce the concentration of DOPAC in rodent brain is a group related to and including amphetamine (Glowinski *et al.* 1966; Roffler-Tarlov *et al.* 1971). Rutledge (1970) concluded from his experiments on the effect of amphetamine on the metabolism of noradrenaline in the brain that amphetamine decreased oxidative deamination by an inhibition of the uptake of the amine into the neuron, thus limiting the access to intraneuronal monoamine oxidase, rather than by inhibition of monoamine oxidase. Deprenyl is related structurally to amphetamine and may possess a similar activity but would be much less potent since a dose of 20 mg/kg of *d*-amphetamine administered intraperitoneally reduced the striatal concentration of DOPAC in the mouse by 55% after two hours (Roffler-Tarlov *et al.* 1971). A possible alternative explanation lies in the suggestion of Roth *et al.* (1973) that the concentration of DOPAC in the brain is related to the impulse flow in the dopamine-containing neurons. The administration of *d*-amphetamine results in a decrease in the firing rate of dopamine-containing nerve cells in the mid-brain (Bunney & Aghajanian 1973) and the response seen in the mouse to (—)-deprenyl might reflect a reduction in impulse traffic in the dopamine-containing neurons.

The present results indicate that in general there appears to be no great distinction between the type or types of monoamine oxidase involved in the formation of the two acidic metabolites of dopamine in the brain of the rodent. In the rat, as described by Yang & Neff (1974), both type A and type B are involved but in the mouse, most of the metabolism of dopamine appears to involve only type A. The results also show that in the mouse, the *in vitro* test would lead to a false picture of the *in vivo* cerebral metabolism of dopamine if it were assumed that the pattern of inhibition of amine metabolism seen in homogenates reflected the enzyme–substrate interactions which take place in the brain. It would thus seem easier to modify the intra- or extra-homoneuronal metabolism of catecholamines in the brain by drugs which act on the processes which are responsible for the apparent separation of the sites at which the two types of metabolite are formed rather than by a direct specific inhibition of the monoamine oxidase activity at the different sites.

The evidence available is clearly not sufficient to conclude with certainty that a change in the formation of a deaminated catecholamine reflects absolutely intra-homoneuronal metabolism or that, in the rat and mouse, the formation of HVA, which can be shown to be transported out of the brain by a probenecid-sensitive active transport system, reflects the extra-homoneuronal metabolism of dopamine. However, the apparent separation, by the action of drugs, of the sites at which the deaminated and the *O*-methylated-deaminated metabolites are formed makes this hypothesis worthy of further investigation.

References

ANDÉN, N.-E., ROOS, B.-E. & WERDINIUS, B. (1964) Effects of chlorpromazine, haloperidol and reserpine on the levels of phenolic acids in rabbit corpus striatum. *Life Sci. (Oxford) 3*, 149-158

BRAESTRUP, C. & NIELSEN, M. (1975) Intra- and extraneuronal formation of the two major noradrenaline metabolites in the CNS of rats. *J. Pharm. Pharmacol. 27*, 413-419

BUNNEY, B. S. & AGHAJANIAN, G. K. (1973) Electrophysiological effects of amphetamine on dopaminergic neurons. In *Frontiers in Catecholamine Research* (Usdin, E. & Snyder, S. H., eds.), pp. 957-962, Pergamon Press, Oxford

CEASAR, P. M., HAGUE, P., SHARMAN, D. F. & WERDINIUS, B. (1974) Studies on the metabolism of catecholamines in the central nervous system of the mouse. *Br. J. Pharmacol. 51*, 187-195

GLOWINSKI, J., AXELROD, J. & IVERSEN, L. L. (1966) Regional studies of catecholamines in the rat brain. IV. Effects of drugs on the disposition and metabolism of H^3-norepinephrine and H^3-dopamine. *J. Pharmacol. Exp. Ther. 153*, 30-41

HUTCHINS, D. A., PEARSON, J. D. M. & SHARMAN, D. F. (1975) Striatal metabolism of dopamine in mice made aggressive by isolation. *J. Neurochem. 24*, 1151-1154

HUTCHINS, D. A., RAYEVSKY, K. & SHARMAN, D. F. (1972) The effect of sodium γ-hydroxybutyrate on the metabolism of dopamine in the brain. *Br. J. Pharmacol. 46*, 409-415

JOHNSTON, J. P. (1968) Some observations upon a new inhibitor of monoamine oxidase in brain tissue. *Biochem. Pharmacol. 17*, 1285-1297

KOPIN, I. J. & GORDON, E. K. (1962) Metabolism of norepinephrine-H^3 released by tyramine and reserpine. *J. Pharmacol. Exp. Ther. 138*, 351-359

KOPIN, I. J. & GORDON, E. K. (1963) Metabolism of administered and drug released norepinephrine-7-H^3 in the rat. *J. Pharmacol. Exp. Ther. 140*, 207-216

LAVERTY, R. & SHARMAN, D. F. (1965) Modification by drugs of the metabolism of 3,4-dihydroxyphenylethylamine, noradrenaline and 5-hydroxytryptamine in the brain. *Br. J. Pharmacol. Chemother. 24*, 759-772

MURPHY, G. F., ROBINSON, D. & SHARMAN, D. F. (1969) The effect of tropolone on the formation of 3,4-dihydroxyphenylacetic acid and 4-hydroxy-3-methoxyphenylacetic acid in the brain of the mouse. *Br. J. Pharmacol. 36*, 107-115

NEFF, N. H. & GORIDIS, C. (1972) Neuronal monoamine oxidase: specific enzyme types and their rates of formation. In *Monoamine Oxidases—New Vistas* (Costa, E. & Sandler, M., eds.) *(Adv. Biochem. Psychopharmacol. 5)*, pp. 307-323, Raven Press, New York and North-Holland, Amsterdam

NEFF, N. H., TOZER, T. N. & BRODIE, B. B. (1967) Application of steady state kinetics to studies of the transfer of 5-hydroxyindoleacetic acid from brain to plasma. *J. Pharmacol. Exp. Ther. 158*, 214-218

ROBINSON, D. S., LOVENBERG, W., KEISER, H. & SJOERDSMA, A. (1968) Effects of drugs on human blood platelet and plasma amine oxidase activity *in vitro* and *in vivo*. *Biochem. Pharmacol. 17*, 109-119

ROFFLER-TARLOV, S., SHARMAN, D. F. & TEGERDINE, P. (1971) 3,4-Dihydroxyphenylacetic acid and 4-hydroxy-3-methoxyphenylacetic acid in the mouse striatum: a reflection of intra- and extra-neuronal metabolism of dopamine? *Br. J. Pharmacol. 42*, 343-351

ROTH, R. H., WALTERS, J. R. & AGHAJANIAN, G. K. (1973) The effect of impulse flow on the release and synthesis of dopamine in the rat striatum. In *Frontiers in Catecholamine Research* (Usdin, E. & Snyder, S., eds.), pp. 567-574, Pergamon Press, Oxford

RUTLEDGE, C. O. (1970) The mechanisms by which amphetamine inhibits oxidative deamination of norepinephrine in brain. *J. Pharmacol. Exp. Ther. 171*, 188-195

SANDLER, M. & YOUDIM, M. B. H. (1972) Multiple forms of monoamine oxidase: functional significance. *Pharmacol. Rev. 24*, 331-348

SHARMAN, D. F. (1966) Changes in the metabolism of 3,4-dihydroxyphenylethylamine (dopamine) in the striatum of the mouse induced by drugs. *Br. J. Pharmacol. Chemother. 28*, 153-163

SHARMAN, D. F. (1967) A discussion of the modes of action of drugs which increase the concentration of 4-hydroxy-3-methoxyphenylacetic acid (homovanillic acid) in the striatum of the mouse. *Br. J. Pharmacol. Chemother. 30*, 620-626

SHARMAN, D. F. (1974) The formation of some acidic metabolites in the brain and their subsequent transport. In *Drugs and Transport Processes* (Callingham, B., ed.), pp. 297-308, Macmillan, London

TOZER, T. N., NEFF, N. H. & BRODIE, B. B. (1966) Application of steady state kinetics to the synthesis rate and turnover time of serotonin in the brain of normal and reserpine-treated rats. *J. Pharmacol. Exp. Ther. 153*, 177-182

WALTERS, J. R. & ROTH, R. H. (1972) Effect of gamma-hydroxybutyrate on dopamine and dopamine metabolites in the rat striatum. *Biochem. Pharmacol. 15*, 2111-2121

YANG, H.-Y. T. & NEFF, N. H. (1974) The monoamine oxidases of brain: selective inhibition with drugs and the consequences for the metabolism of the biogenic amines. *J. Pharmacol. Exp. Ther. 189*, 733-740

YOUDIM, M. B. H. (1974) Heterogeneity of rat brain mitochondrial monoamine oxidase. *Adv. Biochem. Psychopharmacol. 11*, 59-63

YOUDIM, M. B. H., COLLINS, G. G. S., SANDLER, M., BEVAN-JONES, A. B., PARE, C. M. B. & NICHOLSON, W. J. (1972) Human brain monoamine oxidase: multiple forms and selective inhibitors. *Nature (Lond.) 236*, 225-228

Discussion

Fuller: Your data suggesting that dopamine in the mouse may be primarily metabolized by MAO type A fit nicely with the evidence I mentioned earlier (p. 158) that MAO inhibitors of A type are the most potent potentiators of the effects of dopa in mice.

Sharman: I agree. What worries me is the difference between an *in vitro* technique which indicates that dopamine metabolism is inhibited, and the results obtained when you give the drug *in vivo*, when dopamine metabolism is apparently only slightly affected. We shall have to find a way of interpreting the *in vitro* experiments.

Neff: Some of your results might perhaps be explained in other ways, Dr Sharman. For example, you mentioned that tranylcypromine gave a greater slope for decline than pargyline. Tranylcypromine releases catecholamines and it is a partially reversible inhibitor of MAO, and therefore the slope may be greater because the amines are released onto MAO. After pargyline, which does not release amines and is an irreversible inhibitor, the slope would depend solely on transport.

Sharman: My impression is that tranylcypromine inhibits all the MAO within a few minutes, so even if it were releasing dopamine there ought not to be any active enzyme to metabolize the released amine; but that may be wrong. I am again taking the *in vitro* evidence and trying to extrapolate to the whole animal.

Neff: You showed that in some cases HVA levels went up and DOPAC levels didn't, and *vice versa*; however, you were only measuring one particular time interval. If you choose a single time interval you may be led to the false conclusion that one metabolite is not increasing, when in fact it is.

I wonder whether anyone has looked to see if dopamine is metabolized in mouse brain *in vitro* by the B enzyme?

Sharman: I think the only information is from our experiments with deprenyl, which reduced the deamination of dopamine by about 50%.

Neff: So there is no evidence yet that dopamine is metabolized by the B enzyme of mouse brain *in vitro*. In the rat there is evidence for this, as you have shown. It may be that the various species metabolize dopamine and other amines differently.

Sharman: Yes. I suspect that the clue of the inhibition of B enzyme with (—)-deprenyl, causing a small decrease in the concentration of DOPAC in the mouse corpus striatum, means something. Does it represent metabolism of dopamine that is getting somewhere other than the site at which it is required to act?

Neff: You say that the transport of DOPAC is not inhibited by probenecid. Perhaps DOPAC has a higher affinity for the transport system than probenecid and cannot be displaced by probenecid, whereas HVA has a lower affinity and can be displaced.

Sharman: No one has looked at this, but it is unlikely, from experiments on the uptake mechanism of the choroid plexus, where one acid metabolite will interfere with the transport of a second acid metabolite (Pullar 1971). The transport mechanism does not seem to be particularly specific. I agree that these experiments have to be done. I do not know if Moir used DOPAC, but he was able to show cross-interference between acids in their removal from the c.s.f.; he did not study this system in the rodent brain (Moir 1972). It is impossible to study from that point of view; one can only draw conclusions by analogy. I do not think anyone has examined whether DOPAC excretion changes after giving probenecid, but Werdinius (1967) showed that probenecid delayed the removal of DOPAC from the bloodstream.

Maître: As regards Dr Neff's comment on HVA and DOPAC values, we have studied the time-course of the effects of clorgyline and deprenyl on these two metabolites in the rat corpus striatum. Between one and 24 hours the decreases are quite similar (Fig. 1, p. 218).

Neff: We find different results. With doses of deprenyl that do not inhibit the A type enzyme, there is a decrease in DOPAC. We have not studied HVA.

Maître: The dose of deprenyl used (10 mg/kg, subcutaneously) definitely inhibits MAO type A to some extent in the rat corpus striatum, using dopamine and 5-hydroxytryptamine as substrates (see also Fig. 1, p. 26).

Sourkes: Dr Sharman, what was the result with γ-hydroxybutyric acid? And could you comment on species differences in regard to the production of these two acids? When we studied the striatum in monkeys we found very little DOPAC but much HVA (Sharman *et al.* 1967).

Sharman: Of all the species studied, the rat is apparently unusual in its metabolism of dopamine. It appears to make much more DOPAC than HVA. Initially we were unable to detect HVA by our rather crude methods. As you go from rat to mouse to guinea pig the concentrations of DOPAC and HVA in the striatum become about equal, whereas with larger mammals such as dog or sheep there is a predominance of HVA; sometimes twenty times as much as

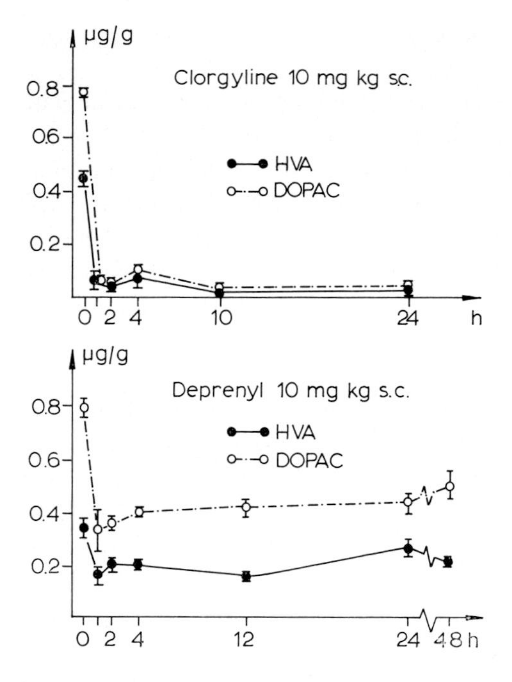

FIG. 1 (Maître). Time-course of the effects of clorgyline and deprenyl on the concentration of homovanillic acid (HVA) and dihydroxyphenylacetic acid (DOPAC) in the rat corpus striatum. The brains were removed at the indicated times after injection of the MAO inhibitors. HVA and DOPAC were extracted and estimated as mentioned in the legend of Fig. 1 (p. 128). Each point represents the mean value ± S.E.M. of four extracts from four corpora striata each.

DOPAC. We analysed the phenolic acids in rat corpus striatum, using gas chromatography (see Pearson & Sharman 1975). We find some differences between strains of rats; one particular strain had about three and a half times as much DOPAC as HVA, a ratio similar to that reported by Guldberg & Broch (1971). We have also used another strain where the concentration of DOPAC is lower than this, but generally in the rat DOPAC is the main metabolite found in the striatum. In the mouse, probenecid does not produce an increase in the striatal concentration of DOPAC. In mouse striatum, the concentrations of HVA and DOPAC are about equal.

With regard to the effect of γ-hydroxybutyric acid or γ-butyrolactone, there is a reduction in the cerebral concentration of HVA in probenecid-treated rats (Roth 1971) and also in the concentration of DOPAC (Roth *et al.* 1973) in the initial stages. We have examined the effects of sodium γ-hydroxybutyrate in mice (Hutchins *et al.* 1972). There was an increase in the concentration of dopamine in the striatum, followed by an increase in the concentration of DOPAC. As the concentration of dopamine began to fall, the concentration of HVA increased. If these effects are due to a build-up of dopamine within the nerve ending and the rate at which the amine can be stored is exceeded, some dopamine would be released within the neuron. If this is correct, one would expect an increase in the concentration of DOPAC to occur before an increase in the concentration of HVA.

Sourkes: What is the time displacement between the dopamine and DOPAC curves?

Sharman: The curves are very close, possibly because the dopamine starts spilling over onto MAO as soon as the concentration increases.

Knoll: The data presented by Dr Sharman seem to support the view that dopamine is mainly oxidized by the MAO A enzyme. We therefore have to explain the beneficial effects of (—)-deprenyl in Parkinson's disease. I would like to refer in this context to the scheme I presented earlier (p. 148), visualizing my working hypothesis of how phenylethylamine facilitates noradrenergic transmission in the guinea pig vas deferens. Phenylethylamine, which induces stereotyped behaviour in the rat, might similarly facilitate dopaminergic transmission in the brain. This approach seems to offer a reasonable explanation for the effects of (—)-deprenyl, which by increasing the brain concentrations of phenylethylamine might activate the dopaminergic system even if dopamine is mainly a substrate of the MAO A enzyme.

Sharman: To go back to my Fig. 5 (p. 211), all it tells us is that the amount of dopamine being metabolized has hardly changed. It does not really tell us whether any enzyme that metabolizes dopamine has been inhibited and that the effect of this inhibition has then been obscured by an increased release of dopamine. In fact, there may be increased release of dopamine in these mice, but the inhibition of dopamine metabolism may have balanced it out and it may be fortuitous that the concentration of HVA has come out at the same value.

van Praag: We have tried to measure DOPAC in human cerebrospinal fluid (unpublished). It was hardly detectable; after giving probenecid it was still very low and after neuroleptic drugs (phenothiazine-type) it remains low. So it seems not to be an important metabolite in man, as far as this is reflected in the c.s.f. I gather that you measured HVA and DOPAC in the corpus striatum; did you analyse whether there was a difference in HVA and DOPAC ratios in the nigro-striatal and mesolimbic dopaminergic systems?

Sharman: In other species there are differences; but it's bad enough trying to do it in the striatum of the mouse! We went on to gas chromatography to try to increase the sensitivity of our measurements. We made some measurements of HVA and DOPAC in the olfactory tubercle of the rat and there were no major differences between the concentrations there and in the striatum (Pearson & Sharman 1975).

Sandler: Whether c.s.f. contains much DOPAC is still an open question but certainly human urine contains appreciable amounts of this acid, up to about 2 mg, half of it free and half conjugated (Weg *et al.* 1975).

van Praag: Westerink & Korf (1975) from our laboratory have developed a sensitive autoanalyser method for measuring HVA and DOPAC, and in human c.s.f. the concentration of DOPAC is low, less than 0.1 μg/ml; whereas HVA is about 0.6 μg/ml.

Sharman: The value for DOPAC is < 0.01 μg/ml and for HVA, 0.06 μg/ml, for lumbar cerebrospinal fluid.

Coppen: Professor van Praag, did you measure c.s.f. metabolites in Parkinsonian patients given L-dopa? We give human subjects large doses of L-dopa for months and years; I wonder what is happening in the central nervous system? Presumably the dopaminergic neurons are filling up with dopamine; and presumably dopamine is being formed at another site and in other neurons.

van Praag: The administration of L-dopa gives rise to a significant increase in the HVA baseline and in post-probenecid levels in c.s.f., both in depressed individuals and in Parkinson patients (Korf *et al.* 1974; van Praag 1974). I have no corresponding data on DOPAC concentrations in c.s.f.

Sharman: If you do not also give a peripheral decarboxylase inhibitor, a lot of dopamine and its metabolites will be formed in blood vessel walls. In mice given L-dopa there are enormous increases in the acid metabolites. You can juggle with catechol *O*-methyltransferase inhibitors and change the concentrations up and down. After a peripheral decarboxylase inhibitor has been given, as Professor Pletscher has done, I'm sure that a lot of excess dopamine is formed, and if it is being made at sites where it is not usually made I would be surprised if MAO enzyme activity were not induced, to get rid of it. I suspect that the so-called type B enzyme might be the one that is induced by this excess dopamine metabolism, purely to stop it reaching places in the brain where it might do harm. Are there enzyme systems in the brain which can be induced if a substance gets to the wrong site?

Coppen: It has always amazed me clinically that there is little dramatic alteration in the behaviour of patients who have been given these unphysiological doses of L-dopa over months and years, where presumably dopamine is being formed in unphysiological sites. Their Parkinsonism may improve but otherwise there is not a great deal of effect.

Sharman: The metabolizing systems have an enormous capacity, and unless they are interfered with, I do not think you will see much effect. One can give

L-dopa to mice and they do not become very excited until you give MAO oxidase inhibitors as well and stop the breakdown of the products formed from L-dopa.

Kety: In Parkinsonism there may be a hypersensitivity of the dopamine receptors in the striate, which would tend to produce a greater effect there than elsewhere.

Sourkes: What really is hypersensitivity?

Kety: Whatever it may be ultimately, it is the operational phenomenon of denervation hypersensitivity. One has always wondered why it is, in Parkinsonism, that L-dopa has such a specific effect on certain areas, and none in many other areas; it may be this hypersensitivity.

Green: The fact that metabolite concentrations increase in c.s.f. doesn't necessarily mean that the receptor has been stimulated, because the intraneuronal MAO present may metabolize amines before their release into the synaptic cleft and the production of a postsynaptic response. This is something to be aware of in quoting amine metabolite concentrations in c.s.f.

Youdim: Professor Sandler and Dr Collins showed that L-dopa treatment of rats caused an increase in uterine MAO activity. We have recently repeated and confirmed this work. We also found an increase in MAO activity in arterial walls as well as in the adrenal gland after three weeks of L-dopa treatment (300 mg/kg per day). We looked at specific areas of brain of the rat and couldn't find any increases, but I believe Dr Oreland has found an increase?

Oreland: Yes. Being aware of the previous negative results, we have been extra careful to foresee pitfalls in this type of experiment, but nevertheless we find it possible to increase MAO activity in the rat brain (A. Wiberg & L. Oreland, in preparation). The increase is in the order of 20% after four weeks of treatment with 40 mg/kg of L-dopa daily.

Sourkes: Dr Andrée Roberge has given L-dopa chronically to animals: she claims that dopa decarboxylase activity increases after some time, so there may be an adaptation even in the synthetic pathway.

Sandler: Other people have found a decrease in dopa decarboxylase activity (Dairman *et al.* 1971; Tate *et al.* 1971).

Sourkes: They were using different experimental conditions, which have to be specified.

Sandler: We have examined platelet MAO in patients treated chronically with L-dopa. We didn't find any evidence of an increase (unpublished).

Dr Collins and Dr Southgate did further experiments (unpublished) in which they noted that the known progesterone-induced increase in dopamine-oxidizing activity in rat uterus (Collins *et al.* 1970) was confined electrophoretically to a specific increase in activity of a previously observed band (Collins *et al.* 1968) migrating from anode to cathode.

Gorkin: Should the increase in uterine MAO activity after the injection of L-dopa be considered in terms of adaptive synthesis of the enzyme? Dr Sandler has published evidence (Davison & Sandler 1956) that in patients with carcinoid there was no adaptive increase in MAO activity. Is it possible that the A type enzyme is adaptively increased and the B type is not?

Youdim: We would like to believe this. If such a thing occurred, it would be an adaptive mechanism for amine catabolism.

Neff: Can one really attribute the increase in MAO to the L-dopa treatment? L-Dopa changes steroid metabolism, which in turn changes the enzymes in the uterus and adrenal gland—areas where one would expect steroids to influence enzyme activity.

Youdim: This is possible. If protein synthesis is increased after giving L-dopa, it could be acting through a steroid mechanism.

Pletscher: It was suggested earlier (p. 179) that chronic administration of irreversible MAO inhibitors might abolish the difference between type A and type B inhibition. Harmaline would fulfil the theoretical criteria for an antidepressant, on the basis of present theories; it is a type A inhibitor and it is reversible, so the 'suicide' type of inhibition is not there. What is the clinical action of harmaline? Does it affect mood?

van Praag: It has been used only on a very small scale. To my knowledge, controlled data are not available.

Sourkes: In the late 20's in Germany it was the wonder drug for Parkinson's disease; it apparently had some stimulatory activity, but it was shown later to increase the tremor. There was greater alertness; the investigators were looking for a decrease in the akinesia of Parkinsonism, and it apparently helped that.

Sandler: What is interesting in this context is the recent suggestion that harmaline-like compounds—tryptolines (carbolines)—might be generated *in vivo* (Barchas *et al.* 1974). There is also much recent information about a formaldehyde-generating system in the brain (see, for example, Leysen & Laduron 1974). This would make possible the generation of three-ring compounds of this type which may inhibit MAO. One wonders whether there are any balance mechanisms here. Perhaps this is what stops us from being depressed, the generation of our own endogenous MAO inhibitors!

Sourkes: Do they really inhibit MAO, or do they just look as though they might?

Sandler: Ho and his colleagues (Ho 1972) have made an extensive study of this action.

in the original publications of Glenner *et al.* (1957) and of Shimizu *et al.* (1959). More recently a similar strategy, but with selective MAO inhibitors, has been adopted by J. E. Gascoigne *et al.* (1975). Therefore if there is inhibition of MAO activity by tetrazolium in the circumstances of the reaction one must assume that it is partial. The apparent selectivity of the stain for central monoamine-containing structures also suggests strongly that the technique is able to discriminate areas of high functional monoamine oxidase activity. It is of course a qualitative technique.

Trendelenburg: In various peripheral tissues there is increasingly good evidence that there is some pre-junctional catechol *O*-methyltransferase activity (COMT). Is there any evidence for the brain?

Sharman: Yes; several studies have shown COMT activity in synaptosomal fractions from brain.

Tipton: One should be cautious in quantitatively interpreting fractionation studies. The fact that catechol *O*-methyltransferase is a soluble enzyme makes it more difficult to localize than those that are particle-bound. In addition we have some evidence that the enzyme is capable of binding to membrane material (R. Singer & K. F. Tipton, unpublished work) and some of the activity that has been found to be associated with the synaptosomes may be bound to the outside of the synaptosomal membrane rather than being located within the nerve ending.

Iversen: Even if there is COMT activity in brain synaptosomes, this does not necessarily imply that it is localized in adrenergic synaptosomes. In the periphery, however, sympathetic denervation of the cat nictitating membrane and rat vas deferens does show the loss of a neuronal component from these tissues (Jarrott 1971; Jarrott & Langer 1971).

Fuller: It is true, as Dr Iversen said, that there may not be adequate biochemical methods for studying different MAO types, but there are pharmacological ways of studying them; and in fact that is how the distinction between A and B arose. In the pharmacological sense we are primarily interested not in the MAO that is certainly present in other cells but the neuronal MAO which is probably important in the metabolism of noradrenaline. Dr Trendelenburg's data support the idea that noradrenaline may be metabolized intra-homoneuronally, to use Dr Sharman's terminology. If uptake is blocked by cocaine, then extra-neuronal deamination of noradrenaline may occur.

Kety: You may differentiate between an A and B type pharmacologically, but how do you know how much of the A type is represented by particular nerve endings and how much by other tissue?

Fuller: My point is not so much that you could know that, but that pharmacologically we are primarily interested in that part of the MAO that actually metabolizes the monoamines. There may be MAO in other cell types which would be capable of metabolizing monoamines and can do so *in vitro*, if one grinds up tissues, but that may not be relevant *in vivo*.

References

BARCHAS, J. D., ELLIOTT, G. R., DO AMARAL, J., ERDELYI, E., O'CONNOR, S., BOWDEN, M., BRODIE, H. J. H., BERGER, P. A., RENSON, J. & WYATT, R. J. (1974) Tryptolines. *Arch. Gen. Psychiatry 31*, 862

BLASCHKO, H. (1974) Introductory remarks on monoamine oxidase. *J. Psychiatr. Res. 11*, 325-327

BOADLE, M. C. & BLOOM, F. E. (1969) A method for the fine structural demonstration of monoamine oxidase activity. *J. Histochem. Cytochem. 17*, 331-340

BRAESTRUP, C. & NIELSEN, M. (1975) Intra- and extraneuronal formation of the two major noradrenaline metabolites in the CNS of rats. *J. Pharm. Pharmacol. 27*, 413-419

COLLINS, G. G. S., YOUDIM, M. B. H. & SANDLER, M. (1968) Isoenzymes of human and rat liver monoamine oxidase. *FEBS Lett. 1*, 215-218

COLLINS, G. G. S., PRYSE-DAVIES, J., SANDLER, M. & SOUTHGATE, J. (1970) Effect of pretreatment with oestradiol, progesterone and dopa on monoamine oxidase activity in the rat. *Nature (Lond.) 226*, 642-643

DAIRMAN, W., CHRISTENSON, J. G. & UDENFRIEND, S. (1971) Decrease in liver aromatic L-amino-acid decarboxylase produced by administration of L-DOPA. *Proc. Natl. Acad. Sci. U.S.A. 68*, 2117-2120

DAVISON, A. N. & SANDLER, M. (1956) Monoamine oxidase activity in the argentaffin carcinoma syndrome. *Clin. Chim. Acta 1*, 450-456

ERÄNKÖ, O. & ERÄNKÖ, L. (1971) Small, intensely fluorescent granule containing cells in the sympathetic ganglion of the rat. *Prog. Brain Res. 34*, 39-51

GASCOIGNE, J. E., WILLIAMS, D. & WILLIAMS, E. D. (1975) Histochemical demonstration of monoamine oxidase of the brain of rodents. *Z. Zellforsch. Mikrosk. Anat. 49*, 389-400

GLENNER, G. G., BURTNER, H. J. & BROWN, G. W. (1957) Histochemical demonstration of monoamine oxidase activity by tetrazolium salts. *J. Histochem. Cytochem. 5*, 591-600

GULDBERG, H. C. & BROCH, O. J., JR (1971) On the mode of action of reserpine on dopamine metabolism in the rat striatum. *Br. J. Pharmacol. 13*, 155-167

HO, B. T. (1972) Monoamine oxidase inhibitors. *J. Pharm. Sci. 61*, 821

HUTCHINS, D. A., RAYEVSKY, K. & SHARMAN, D. F. (1972) The effect of sodium γ-hydroxybutyrate on the metabolism of dopamine in the brain. *Br. J. Pharmacol. 46*, 409-415

JARROTT, B. (1971) Occurrence and properties of catechol-*O*-methyl transferase in adrenergic neurons. *J. Neurochem. 18*, 17-27

JARROTT, B. & LANGER, S. Z. (1971) Changes in monoamine oxidase and catechol-*O*-methyl transferase activities after denervation of the nictitating membrane of the cat. *J. Physiol. (Lond.) 212*, 549-559

KORF, J., VAN PRAAG, H. M., SCHUT, T., NIENHUIS, R. J. & LAKKE, J. P. W. F. (1974) Parkinson's disease and amine metabolites in cerebrospinal fluid: implications for L-DOPA therapy. *Eur. Neurol. 12*, 340-350

LAGNADO, J. R., OKAMOTO, M. & YOUDIM, M. B. H. (1971) The effect of tetrazolium salts on monoamine oxidase activity. *FEBS Lett. 17*, 117-120

LEYSEN, J. & LADURON, P. (1974) Characterization of an enzyme yielding formaldehyde from 5-methyltetrahydrofolic acid. *FEBS Lett. 47*, 299

LINDVALL, O. & BJÖRKLUND, A. (1974) The organization of the ascending catecholamine neuron systems in the rat brain as revealed by the glyoxylic acid fluorescence method. *Acta Physiol. Scand.* Suppl. 412, 1-48

MOIR, A. T. B. (1972) Interaction in the cerebral metabolism of the biogenic amines. Effects of phenelzine on the cerebral metabolism of the 5-hydroxyindoles in dog brain. *Br. J. Pharmacol. 45*, 249-264

PEARSON, J. D. M. & SHARMAN, D. F. (1974) Increased concentrations of acidic metabolites of dopamine in the superior cervical ganglion following preganglionic stimulation *in vivo*. *J. Neurochem. 22*, 547-550

PEARSON, J. D. M. & SHARMAN, D. F. (1975) The estimation of 3,4-dihydroxyphenylacetic acid, homovanillic acid and homoisovanillic acid in nervous tissue by gas-liquid chromatography and electron capture detection. *Br. J. Pharmacol. 53*, 143-148

PULLAR, I. A. (1971) The accumulation of [^{14}C]5-hydroxyindolyl-3-acetic acid by the rabbit choroid plexus *in vitro*. *J. Physiol. (Lond.) 216*, 201-211

ROTH, R. H. (1971) Effect of anesthetic doses of γ-hydroxybutyrate on subcortical concentrations of homovanillic acid. *Eur. J. Pharmacol. 15*, 52-59

ROTH, R. H., WALTERS, J. R. & AGHAJANIAN, G. K. (1973) The effect of impulse flow on release and synthesis of dopamine in the rat striatum, in *Frontiers in Catecholamine Research* (Usdin, E. & Snyder, S. H., eds.), pp. 567-574, Pergamon Press, Oxford

SHANNON, W. A., WASSERKRUG, H. L. & SELIGMAN, A. M. (1974) The ultrastructural localization of monoamine oxidase (MAO) with tryptamine and a new tetrazolium salt, 2-(2′-benzothiazolyl)-5-styryl-3-(4′-phthalhydrazidyl) tetrazolium chloride (BSPT). *J. Histochem. Cytochem. 22*, 170-182

SHARMAN, D. F., POIRIER, L. J., MURPHY, G. F. & SOURKES, T. L. (1967) Homovanillic acid and dihydroxyphenylacetic acid in the striatum of monkeys with brain lesions. *Can. J. Physiol. Pharmacol. 45*, 57-62

SHIMIZU, N., MORIKAWA, N. & OKADA, M. (1959) Histochemical studies of monoamine oxidase of the brain of rodents. *Z. Zellforsch. Mikrosk. Anat. 49*, 389-400

TATE, S. S., SWEET, R., McDOWELL, F. H. & MEISTER, A. (1971) Decrease of the 3,4-dihydroxyphenylalanine (DOPA) decarboxylase activities in human erythrocytes and mouse tissues after administration of DOPA. *Proc. Natl. Acad. Sci. U.S.A. 68*, 2121-2123

URETSKY, N. J. & IVERSEN, L. L. (1970) Effects of 6-hydroxydopamine on catecholamine containing neurones in the rat brain. *J. Neurochem. 17*, 269-278

VAN PRAAG, H. M. (1974) Towards a biochemical typology of depressions? *Pharmacopsychiat. 7*, 281-292

WEG, M. W., RUTHVEN, C. R. J., GOODWIN, B. L. & SANDLER, M. (1975) Specific gas chromatographic measurement of urinary 3,4-dihydroxyphenylacetic acid. *Clin. Chim. Acta 59*, 249-251

WERDINIUS, B. (1967) Elimination of 3,4-dihydroxyphenylacetic acid from the blood. *Acta Pharmacol. Toxicol. 25*, 9-17

WESTERINK, B. H. C. & KORF, J. (1975) Determination of nanogram amounts of homovanillic acid in the central nervous system with a rapid semi-automated fluorometric method. *Biochem. Med. 12*, 106-114

YOUDIM, M. B. H. & LAGNADO, J. R. (1972) Limitation in the use of tetrazolium salts for the detection of multiple forms of monoamine oxidase, in *Monoamine Oxidase—New Vistas* (Costa, E. & Sandler, M., eds.) *(Adv. Biochem. Psychopharmacol. 5)*, pp. 289-292, Raven Press, New York and North-Holland, Amsterdam

Use of a behavioural model to study the action of monoamine oxidase inhibition *in vivo*

A. RICHARD GREEN and MOUSSA B. H. YOUDIM

Medical Research Council Unit and University Department of Clinical Pharmacology, Radcliffe Infirmary, Oxford

Abstract When rats are given the monoamine oxidase (MAO) inhibitor tranylcypromine (20 mg/kg) followed by L-tryptophan (100 mg/kg) they display a characteristic hyperactivity syndrome which appears to be due to increased synthesis and release of 5-hydroxytryptamine (5HT). Administration of either of the 'selective' inhibitors clorgyline or deprenyl, even at high doses (10 mg/kg), in place of tranylcypromine does not result in hyperactivity. Furthermore, brain 5HT does not rise to the same degree as seen after tranylcypromine. When clorgyline and deprenyl (2.5 mg/kg of each) are given together, brain 5HT shows the rise observed after tranylcypromine and the animals become hyperactive. Only when both inhibitors are given is MAO 'type A' and 'type B' totally inhibited, suggesting that when 5HT synthesis is increased the 5HT formed can still be metabolized by 'type B' when 'type A' is inhibited by clorgyline. Administration to rats of the 5HT agonist quipazine produces an identical hyperactivity syndrome to that which follows tranylcypromine and tryptophan. The syndrome is enhanced by deprenyl administration but not by clorgyline. Since it has been shown that a dopaminergic system is involved in the production of hyperactivity following increased 5HT synthesis and release, it seems possible that MAO inhibition by deprenyl in those neurons increases the behavioural response. These data indicate possible limitations of using 'selective' inhibitors for changing the functional activity of amine transmitters in the brain.

A major problem in assessing the functional activity of the biogenic amines is the control of amine degradation by monoamine oxidase (amine: oxygen oxidoreductase [deaminating] [flavin-containing]; EC 1.4.3.4) (MAO). This is because MAO is present both intraneuronally and extraneuronally. The experiments of Grahame-Smith (1971) showed that 5-hydroxytryptamine (5HT) synthesis could be increased considerably in rat brain without any overt behavioural changes by giving L-tryptophan, whereas a characteristic hyperactivity syndrome occurred if a MAO inhibitor was also given. He concluded that this demonstrated the ability of intraneuronal MAO to metabolize the increased 5HT being synthesized, thereby preventing the amine being released into the synaptic cleft, stimulating the receptor and producing the behavioural change. When an MAO inhibitor was given this intraneuronal metabolism was blocked and the amine 'spilt over' onto the receptor. Using this behavioural model together with biochemical measurements, we have investigated the action of various drugs on the functional activity of 5HT (Grahame-Smith & Green 1974; Green & Grahame-Smith 1974*a*, 1975).

We have now used the hyperactivity syndrome in another way, investigating the mode of action of the MAO inhibitors clorgyline, deprenyl and tranylcypromine in altering 5HT metabolism and functional activity. As reported elsewhere in this book (Knoll, pp. 135–155), clorgyline and deprenyl are selective inhibitors of MAO, clorgyline inhibiting type A MAO which oxidatively deaminates 5-hydroxytryptamine and dopamine and deprenyl inhibiting type B MAO which is relatively resistant to clorgyline and metabolizes phenylethylamine and dopamine.

METHODS

Adult male Sprague-Dawley rats weighing 150–200 g were used in all experiments. Brain 5HT was measured by the method of Curzon & Green (1970) in one half of the brain and MAO activity was measured in the other half of the brain using radioactive substrates, by the method of Southgate & Collins (1969). Activity measurements were made on groups of three animals using Animex activity meters, as described elsewhere (Grahame-Smith 1971; Green & Grahame-Smith 1974*b*).

RESULTS

Studies on the effect of L-*tryptophan administration to rats pretreated with tranylcypromine, clorgyline or deprenyl*

Initial experiments were performed to see whether L-tryptophan administration altered MAO activity.

Rats were injected with saline and 30 min later given either saline or L-tryptophan (100 mg/kg). After a further 90 min they were killed and brain 5HT, tryptophan and MAO activity were measured. After L-tryptophan there was an increase in the concentration of both tryptophan and 5HT (Table 1). Tryptophan did not inhibit MAO activity to any of the substrates examined and no behavioural activity changes were observed.

TABLE 1

Effect of L-tryptophan (100 mg/kg) on brain tryptophan and 5-hydroxytryptamine (5HT) concentrations 90 min later

Injected	*Brain tryptophan (μg tryptophan/g, wet wt)*	*Brain 5HT (μg 5HT/g brain, wet wt)*
Saline	4.71 ± 0.38 (3)	0.49 ± 0.01 (10)
L-Tryptophan (100 mg/kg)	35.0 ± 0.59 (3)	0.81 ± 0.05 (3)

Brain 5HT and tryptophan concentrations are expressed as mean ± 1 S.E.M. with number of determinations in brackets.

The next experiments examined the effects of giving various doses of the inhibitors tranylcypromine, clorgyline and deprenyl before L-tryptophan in producing the hyperactivity syndrome. Rats were injected with 1.0, 2.5, 5.0 or 10.0 mg/kg tranylcypromine, clorgyline or deprenyl followed 30 min later by saline or L-tryptophan (100 mg/kg). Activity was measured over the next 90 min, after which time the rats were killed and 5HT and MAO activity was measured in the brain.

The administration of L-tryptophan to a rat pretreated with tranylcypromine (1.0 mg/kg) did not result in hyperactivity. However, when tryptophan was given to rats pretreated with any of the higher doses of tranylcypromine (2.5–10 mg/kg) the animals became hyperactive and the degree of hyperactivity was the same in all cases (Fig. 1). Tranylcypromine (1 mg/kg) produced a slight rise of 5HT. When L-tryptophan was also administered the concentration of brain 5HT did not increase further than when L-tryptophan alone had been given (Table 2). Higher doses of tranylcypromine produced similar rises in brain 5HT concentrations when the inhibitor alone was given, with a further increase (which was the same after any dose of tranylcypromine from 2.5 mg/kg to 10 mg/kg) if tryptophan had also been given (Table 2). Tranylcypromine (1 mg/kg) did not totally inhibit MAO activity towards either 5HT (a 'type A' MAO substrate) or phenylethylamine (a 'type B' MAO substrate) but all higher doses inhibited both forms of the enzyme (Table 2).

When L-tryptophan was administered to rats given the type A MAO inhibitor clorgyline in place of tranylcypromine no hyperactivity was observed even after the highest dose of inhibitor (Fig. 2). However, it was found that brain 5HT did not rise to the concentration seen after tranylcypromine (2.5 mg/kg), even after 10 mg/kg of clorgyline. Furthermore, when tryptophan was given after clorgyline, the accumulation of brain 5HT over the next 90 min was less than when tryptophan was given after tranylcypromine (2.5 mg/kg) (Table 3). The measurement of MAO activity in these animals showed that

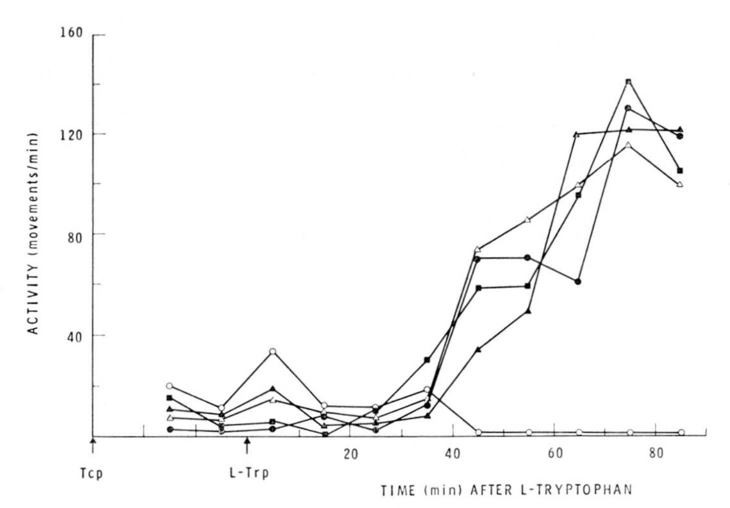

FIG. 1. The effect of injecting L-tryptophan (100 mg/kg) after various doses of tranylcypromine on hyperactivity. Rats were injected with tranylcypromine and then with L-tryptophan (100 mg/kg) 30 min later. Activity was measured as movements/min. Tranylcypromine doses: 1.0 mg/kg (○—○), 2.5 mg/kg (●—●), 5.0 mg/kg (△—△), 10.0 mg/kg (▲—▲), and 20.0 mg/kg (■—■).

TABLE 2

Effect of various doses of tranylcypromine with or without L-tryptophan injection (100 mg/kg) 30 min later on brain 5-hydroxytryptamine (5HT) concentrations and MAO activity to 5HT, dopamine and phenylethylamine 120 min after injection

Dose of tranylcypromine (mg/kg)	*Brain 5HT (μg 5HT/g brain, wet wt)*		*% inhibition of MAO activity to substrates*		
	Saline	*L-Tryptophan*	*5HT*	*Dopamine*	*Phenyl-ethylamine*
1.0	0.60 ± 0.03 (3)	0.75 ± 0.14 (3)	86 ± 5 (6)	85 ± 1 (6)	86 ± 1 (3)
2.5	0.78 ± 0.02 (3)	1.12 ± 0.14 (3)	100	100	100
5.0	0.77 ± 0.02 (3)	1.16 ± 0.06 (3)	100	100	100
10.0	0.79 ± 0.05 (3)	1.18 ± 0.07 (3)	100	100	N.D.

Brain 5HT concentrations are shown 120 min after injection of inhibitor when saline or L-tryptophan (100 mg/kg) had also been given, 30 min after tranylcypromine. The percentage inhibition of MAO activity is also shown 120 min after the injection of tranylcypromine. Results expressed as mean ± 1 S.E.M. with number of determinations in brackets. N.D.: not determined.

even when the oxidative deamination of 5HT was totally inhibited after the injection of clorgyline (10 mg/kg), phenylethylamine oxidation was inhibited by only 67% (Table 3).

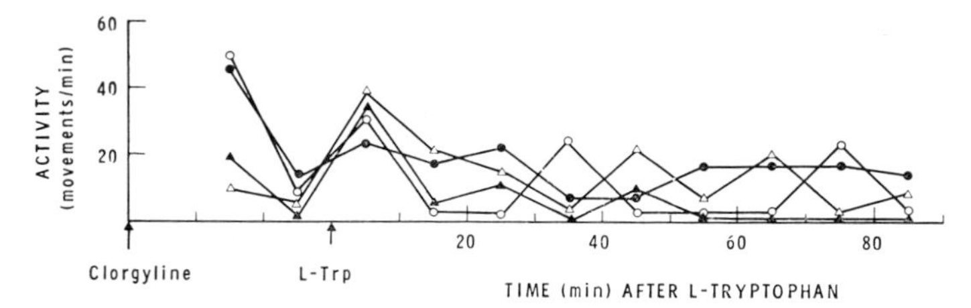

FIG. 2. The effect of injecting L-tryptophan (100 mg/kg) after various doses of clorgyline on hyperactivity. Rats were injected with L-tryptophan (100 mg/kg) 30 min after clorgyline. Activity was measured as movements/min. Clorgyline doses: 1.0 mg/kg (○—○), 2.5 mg/kg (●—●), 5.0 mg/kg (△—△), and 10.0 mg/kg (▲—▲).

TABLE 3

Effect of various doses of clorgyline or deprenyl with or without L-tryptophan injection (100 mg/kg) 30 min later on brain 5-hydroxytryptamine (5HT) concentrations and MAO activity to 5HT, dopamine and phenylethylamine 120 min after injection

Injected	*Dose of deprenyl or clorgyline (mg/kg)*	*Brain 5HT (µg 5HT/g brain wet wt)*		*% inhibition of MAO activity to substrates*		
		Saline	*L-tryptophan*	*5HT*	*Dopamine*	*Phenyl-ethylamine*
Clorgyline	1.0	0.60 ± 0.06 (9)	0.90 ± 0.10 (6)	61 ± 7 (11)	51 ± 6 (12)	19 ± 7 (3)
	10.0	0.69 ± 0.03 (6)	0.90 ± 0.09 (6)	99 ± 2 (9)	96 ± 18 (9)	67 ± 8 (3)
Deprenyl	1.0	0.51 ± 0.01 (9)	0.73 ± 0.09 (6)	23 ± 3 (12)	28 ± 5 (9)	58 ± 5 (3)
	10.0	0.57 ± 0.02 (6)	0.78 ± 0.12 (6)	44 ± 8 (9)	57 ± 5 (8)	94 ± 5 (3)

Brain 5HT concentrations are shown 120 min after injection of inhibitor when saline or L-tryptophan (100 mg/kg) had also been given, 30 min after the inhibitor. The percentage inhibition of MAO activity is also shown 120 min after the injection of the inhibitor. Results expressed as mean ± 1 S.E.M. with number of determinations in brackets.

When deprenyl was given before L-tryptophan it was again found that no dose of the inhibitor (1.0–10.0 mg/kg) resulted in the appearance of the hyperactivity syndrome although the administration of 10 mg/kg did produce amphetamine-like locomotor activity. This was behaviourally unlike the hyperactivity syndrome, however, and was not altered by subsequent administration of tryptophan. No dose of deprenyl produced the rise of brain 5HT seen after tranylcypromine (2.5 mg/kg) and when L-tryptophan had also been administered brain 5HT concentrations did not rise above the concentration seen when the amino acid was given alone (Table 3). The oxidation of phenylethylamine was almost totally inhibited after higher doses of deprenyl, while the degradation of 5HT was not inhibited by more than 50% even after administration of 10 mg/kg of deprenyl (Table 3).

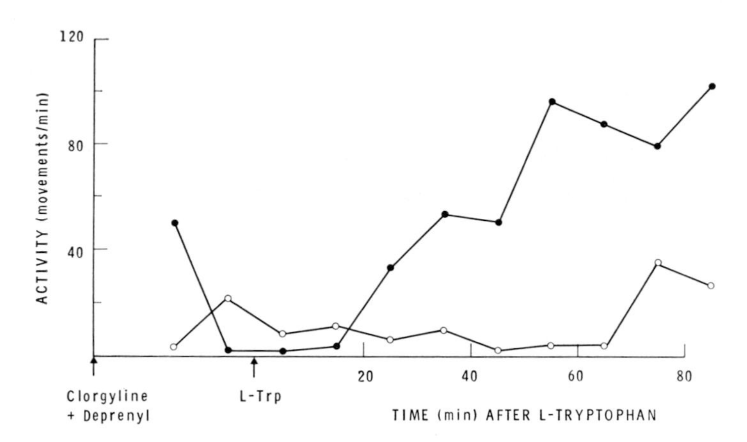

FIG. 3. The effect of injecting L-tryptophan (100 mg/kg) after various doses of clorgyline plus deprenyl on hyperactivity. Rats were injected with clorgyline plus deprenyl, with L-tryptophan (100 mg/kg) 30 min later. Activity was measured as movements/min. Clorgyline (1.0 mg/kg) + deprenyl (1.0 mg/kg) (○—○); clorgyline (2.5 mg/kg) + deprenyl (2.5 mg/kg) (●—●).

When clorgyline (2.5 mg/kg) and deprenyl (2.5 mg/kg) were given together before L-tryptophan, the oxidation of both 5HT and phenylethylamine was inhibited by about 90% (Table 4), brain 5HT concentrations rose to the value seen after tranylcypromine (2.5 mg/kg) had been given, and the animals became hyperactive when tryptophan was also given (Fig. 3).

Effect of pretreatment with either clorgyline or deprenyl on the hyperactivity produced by quipazine

Quipazine (2-(1-piperazinyl) quinoline maleate) has been reported to be a 5-HT agonist both peripherally (Hong *et al.* 1969) and in the brain (Rodriguez *et al.* 1973). Recent studies in our laboratory have shown that in rats it produces behavioural changes, including hyperactivity, identical to those following tranylcypromine and L-tryptophan. This suggests that quipazine stimulates central 5HT receptors, and other studies have strengthened this view (Green *et al.* 1976). In the course of these investigations we observed that the hyperactivity produced by quipazine was enhanced if the animals were pretreated with tranylcypromine (20 mg/kg). If quipazine is a 5HT agonist then MAO inhibition would not be expected to alter its behavioural effects. Further investigations were therefore made on the effect of MAO inhibition on the hyperactivity produced by quipazine.

Administration of clorgyline (5.0 mg/kg) 30 minutes before the injection

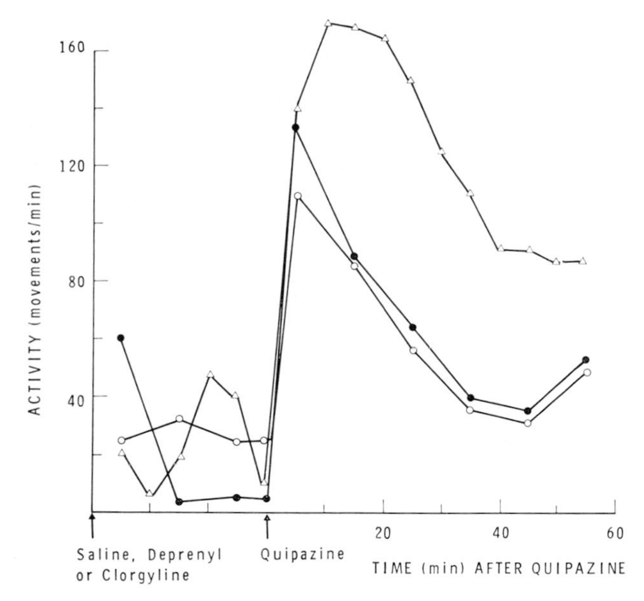

FIG. 4. The effect of pretreatment with deprenyl or clorgyline on the activity following the injection of quipazine (25 mg/kg) to rats. Rats were given either saline (○), clorgyline 5 mg/kg (●) or deprenyl 5 mg/kg (△), followed 30 min later by quipazine (25 mg/kg), and their activity was measured.

TABLE 4

Effect of various doses of clorgyline plus deprenyl with or without L-tryptophan injection (100 mg/kg) 30 min later on brain 5-hydroxytryptamine (5HT) concentrations and MAO activity to 5HT, dopamine and phenylethylamine 120 min after injection

Dose of clorgyline plus deprenyl (mg/kg of each)	*Brain 5HT (μg 5HT/g brain wet wt)*		*% inhibition of MAO activity to substrates*		
	Saline	L-*tryptophan*	*5HT*	*Dopamine*	*Phenyl-ethylamine*
1.0 + 1.0	0.68 ± 0.03 (9)	0.80 ± 0.08 (3)	84 ± 7 (10)	83 ± 4 (12)	76 ± 20 (3)
2.5 + 2.5	0.76 ± 0.05 (8)	1.04 ± 0.14 (3)	97 ± 14 (12)	99 ± 2 (12)	90 ± 13 (3)

Brain 5HT concentrations are shown 120 min after injection of inhibitor when saline or L-tryptophan (100 mg/kg) had also been given, 30 min after the inhibitor. The percentage inhibition of MAO activity is also shown 120 min after the injection of the inhibitor. Results expressed as mean ± 1 S.E.M. with number of determinations in brackets.

of quipazine (25 mg/kg) did not alter the hyperactivity resulting from the quipazine injection (Fig. 4). However, when deprenyl (5.0 mg/kg) was given 30 min before quipazine (25 mg/kg), the animals displayed enhanced hyperactivity (Fig. 4).

DISCUSSION

Both investigations reported in this paper raise interesting points on the role of the specific MAO inhibitors in altering the functional activity of 5-hydroxytryptamine in the rat brain. In agreement with Squires & Buus Lassen (1975), we found that the administration of tryptophan to rats pretreated with either clorgyline or deprenyl alone did not produce hyperactivity, while hyperactivity did occur when both inhibitors were given before tryptophan. Squires (1975) suggested that hyperactivity is due to the formation of an *N*-substituted derivative of 5HT which is deaminated by type B MAO to 5HIAA. Our results suggest a quite different conclusion.

Pretreatment of rats with tranylcypromine (2.5 mg/kg or more) totally inhibits both type A and type B MAO. In contrast, as would be expected with selective inhibitors, neither clorgyline nor deprenyl inhibited both type A and type B MAO, even after high doses of the inhibitor had been given. Administration of either of the drugs did not produce an increase in brain 5HT as large as that seen after tranylcypromine (2.5 mg/kg) and the accumulation of 5HT seen when L-tryptophan had been given after clorgyline or deprenyl was smaller than when tranylcypromine (2.5 mg/kg) was the inhibitor used. However, when deprenyl and clorgyline were given together at a dose of 2.5 mg/kg of each, both forms of the enzyme were inhibited and brain 5HT levels rose to the concentrations seen after tranylcypromine (2.5 mg/kg). After the subsequent administration of L-tryptophan, 5HT accumulated to almost the same amount seen when tranylcypromine was the inhibitor and the animals became hyperactive.

These results suggest that while *in vivo* 5HT may normally be metabolized by 'type A' MAO, nevertheless when this functional form is inhibited (as occurs when clorgyline has been given), 'type B' MAO can continue to metabolize the amine. When deprenyl is given, 'type A' MAO is only partly inhibited. Thus when either of these inhibitors is given, brain 5HT concentrations do not rise to the value seen when the non-specific inhibitor tranylcypromine is given or when both clorgyline and deprenyl are injected together. The results on hyperactivity complement this interpretation of the results, since they also indicate that the functional activity of 5HT is increased only when both forms of the enzyme are inhibited. Presumably unless both forms of the enzyme are inhibited in the presynaptic nerve ending, the increased 5HT being made after the L-tryptophan administration can be metabolized by intraneuronal MAO and thus does not 'spill over' into functional activity, producing hyperactivity. A further argument against the involvement of any other amine is that 5HT did not increase after clorgyline alone to the concentration seen after tranylcypromine, indicating incomplete inhibition of 5HT degradation. The rise in 5HT concentrations seen after the injection of either clorgyline plus deprenyl (2.5 mg/kg of each) or tranylcypromine (2.5 mg/kg) does not appear to be due to the formation of a 5HT derivative, as the values obtained when calculated as rates of 5HT synthesis are in good agreement with the synthesis rate measured by other techniques not involving MAO inhibition (Neff & Tozer 1968). The observation that hyperactivity did not occur when tryptophan was given after tranylcypromine (1.0 mg/kg), when the activity of both type A and type B MAO was inhibited by about 85%, suggests that MAO is present in the brain grossly in excess of normal requirements. This may be the reason why successful therapy of psychiatric conditions by MAO inhibitors is difficult to achieve. Previous studies on human brain by Youdim *et al.* (1972) indicated that MAO activity to various substrates is not normally inhibited above 70% in patients who had been treated with various inhibitors. The degree of inhibition of dopamine oxidation by MAO after either of the selective inhibitors suggests that dopamine has access to both type A and type B MAO and is oxidized by both forms, in agreement with Yang & Neff (1974).

The experiments with quipazine and the MAO inhibitors produced some intriguing observations. Quipazine does have some MAO-inhibiting actions and appears to be exclusively a type A MAO inhibitor (at the dose of 25 mg/kg). When deprenyl is also given, 5HT oxidation was inhibited by 85 $\pm$ 2% ($n = 4$) and phenylethylamine oxidation by 100%. If the interpretation of other observations is correct and quipazine is a 5HT agonist, then MAO inhibition should not alter its action. However, hyperactivity was, in fact, enhanced. It has previously been demonstrated that dopamine is involved in the hyperactivity syndrome, postsynaptically to the 5HT neurons initiating the behavioural response (Green & Grahame-Smith 1974*b*). Since Yang & Neff (1974) have reported that dopamine is metabolized by both forms of the MAO enzyme, one interpretation of these results is that when dopamine oxidation is inhibited in this postsynaptic dopaminergic system, hyperactivity is enhanced. If this interpretation is correct, it suggests that these dopaminergic neurons are to some degree able to control the magnitude of the total neuronal responses after stimulation of 5HT receptors.

Our results indicate, therefore, that the functional activity of 5HT may not necessarily be increased when the selective MAO inhibitor clorgyline is increased. Furthermore, it is possible that 5HT functional activity may be altered by the inhibition of monoamine oxidase in dopaminergic neurons. Together these results indicate possible difficulties in the use of these inhibitors in therapeutic situations where it is hoped to increase selectively the metabolism and function of 5-hydroxytryptamine in the brain.

References

CURZON, G. & GREEN, A. R. (1970) Rapid method for the determination of 5-hydroxytryptamine and 5-hydroxyindoleacetic acid in small regions of rat brain. *Br. J. Pharmacol. 39*, 653-655

GRAHAME-SMITH, D. G. (1971) Studies *in vivo* on the relationship between brain tryptophan, brain 5-HT synthesis and hyperactivity in rats treated with a monoamine oxidase inhibitor and L-tryptophan. *J. Neurochem. 18*, 1053-1066

GRAHAME-SMITH, D. G. & GREEN, A. R. (1974) The role of brain 5-hydroxytryptamine in the hyperactivity produced in rats by lithium and monoamine oxidase inhibition. *Br. J. Pharmacol. 52*, 19-26

GREEN, A. R. & GRAHAME-SMITH, D. G. (1974*a*) TRH potentiates the behavioural changes following increased 5-hydroxytryptamine accumulation in rats. *Nature (Lond.) 251*, 524-526

GREEN, A. R. & GRAHAME-SMITH, D. G. (1974*b*) The role of brain dopamine in the hyperactivity syndrome produced by increased 5-hydroxytryptamine synthesis in rats. *Neuropharmacology 13*, 949-959

GREEN, A. R. & GRAHAME-SMITH, D. G. (1975) The effect of diphenylhydantoin on brain 5-hydroxytryptamine metabolism and function. *Neuropharmacology 14*, 107-113

GREEN, A. R., YOUDIM, M. B. H. & GRAHAME-SMITH, D. G. (1976) Quipazine: its effects on rat brain 5-hydroxytryptamine metabolism, monoamine oxidase activity and behaviour. *Neuropharmacology*' in press

HONG, E., SANCILIO, L. F., VARGAS, R. & PARDO, E. G. (1969) Similarities between the pharmacological actions of quipazine and serotonin. *Eur. J. Pharmacol. 6*, 274-280

KNOLL, J. (1976) This volume, pp. 135-155

NEFF, N. H. & TOZER, T. N. (1968) *In vivo* measurement of brain serotonin turnover. *Adv. Pharmacol. 6A*, 97-109

RODRIGUEZ, R., ROJAS-RAMIREZ, J. A. & DRUCKER-COLIN, R. R. (1973) Serotonin-like actions of quipazine on the central nervous system. *Eur. J. Pharmacol. 24*, 164-171

SOUTHGATE, J. & COLLINS, G. G. S. (1969) The estimation of monoamine oxidase using ^{14}C-labelled substrates. *Biochem. Pharmacol. 18*, 2285-2287

SQUIRES, R. F. (1975) Evidence that 5-methoxy, *N*, *N*-dimethyltryptamine is a specific substrate for MAO - A in the rat: implications for the indoleamine dependent behavioural syndrome. *J. Neurochem. 24*, 47-50

SQUIRES, R. F. & BUUS LASSEN, J. (1975) The inhibition of A and B forms of MAO in the production of a characteristic behavioural syndrome in rats after L-tryptophan loading. *Psychopharmacologia 41*, 145-151

YANG, H.-Y. T. & NEFF, N. H. (1974) The monoamine oxidases of brain; selective inhibition with drugs and the consequences for the metabolism of the biogenic amines. *J. Pharmacol. Exp. Ther. 189*, 733-740

YOUDIM, M. B. H., COLLINS, G. G. S., SANDLER, M., BEVAN-JONES, A. B., PARE, C. M. B. & NICHOLSON, W. J. (1972) Human brain monoamine oxidase: multiple forms and selective inhibitors. *Nature (Lond.) 236*, 225-228

Discussion

Fuller: Does α-methyl-*p*-tyrosine block the effect of quipazine?

Green: It reduces hyperactivity by 60%, which is less than blockade of the response to tranylcypromine and tryptophan.

Crow: Have you thought of combining quipazine with apomorphine or amphetamine?

Green: No.

Sourkes: Why doesn't apomorphine (given alone) reproduce the behavioural syndrome, if it is being mediated through dopamine?

Green: Apomorphine produces behavioural changes but they are not quite the same as those seen after giving quipazine. You don't get the aggression with quipazine that is seen after tranylcypromine and L-dopa. The rats circle and pad, but we don't see rearing and aggression. But in a situation like this we are probably exciting many systems, some of them perhaps being mediated by dopamine and some not.

Crow: You use tranylcypromine as the reference compound in all your experiments. Is the syndrome really identical to that found with pargyline?

Green: It is the same. I use tranylcypromine because it was used in the original experiments of Grahame-Smith (1971) and I did not want to alter the conditions. Nevertheless, we have used pargyline and iproniazid and the response is the same.

Coppen: Some of our clinical work parallels your animal work. In the 1960s, before the selective inhibitors were available, we were interested in whether 5-hydroxytryptamine (5HT) was more important in ameliorating depressive illness than noradrenaline and so on. We gave a group of depressed patients, who were probably not normal from the point of view of MAO and perhaps other points of view, 30 mg/day of tranylcypromine, which is a low dose compared to your rats, so one was probably getting only 70% inhibition of MAO (Youdim *et al.* 1972). Some patients also received tryptophan; others received a placebo. We studied them over a four-week period. The patients given tryptophan in addition to the MAO inhibitor were much less depressed after four weeks than those receiving the MAO inhibitor alone (Coppen *et al.* 1963). This experiment has been repeated four or five times now, under double-blind conditions, and everybody has found a similar picture, namely potentiation (Pare 1963; Glassman & Platman 1969; Ayuso Gutierrez & López-Ibor Aliño 1971). I have never seen evidence of hyperactivity, or hypomania or mania, under these conditions—that is, anything on a human level resembling the hyperactivity syndrome in animals. This is probably because we were using relatively low doses of tranylcypromine.

Green: I agree. We are not saying that hyperactivity is anything like a normal behavioural change: perhaps the best description of what we are doing is a 'whole-animal bioassay system'. Without total MAO inhibition and the hyperactivity appearing you can still get a rise in 5HT, as we have shown in this paper.

Coppen: When we compared the effects of various agents we found that tryptophan has the same sort of activity as imipramine in doses of 150 mg/day. Tryptophan plus MAO inhibitor was the best antidepressant.

We have examined this further. We have been looking at the precursors of 5HT in blood, particularly at free tryptophan, as we think it may be related to brain tryptophan (Coppen *et al.* 1973). We measured plasma tryptophan under standardized control conditions in depressives. Although there is no difference from controls in total plasma tryptophan, there is evidence of a considerable decrease in free tryptophan in depressives. If this has any relation to the concentration of tryptophan in the brain, it may account for the decreased synthesis of 5HT that has been described by Dr van Praag in his work with probenecid (van Praag *et al.* 1970). We did find a tendency for an increase in free tryptophan after recovery.

Green: If we increase tryptophan pyrrolase by injecting hydrocortisone into rats there is a decrease in free and bound plasma tryptophan six hours after injection. Liver and brain tryptophan concentrations decrease, as do brain 5HT and 5HIAA concentrations (Green *et al.* 1975). However, this is an acute situation and may not be comparable.

There is some other work on deprenyl, clorgyline and hyperactivity by Squires (1975; Squires & Buus Lassen 1975), who also found that a combination of deprenyl and clorgyline produces hyperactivity whereas separately they do not. He suggested that hyperactivity was due to an *N*-methylated 5HT derivative which was metabolized by the type B enzyme. We think this is unlikely for two reasons: (1) Measurement of 5HT in brain after giving an MAO inhibitor and L-tryptophan, by either the specific ninhydrin method (Snyder *et al.* 1965) or the *O*-phthalaldehyde method (Curzon & Green 1970), which measures other 5-substituted indoles, results in identical values. With the *O*-phthalaldehyde method one would expect to detect *N*-methyl derivatives (Green & Youdim 1975). (2) The increase in brain 5HT after the injection of clorgyline plus deprenyl (2.5 mg/kg of each) or tranylcypromine (20 mg/kg) produces a value which when translated into a rate of 5HT synthesis agrees well with rates obtained using other methods not requiring the injection of an MAO inhibitor (Neff & Tozer 1968).

van Praag: If you give 5-hydroxytryptophan (5HTP) instead of tryptophan, in combination with an MAO inhibitor, do you see the same behavioural syndrome in your rats?

Green: We see a similar hyperactivity syndrome but with a more rapid onset, presumably because tryptophan hydroxylase is the rate-limiting step; 5HTP can be decarboxylated very rapidly. We find a bigger response than with tryptophan, dose for dose, presumably because 5HTP is also being decarboxylated in dopamine neurons.

van Praag: I have been interested for several years in 5HTP in man, and especially in depressive behaviour (van Praag *et al.* 1972). We tried to discover, first, whether 5HTP has any psychopharmacological activity and, secondly, whether patients with low c.s.f. 5HIAA responses to probenecid are more responsive towards 5HTP than depressives with normal 5HT metabolism, as measured in c.s.f. (van Praag *et al.* 1973). 5HTP is pharmacologically active and has some antidepressant activity in depressed patients. If the effect has anything to do with increased availability of 5HT at the relevant receptor sites in the brain, one would expect to obtain the potentiation of the antidepressant effect by combining 5HTP with a drug which, along another channels, increases available 5HT near the receptor sites. We therefore combined 5HTP with the tricyclic antidepressant clomipramine (a more or less selective 5HT re-uptake inhibitor) in a pilot study that is now being repeated in a double-blind manner.

We did the following experiment (van Praag *et al.* 1974). We first treated a series of very serious, therapy-resistant depressions with a combination of 5HTP and a peripheral decarboxylase inhibitor. Afterwards we introduced low doses of clomipramine, because we wanted to minimize the intrinsic antidepressant effect of the tricyclic drug. All these patients had been depressed for several years and had had all kinds of treatment—psychotherapy, drug treatment, and ECT—with little or no effect. After two weeks on 5HTP there was some effect in these patients. With a combination of 5HTP plus the decarboxylase inhibitor, which gives rise to an enormous increase of 5HTP in the blood, plus clomipramine, there was a worthwhile effect in several patients. Some of them were discharged and are still receiving 5HTP, in combination with clomipramine, and are doing well.

Our conclusion so far is that 5HTP can be of value in the treatment of depression and in particular in combination with a tricyclic drug which is supposed to have an influence on 5HT in central synapses.

We haven't tried the combination of 5HTP and an MAO inhibitor.

Green: We used clomipramine in the hyperactivity model. With tranylcypromine and tryptophan we get a hyperactivity response. If we pretreat the rat with clomipramine, the hyperactivity starts earlier (Green & Grahame-Smith 1975). When synthesis is increased, 5HT is released into the synaptic cleft and with re-uptake blocked the receptor is stimulated sooner. If you just give clomipramine and tryptophan, there is no hyperactivity. We haven't tried 5HTP and clomipramine; it would be interesting to do so because 5HTP is decarboxylated rapidly, so we might see an effect.

van Praag: Modigh (1973) in Göteborg described the same syndrome as you find with MAO inhibitors and tryptophan. He used clomipramine, 5HTP and a peripheral decarboxylase inhibitor.

Fuller: Does clomipramine have any effect on the action of quipazine?

Green: I don't know; it's an interesting thought.

Fuller: Malick *et al.* (1975) have reported that MAO inhibitors potentiate the quipazine effect (they used pargyline), and suggested that quipazine was acting in part by releasing 5HT onto the receptor. They supported that theory by the finding that *p*-chlorophenylalanine, which didn't alter the action of quipazine itself, did prevent the potentiation of quipazine by pargyline.

Green: We find that in rats pretreated with *p*-chlorophenylalanine quipazine produces more hyperactivity then in control animals; this effect is small but repeatable (Green *et al.* 1976). Grabowska and colleagues have also reported this (Grabowska *et al.* 1974). They also find that in rats in which 5HT is severely depleted with reserpine you still see the effect of quipazine. We have simply produced one possible explanation for the enhanced response to quipazine seen after MAO inhibition and I wouldn't rule another out. We have to see whether quipazine releases 5HT in brain slices; this hasn't been done, to my knowledge.

Coppen: Professor Carlsson and Dr Roos have reported that tryptophan potentiates the beneficial action of clomipramine in depression, in a rather similar way to what we found with tryptophan and tranylcypromine (Walinder *et al.* 1975).

References

Ayuso Gutierrez, J. L. & López-Ibor Aliño, J. J. (1971) Tryptophan and an MAOI (nialamide) in the treatment of depression. *Int. Pharmacopsychiatry 6*, 92-97

Coppen, A., Shaw, D. M. & Farrell, J. P. (1963) Potentiation of the antidepressive effect of a monoamine-oxidase inhibitor by tryptophan. *Lancet 1*, 79-81

Coppen, A., Eccleston, E. G. & Peet, M. (1973) Total and free tryptophan concentration in the plasma of depressive patients. *Lancet 2*, 60-63

Curzon, G. & Green, A. R. (1970) Rapid method for the determination of 5-hydroxytryptamine and 5-hydroxyindoleacetic acid in small regions of rat brain. *Br. J. Pharmacol. 39*, 653-655

Glassman, A. M. & Platman, S. R. (1969) Potentiation of a monoamine oxidase inhibitor by tryptophan. *J. Psychiatr. Res. 7*, 83.-88

Grahame-Smith, D. G. (1971) Studies *in vivo* on the relationship between brain tryptophan, brain 5-HT synthesis and hyperactivity in rats treated with a monoamine oxidase inhibitor and L-tryptophan. *J. Neurochem. 18*, 1053-1066

Grabowska, M., Antkiewicz, L. & Michaluk, J. (1974) A possible interaction of quipazine with central dopamine structures. *J. Pharm. Pharmacol. 26*, 74-76

Green, A. R. & Grahame-Smith, D. G. (1975) The effect of diphenylhydantoin on brain 5-hydroxytryptamine metabolism and function. *Neuropharmacology 14*, 107-113

Green, A. R. & Youdim, M. B. H. (1975) Effects of monoamine oxidase inhibition by clorgyline, deprenil or tranylcypromine on 5-hydroxytryptamine concentrations in rat brain and hyperactivity following subsequent tryptophan administration. *Br. J. Pharmacol. 55*, 415-422

GREEN, A. R., WOODS, H. F., KNOTT, P. J. & CURZON, G. (1975) Factors influencing the effect of hydrocortisone on rat brain tryptophan metabolism. *Nature (Lond.) 255*, 170

GREEN, A. R., YOUDIM, M. B. H. & GRAHAME-SMITH, D. G. (1976) Quipazine: its effects on rat brain 5-hydroxytryptamine, monoamine oxidase activity and behaviour. *Neuropharmacology* in press

MALICK, J. B., DOREN, E. & BARNETT, A. (1975) Quipazine-induced head-twitch produced by serotonin receptor activation in mice. *Fed. Proc. 34*, 801 (abstr.)

MODIGH, K. (1973) Effects of chloroimipramine and protriptyline on the hyperactivity induced by 5-hydroxytryptophan after peripheral decarboxylase inhibition in mice. *J. Neural Transm. 34*, 101-109

NEFF, N. H. & TOZER, J. N. (1968) *In vivo* measurement of brain serotonin turnover. *Adv. Pharmacol. 6A*, 97-109

PARE, C. M. B. (1963) Potentiation of monoamine-oxidase inhibitors by tryptophan. *Lancet 2*, 527-528

SNYDER, S. H., AXELROD, J. & ZWEIG, M. (1965) A sensitive and specific fluorescence assay for tissue serotonin. *Biochem. Pharmacol. 14*, 831-835

SQUIRES, R. F. (1975) Evidence that 5-methoxy, *N*, *N*-dimethyltryptamine is a specific substrate for MAO A in the rat: implications for the indoleamine dependent behavioural syndrome. *J. Neurochem. 24*, 47-50

SQUIRES, R. F. & BUUS LASSEN, J. (1975) The inhibition of A and B forms of MAO in the production of a characteristic behavioural syndrome in rats after L-tryptophan loading. *Psychopharmacologia 41*, 145-151

VAN PRAAG, H. M., KORF, T. & PUITE, J. (1970) 5-Hydroxyindoleacetic acid levels in the cerebrospinal fluid of depressed patients treated with probenecid. *Nature (Lond.) 225*, 1259-1260

VAN PRAAG, H. M., KORF, J., DOLS, L. C. W. & SCHUT, T. (1972) A pilot study of the predictive value of the probenecid test in application of 5-hydroxytryptophan as antidepressant. *Psychopharmacologia 25*, 14-21

VAN PRAAG, H. M., KORF, J. & SCHUT, T. (1973) Cerebral monoamines and depression. An investigation with the probenecid technique. *Arch. Gen. Psychiatry 28*, 827-831

VAN PRAAG, H. M., BURG, W. VAN DEN, BOS, E. R. H. & DOLS, L. C. W. (1974) 5-Hydroxytryptophan in combination with clomipramine in 'therapy-resistant' depression. *Psychopharmacologia 38*, 267-269

WALINDER, J., SKOTT, A., NAGY, A., CARLSSON, A. & ROOS, B.-E. (1975) Potentiation of antidepressant action of clomipramine by tryptophan. *Lancet 1*, 984

YOUDIM, M. B. H., COLLINS, G. G. S., SANDLER, M., BEVAN-JONES, A. B., PARE, C. M. B. & NICHOLSON, W. J. (1972) Human brain monoamine oxidase: multiple forms and selective inhibitors. *Nature (Lond.) 236*, 225-228

Relations between the degree of monoamine oxidase inhibition and some psychopharmacological responses to monoamine oxidase inhibitors in rats

L. MAÎTRE, A. DELINI-STULA and P. C. WALDMEIER

Research Department, Pharmaceuticals Division, CIBA-GEIGY Limited, Basle

Abstract Monoamine oxidase (MAO) inhibitors antagonize tetrabenazine-induced catalepsy and when combined with dopa or 5-hydroxytryptophan (5HTP) increase locomotor activity and body temperature. These effects are of shorter duration than the inhibition of MAO activities in central or peripheral tissues, or the reduction of striatal homovanillic acid (HVA) and 3,4-dihydroxyphenylacetic acid (DOPAC) levels.

After the injection of ^{3}H-labelled dopa into rats pretreated with MAO inhibitors, [^{3}H]dopamine and [^{3}H]methoxytyramine accumulate in the brain in much greater amounts than in control rats. This effect, however, is of considerably shorter duration (half-life: 12–18 hours, depending on the inhibitor used) than MAO inhibition in the brain (half-life: 2.5–12 days) or the time required for the levels of endogenous HVA and DOPAC (half-life: 3–15 days) in the corpus striatum to revert to normal. After the injection of [^{3}H]5HTP into rats pretreated with pargyline, [^{3}H]5-hydroxytryptamine also accumulates in the brain, but in quantities approximately eight times less than the ^{3}H-labelled amines derived from [^{3}H]dopa. The duration of this effect of pargyline is also relatively short (half-life: 18 hours).

These data suggest the presence of at least two MAO pools in the rat brain. It may be assumed that besides bulk MAO, there is a small pool characterized by relative resistance to inhibition, rapid turnover and preference for dopamine deamination. The inhibition of this small pool seems to be more directly responsible for several of the pharmacological actions of MAO inhibitors than the inhibition of bulk MAO.

Irrespective of their chemical structure, all the widely used monoamine oxidase (MAO) inhibitors are characterized by the prolonged suppressant effect they exert on the activity of this enzyme. The same compounds induce a variety of central pharmacological effects, especially on interaction with other substances or amine precursors. Attempts have been made to correlate these effects with their effects on brain amine concentrations. Spector *et al.* (1963), for instance, have found that in laboratory animals there is a clear-cut relation between the reversing action of pargyline on reserpine-induced sedation and the increase it produces in the noradrenaline concentration. In their investigations of brain amine concentrations and motor hyperactivity in animals treated with pargyline and L-dopa, Wiegand & Perry (1961) were unable to determine the extent to which each of the increased amines contributed towards the pharmacological effect. Nevertheless, they found that after the administration of these two compounds the levels of noradrenaline and 5-hydroxytryptamine (serotonin, 5HT) remained elevated for more than 48 hours, whereas the dopamine levels returned to normal within that time. On the basis of these observations, they suggested that the effects associated with the prolonged inhibition of MAO produced *in vivo* by pargyline may be due to noradrenaline and 5HT rather than to dopamine. To the best of our knowledge, however, no adequate studies have been made to compare the time-course of the pharmacological actions of either pargyline or other MAO inhibitors with the time-course of their biochemical effects. The present experiments were carried out in an attempt to correlate these two aspects of the activity of different types of MAO inhibitor. Particular attention was paid to their duration of action.

METHODS

MAO activity in the whole brain of rats was determined by radioassay, essentially as described by Wurtman & Axelrod (1963). ^{14}C-labelled 5-hydroxytryptamine binoxalate (5HT, 26.7 mCi/mM, New England Nuclear, Boston, Mass.), [^{3}H]dopamine (3.8 Ci/mM, Radiochemical Centre, Amersham) and [^{14}C]phenethylamine HCl (PEA, 7 mCi/mM, New England Nuclear) were used as substrates at a concentration of 6.25 nM (5HT and phenethylamine: 10 nCi; dopamine: 20 nCi) for a final incubation mixture of 0.3 ml. The reaction was carried out with 2.5 mg of brain tissue. The ^{14}C- or ^{3}H-labelled deaminated material formed after an incubation period of 20 minutes at 37 °C was extracted with 6 ml of ethyl acetate. Aliquots of 4 ml were counted together with 1 ml ethanol and 10 ml scintillator. The estimations were made in groups of four rats. Homovanillic acid (HVA) and 3,4-dihydroxyphenylacetic acid (DOPAC) were isolated from rat corpus striatum according to the technique used by Murphy *et al.* (1969). HVA was determined fluorometrically by an automated procedure based on the method of Andén *et al.* (1963). DOPAC was determined fluorometrically as described by Sharman (1971). All estimations were made in four samples, each comprising four striata (controls, $n = 6$). At various times after treatment with MAO inhibitors, rats were injected with [^{3}H]L-dopa (350 μCi/kg, i.v.; specific activity, 5.92 Ci/mM, The

Radiochemical Centre, Amersham, England). Brains were removed *in each case* 30 minutes after the injection of [^{3}H]dopa. A fraction containing [^{3}H]-dopamine + [^{3}H]3-methoxytyramine (MT) was separated on DOWEX 50 WX 4 columns, essentially according to the procedure of Atack & Magnusson (1970). Details of the method will be described elsewhere (P. C. Waldmeier & L. Maître, in preparation).

Analogous experiments have been carried out using [^{3}H]5-hydroxytryptophan (5HTP, 350 μCi/kg, i.v.; specific activity 3.3 Ci/mM, The Radiochemical Centre, Amersham). Again, the rats were invariably decapitated 30 minutes after the administration of the labelled amino acid. [^{3}H]5HT and [^{3}H]5-hydroxyindoleacetic acid (5HIAA) were separated on DOWEX 50 WX 4 according to the same procedure as was used in the experiments with [^{3}H]dopa. The identity of [^{3}H]5HIAA was checked by thin layer chromatography of the column eluates.

In pharmacological experiments, the interactions between MAO inhibitors and tetrabenazine, L-dopa and 5HTP were studied. In each case, a different parameter was investigated, these being, respectively, cataleptic stance, locomotor activity and rectal temperature.

The cataleptic stance produced by tetrabenazine (20 mg/kg, i.p.) was evaluated by reference to the Stages III and IV defined by Wirth (Wirth *et al.* 1958). The MAO inhibitors were given at different times before the injection of tetrabenazine, and catalepsy was assessed intermittently, either until six hours thereafter or until the animals' condition had become normal again. The effect of the drugs was expressed in terms of inhibition, as a percentage of the response shown by the controls. Each group of animals consisted of 12 rats. Control determinations were made simultaneously.

Motor activity of rats treated with MAO inhibitors and L-dopa (250 mg/kg, i.p.) was evaluated by means of a Motron Motilimeter device. As previously, MAO inhibitors were given at different times before the injection of L-dopa. Locomotor activity was recorded at 15-minute intervals over a period of one hour in groups of three treated rats, and a corresponding control group.

The effects of pretreatment with MAO inhibitors on body temperature after the administration of 30 mg/kg (i.p.) 5HTP were recorded by means of an electric thermometer (Ellab) fitted with a rectal probe. The inhibitors were injected at various intervals before 5HTP, and body temperature was measured before each injection as well as 30 minutes and one and two hours after 5HTP. Six to 12 rats per group were used, and corresponding control determinations were made at the same times.

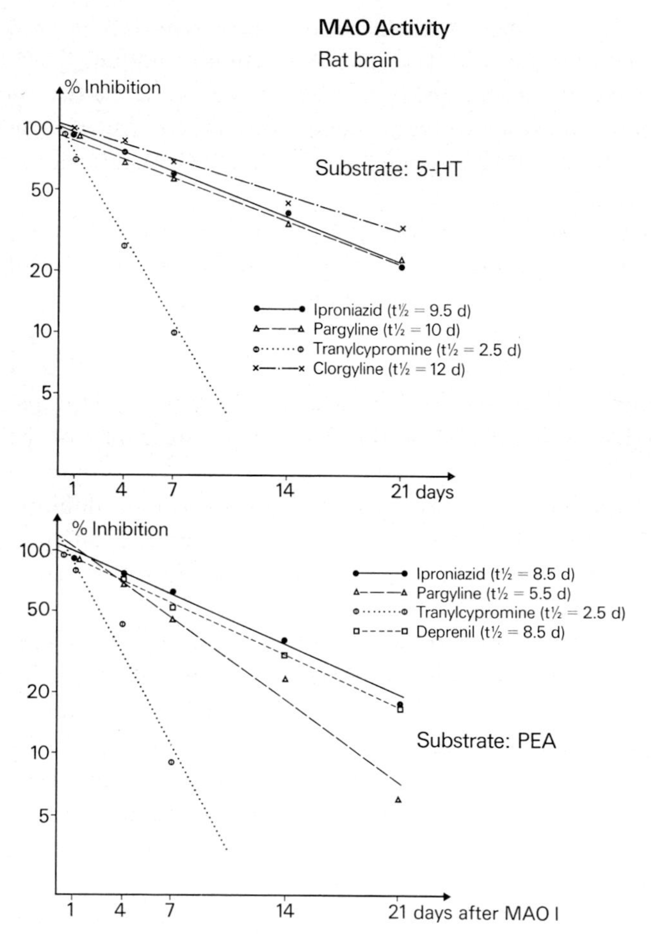

FIG. 1. Half-lives of the disappearance of the inhibitory effect of drugs on monoamine oxidase (MAO) activity in rat whole brain. The extent of MAO inhibition at different times up to 21 days after single administrations of iproniazid (100 mg/kg, s.c.), pargyline (50 mg/kg, s.c.), tranylcypromine (12 mg/kg, s.c.), clorgyline (10 mg/kg, s.c.) or deprenyl (10 mg/kg, s.c.) was plotted semilogarithmically against time. The upper part of the graph shows the data obtained when 5-hydroxytryptamine (5HT) was used as substrate, and the lower shows the results with phenylethylamine (PEA). Each point represents the average of four determinations.

RESULTS AND DISCUSSION

Pargyline, iproniazid, tranylcypromine and deprenyl exert a persistent inhibitory action on brain MAO (Fig. 1). The half-life of the effect of tranylcypromine (2.5 days) was shorter than that of the other MAO inhibitors (5.5–10 days). A similar difference between tranylcypromine and pargyline has been reported by Planz *et al.* (1972). This is consistent with the assumption that MAO inhibition by pargyline, iproniazid and deprenyl is irreversible, and that the resumption of the activity of the enzyme is due to neosynthesis. Tranylcypromine, however, seems to produce a partially reversible MAO inhibition (Planz *et al.* 1972; Neff & Goridis 1972). Clorgyline, a relatively selective inhibitor of type A MAO (Johnston 1968), also belongs to the class of compounds causing irreversible inhibition (Fig. 1).

These results relate to whole-brain homogenates. The effects of the same MAO inhibitors in different regions of the brain (corpus striatum, cortex, hypothalamus and brainstem) were qualitatively and quantitatively very similar to those seen in the whole brain (L. Maître, R. A. Felner & P. C. Waldmeier, in preparation).

If dopamine is used as substrate in either the whole brain or the corpus striatum, both the dose–response relations and the time-courses closely resemble those observed when 5HT is used as substrate (Waldmeier & Maître 1975; L. Maître, R. A. Felner & P. C. Waldmeier, in preparation).

A typical effect of MAO inhibitors on dopamine metabolism is the decrease in the concentrations of the main metabolites, HVA and DOPAC, in the corpus striatum. In analogous experiments to those shown in Fig. 1, the time-courses of the effects of MAO inhibitors on the endogenous levels of HVA and DOPAC in the corpus striatum were determined. From the results obtained, the rates of recovery were calculated (Fig. 2). The normalization rate of HVA corresponds closely to that of MAO A activity. DOPAC levels, however, have a tendency to revert to normal at a slightly slower rate.

These results indicate that the levels of the deaminated acidic dopamine metabolites reflect the overall MAO activity in the same tissue. Interestingly enough, the effects of MAO inhibitors on 5HT metabolism are not strictly comparable with their effects on dopamine metabolism. The decreases in 5HIAA concentrations hardly exceed 50% under conditions in which MAO inhibition is nearly complete (L. Maître, R. A. Felner & P. C. Waldmeier, in preparation).

In pharmacological experiments, it has been demonstrated that MAO inhibitors inhibit tetrabenazine-induced catalepsy. However, in contrast to their prolonged inhibitory effect on MAO activity in general, the duration of their anticataleptic action was short. Figs. 3 and 4 illustrate the antagonistic actions of pargyline and tranylcypromine. Pargyline completely suppresses catalepsy if it is given two hours before tetrabenazine. The inhibitory effect is, however, less marked if it is given 24 hours before tetrabenazine, and only a very slight antagonistic activity persists after four days. Like pargyline, tranylcypromine completely prevented the development of catalepsy in response

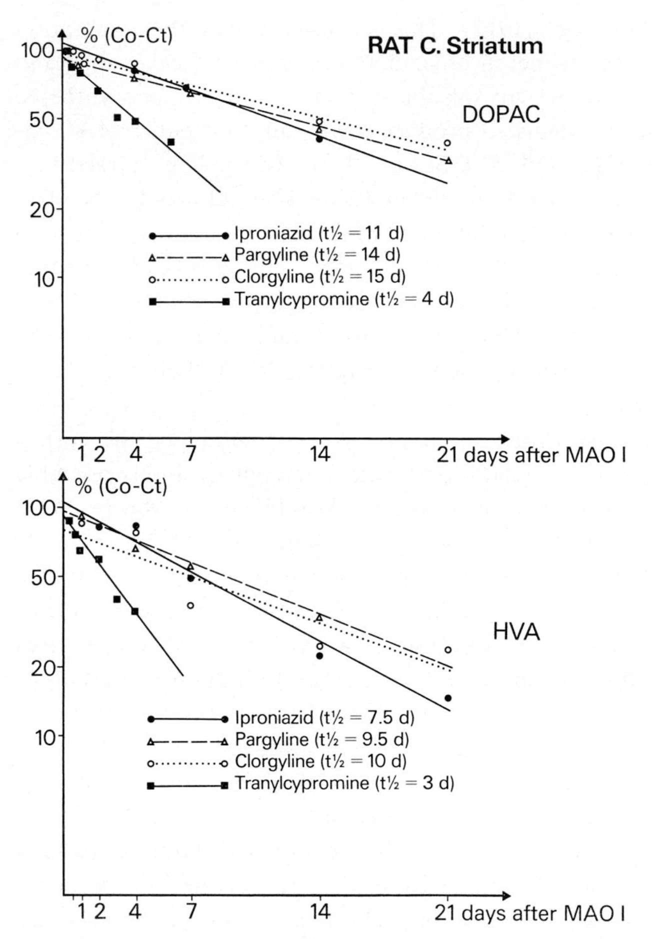

FIG. 2. Half-lives of the disappearance of the 3,4-dihydroxyphenylacetic acid (DOPAC) (*upper part*) and homovanillic acid (HVA) lowering (*lower part*) effects of monoamine oxidase inhibitors in rat corpus striatum. The doses and schedule of treatment were the same as in the experiments illustrated in Fig. 1. The differences between control levels of HVA and DOPAC and the levels at a given time after the administration of the drugs were plotted semi-logarithmically against time. Control levels for HVA ranged from 0.373 ± 0.011 to 0.453 ± 0.032 μg/g and those of DOPAC from 0.719 ± 0.023 to 0.901 ± 0.019 μg/g. Each point represents the average of four determinations.

to the administration of tetrabenazine two hours later; but two days after pretreatment with tranylcypromine tetrabenazine produced a full cataleptic response in the animals, indicating that the effect of the MAO inhibitor had subsided.

The duration of the anticataleptic action of deprenyl is likewise short. This

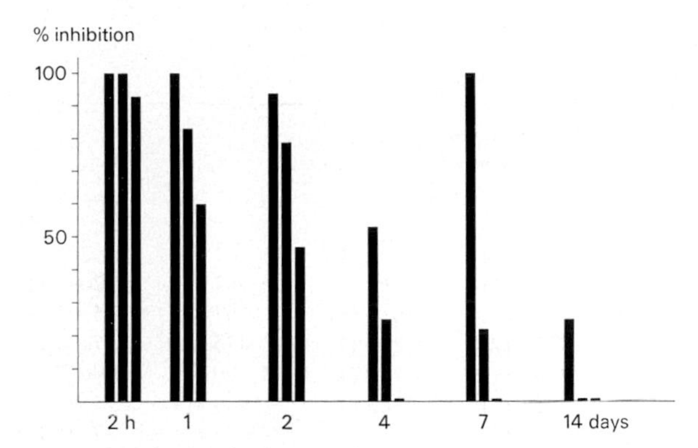

FIG. 3. Antagonism of tetrabenazine-induced catalepsy by pargyline (50 mg/kg, s.c.). Each vertical column represents the average cataleptic response of 10 rats expressed as percentage inhibition in comparison with the controls. In the groups of three columns, the first represents the values obtained 30 minutes, the second one hour and the third two hours after tetrabenazine. The times of pretreatment with pargyline are given on the abscissa.

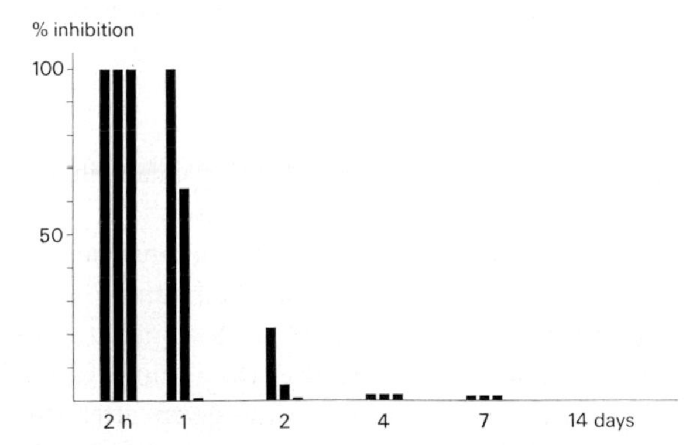

FIG. 4. Antagonism of tetrabenazine-induced catalepsy by tranylcypromine (12 mg/kg, s.c.); for details see legend to Fig. 3.

compound completely inhibited catalepsy when given 30 minutes before tetrabenazine, but its effect was less pronounced after one hour and only marginal after two hours.

Animals treated with pargyline or clorgyline and subsequently given an injection of L-dopa displayed a marked increase in motor activity accompanied by pronounced signs of peripheral sympathomimetic stimulation. Depending on the length of time elapsing between the doses of the two substances, the combination proved more or less toxic, and the animals died within a few hours.

TABLE 1

Interaction of monoamine oxidase inhibitors with L-dopa

Treatment	*Dose (mg/kg)*	*Activity counts (pretreatment time in hours)*			
		2	*8*	*24*	*48*
Pargyline + L-dopa	50 s.c. + 250 i.p.	15 547 (2)	16 829 (1)[a]	16 263 (1)	11 206 (2)
Pargyline + saline	50 s.c.	3572 (1)	2819 (1)	15 583 (1)	7090 (1)
Saline + L-dopa	250 i.p.	4711 (1)	4635 (1)	4867 (1)	5201 (1)
Clorgyline + L-dopa	10 s.c. + 250 i.p.	7978 (2)	15 233 (1)[b]	15 425 (1)[a]	7044 (2)
Clorgyline + saline	10 s.c.	3589 (1)	5740 (1)[b]	11 390 (1)	9657 (1)
Saline + L-dopa	250 i.p.	3229 (1)	4166 (1)[b]	5949 (1)	5583 (1)
Deprenyl + L-dopa	10 s.c. + 250 i.p.	8767 (2)		7084 (2)	7936 (2)
Deprenyl + saline	10 s.c.	19 003 (1)		13 022 (1)	6021 (1)
Saline + L-dopa	250 i.p.	3021 (1)		4189 (1)	4998 (1)

Unless otherwise stated, the values in the table represent the total number of counts effected by movements of rats (three in a group) during a period of one hour after L-dopa injection. Number of experiments performed is given in brackets.

[a] One of the three animals died within the first hour. The dead animal was not replaced and not removed from the cage until the end of the test session.

[b] The activity was only recorded for 45 minutes because the clorgyline-treated animals died 45–60 minutes after L-dopa injection.

For this reason the measurements of motor activity were only made in the first hour after the injection of L-dopa.

The maximum degree of hyperactivity resulting from the combined action of pargyline and L-dopa was observed when the MAO inhibitor was administered two and eight hours before the injection of L-dopa (Table 1). Since a considerable increase in motor activity was noted 24 hours after the administration of pargyline alone, the increase observed when L-dopa was given after this interval could not definitely be regarded as being due to the combination. Forty-eight hours after the administration of pargyline, its effect had diminished distinctly; the injection of L-dopa at this time evoked a relatively weak response. In general, the pattern of effect produced by clorgyline and L-dopa was the same as that produced by pargyline and L-dopa. However, the initial effects (two hours after pretreatment with the inhibitor) were less marked than those observed when the substances were administered at an interval of eight hours; under these latter conditions, the toxicity of the combination was also much greater, resulting in the death of the animals shortly after L-dopa injection. Like pargyline, clorgyline alone showed an overt stimulant effect 24 hours after its administration; no explanation for this phenomenon can be given at present.

In contrast to pargyline and clorgyline, deprenyl did not augment the behavioural response to L-dopa, even though it produced marked stimulation when given alone. Interestingly enough, the combination had less effect than deprenyl alone.

The time-course of the hyperthermic response caused by pargyline and clorgyline in combination with 5HTP paralleled the time-course of the effects resulting from interaction with L-dopa (Table 2). It seems, however, that the changes in temperature produced by pargyline and 5HTP are somewhat shorter-lasting than those produced by clorgyline and 5HTP in combination. This is also evident if the mortality rates after the administration of the substances at different intervals are taken into account.

In this test-system deprenyl was ineffective.

The behavioural and autonomic responses described here are conventionally ascribed to the inhibitory action of these compounds on MAO (Carlsson *et al.* 1957; Udenfriend *et al.* 1957; Everett *et al.* 1959; Everett 1961; Blaschko & Chrusciel 1960), which is of long duration. However, in the present study, the pharmacological effects of the MAO inhibitors tested were relatively short-lived. Antagonism against tetrabenazine had almost completely subsided four days after administration in the case of pargyline and two days thereafter in the case of tranylcypromine. Deprenyl was only active during the first hour after its injection.

The duration of action of pargyline and clorgyline in combination with amine precursors did not exceed 48 hours. No effects of deprenyl could have been demonstrated in this test system.

The short duration of action of pargyline and other MAO inhibitors on the behavioural and somatic responses to dopa or 5HTP prompted us to study the influence of the inhibitors on the accumulation of ^{3}H-labelled amines and some of their metabolites after the injection of [^{3}H]dopa or [^{3}H]5HTP.

The accumulation of ^{3}H-labelled amines and metabolites was assessed by measuring [^{3}H]dopamine + [^{3}H]methoxytyramine (MT) 30 minutes after the administration of [^{3}H]dopa, or measuring [^{3}H]5HT and [^{3}H]5HIAA 30 minutes after the injection of [^{3}H]5HTP. Pargyline, iproniazid, tranylcypromine, and clorgyline given 12 hours before sacrifice increased [^{3}H]dopamine + [^{3}H]MT 30–80-fold (Table 3). Pargyline also augmented the accumulation of [^{3}H]5HT, but to a lesser extent (five-fold; Table 3). In contrast, deprenyl led to only a slight rise in [^{3}H]dopamine + [^{3}H]MT accumulation and an even smaller one in the accumulation of [^{3}H]5HT.

In a subsequent experiment, the interval after treatment with MAO inhibitors was varied (without changing the interval between the injection of the ^{3}H-labelled amino acids and the removal of the brains) to allow the time-

TABLE 2

Interaction of monoamine oxidase inhibitors with 5-hydroxytryptophan (5HTP)

Treatment	*Initial temperature (°C)*	*Temperature (°C)*[a] *(pretreatment time in hours)*				*N*
		2	*16*	*24*	*48*	
Pargyline + 5HTP	37.5 ± 0.19	39.8 ± 0.27 (12/12)	40.7 ± 0.23 (5/12)	38.4 ± 0.30 (0/12)	35.8 ± 0.14 (0/12)	12
Pargyline	37.5 ± 0.27	37.0 ± 0.16	37.5 ± 0.03	37.6 ± 0.11	37.0 ± 0.05	6
5HTP	37.6 ± 0.16	37.1 ± 0.07	37.4 ± 0.08	37.3 ± 0.08	37.0 ± 0.06	6
Saline	37.1 ± 0.10	36.7 ± 0.10	37.3 ± 0.06	37.1 ± 0.12	37.4 ± 0.08	6
Clorgyline + 5HTP	38.1 ± 0.09	40.6 ± 0.15 (12/12)	40.1 ± 0.27 (9/12)	40.2 ± 0.26 (12/12)[b]	38.5 ± 0.38 (0/12)	12
5HTP	38.1 ± 0.16	37.7 ± 0.13	37.6 ± 0.04	37.9 ± 0.13[b]	37.1 ± 0.07	6
Clorgyline	38.3 ± 0.06	37.1 ± 0.11	37.1 ± 0.10	37.8 ± 0.14[b]	37.3 ± 0.09	6
Saline	37.9 ± 0.12	37.0 ± 0.15	37.1 ± 0.07	37.9 ± 0.10[b]	36.9 ± 0.10	6
Deprenyl + 5HTP	37.7 ± 0.11	37.4 ± 0.08 (0/12)	37.3 ± 0.09 (0/12)	37.4 ± 0.10 (0/12)	37.2 ± 0.09 (0/12)	12
5HTP	37.8 ± 0.12	37.3 ± 0.08	37.4 ± 0.07	37.6 ± 0.06	37.5 ± 0.10	6
Deprenyl	37.7 ± 0.18	37.7 ± 0.16	37.6 ± 0.16	37.4 ± 0.27	37.4 ± 0.13	6
Saline	37.9 ± 0.07	37.2 ± 0.05	37.4 ± 0.10	37.6 ± 0.20	37.5 ± 0.10	6

[a] Values in the table represent rectal temperature ± S.E.M. recorded one hour after 5HTP injection. The number of deaths occurring two hours after the 5HTP load is given in brackets.
[b] The temperature in this case was recorded 30 min after 5HTP because the animals treated with the combination died 30–60 min thereafter.

TABLE 3

Effect of monoamine oxidase inhibitors on the accumulation of ^{3}H-labelled amines formed from [^{3}H]dopa or [^{3}H]5-hydroxytryptophan in the rat brain

Drug	*Dose (mg/kg, s.c.)*	[^{3}*H*]dopamine + [^{3}*H*]*MT* (*c.p.m./g*)	%	[^{3}*H*]*5HT*	%
Control		760 ± 50	100	340 ± 20	100
Pargyline	50	35 160 ± 2000	4600	1750 ± 90	520
Control		630 ± 50	100		
Iproniazid	100	49 250 ± 2430	7807		
Control		1330 ± 90	100		
Tranylcypromine	10	42 510 ± 1590	3190		
Control		980 ± 60	100	390 ± 10	100
Deprenyl	10	1790 ± 170	180	390 ± 10	100
Control		1090 ± 100	100		
Clorgyline	10	43 080 ± 2900	3960		
Control		1230 ± 100	100		
Clorgyline	1	12 350 ± 580	1000		

Groups of five animals were treated with MAO inhibitors; 11.5 h later a tracer dose of [^{3}H]dopa or [^{3}H]5HTP was injected intravenously. The brains were removed for analysis 30 min after the injection of the labelled precursors. Figures represent $\bar{x} \pm S_{\bar{x}}$.

course of the effect of the inhibitors to be followed. The results of measurements made between two and 48 hours after pretreatment with MAO inhibitors are shown in Fig. 5.

The effect of pargyline was greatest at the earliest determination time—that is, two hours after treatment. It decreased linearly over the period of 48 hours when plotted semilogarithmically against time. The half-life of this decline was 15 hours. The onset of action of iproniazid was slower. The maximum accumulation occurred between 6 and 12 hours. Between 12 and 36 hours the effect again declined linearly with a half-life of 16.5 hours, similar to that of pargyline. Tranylcypromine induced the highest initial rise in [^{3}H]dopamine + [^{3}H]MT accumulation. The intensity of this effect diminished relatively rapidly from 2 to 18 hours by a factor of 5–10. From 18 to 48 hours it decreased linearly with a half-life of 12 hours.

The rise produced by deprenyl was small compared with those caused by the other MAO inhibitors. This is also reflected in the relatively wide scatter (Fig. 5). The half-life of the disappearance of this effect between 6 and 48 hours was approximately 18 hours.

The [^{3}H]dopamine + [^{3}H]MT fraction consists of 50–70% [^{3}H]dopamine and 30–50% [^{3}H]MT in the controls. This relation was markedly shifted in favour of [^{3}H]MT by pretreatment with either pargyline or tranylcypromine

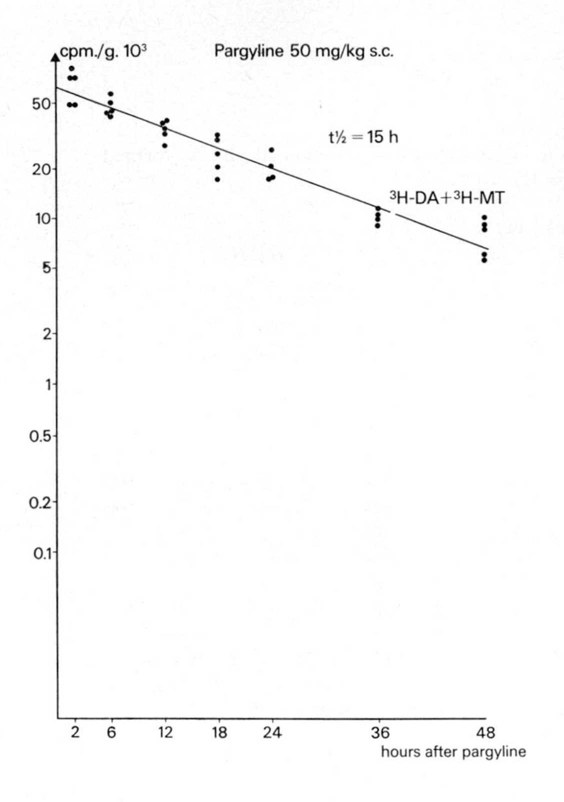
cpm./g. 10^3
Pargyline 50 mg/kg s.c.
t½ = 15 h
^{3}H-DA+^{3}H-MT
50
20
10
5
2
1
0.5
0.2
0.1
2 6 12 18 24 36 48
hours after pargyline

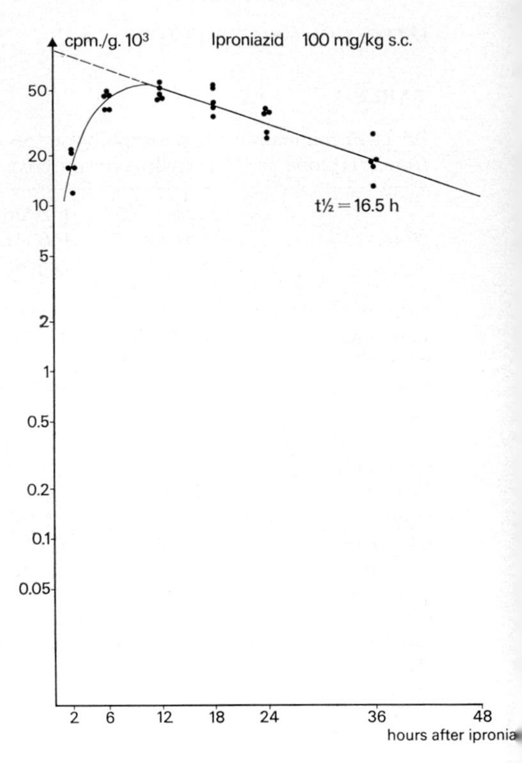
cpm./g. 10^3
Iproniazid 100 mg/kg s.c.
t½ = 16.5 h
50
20
10
5
2
1
0.5
0.2
0.1
0.05
2 6 12 18 24 36 48
hours after ipronia

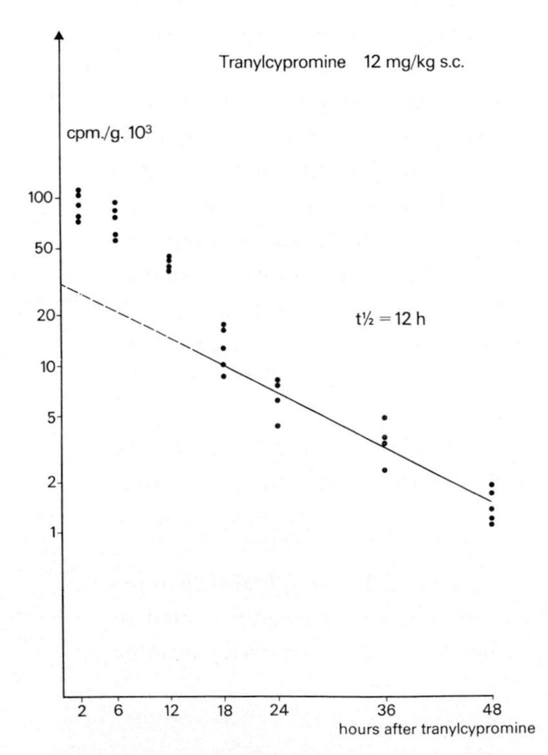
Tranylcypromine 12 mg/kg s.c.
cpm./g. 10^3
t½ = 12 h
100
50
20
10
5
2
1
2 6 12 18 24 36 48
hours after tranylcypromine

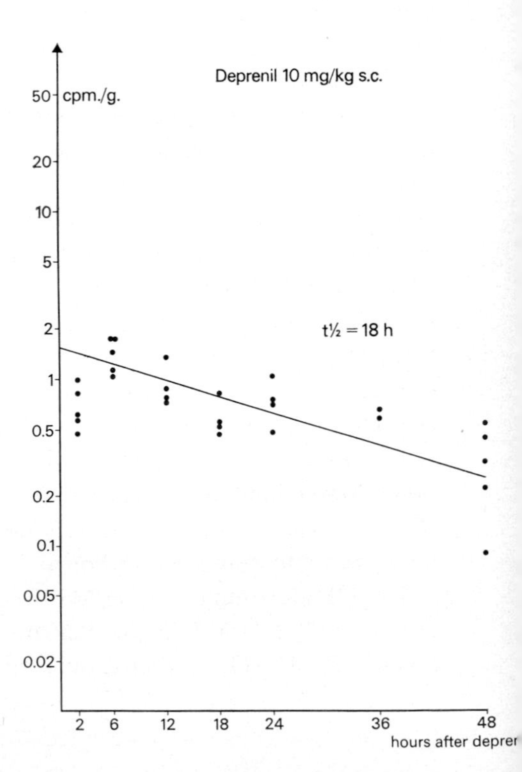
Deprenil 10 mg/kg s.c.
cpm./g.
t½ = 18 h
50
20
10
5
2
1
0.5
0.2
0.1
0.05
0.02
2 6 12 18 24 36 48
hours after depren

TABLE 4

Effects of pargyline and tranylcypromine on the proportions of [^{3}H]dopamine and [^{3}H]methoxytyramine (MT) found in the [^{3}H]dopamine + [^{3}H]MT fraction

Drug	*Dose (mg/kg, s.c.)*	*[^{3}H]dopamine (%)*	*[^{3}H]MT (%)*
Control		70	30
Pargyline 2 h	50	5	95
Pargyline 24 h	50	8	92
Control		51	49
Tranylcypromine 2 h	12	3	97
Tranylcypromine 30 h	12	11	89

Each value represents the mean of 4–5 determinations.

(Table 4). The effect of iproniazid and deprenyl on this ratio has not been studied. In Table 3 and Fig. 5 the effects of the inhibitors on the accumulation of [^{3}H]noradrenaline + [^{3}H]normetanephrine are not shown. They were slight, as will be reported in respect of pargyline (P. C. Waldmeier & L. Maître, in preparation). The greatest rise after 12 hours (approx. eight-fold) was induced by iproniazid.

The short duration of this effect of MAO inhibitors is in marked contrast to the long-lasting blockade of overall MAO activity that they produce. It seems reasonable to assume that the effect is related to their MAO-inhibiting properties for the following reasons: firstly, the four inhibitors belong to different chemical classes; secondly, on replacement of [^{3}H]dopa by [^{3}H]α-methyldopa in the 'time-course' experiment with pargyline, no change in the accumulation of [^{3}H]α-methyldopamine + [^{3}H]α-methylmethoxytyramine in comparison with the controls was observed (P. C. Waldmeier & L. Maître, in preparation), presumably because α-methylated amines are not deaminated by MAO (Blaschko *et al.* 1937).

On the assumption that the observed effect is indeed related to the MAO-inhibiting properties of the drugs tested, their respective potencies with regard to the inhibition of MAO and the increase in [^{3}H]dopamine + [^{3}H]MT accumulation from [^{3}H]dopa were determined. For this purpose, MAO activity was assayed with dopamine as substrate. The doses of pargyline,

← FIG. 5. Time-course of the effects of monoamine oxidase inhibitors on the accumulation of [^{3}H]dopamine + [^{3}H]methoxytyramine (MT) 30 minutes after [^{3}H]dopa. Points represent values obtained by subtraction of the average of the control values from the values of individual treated animals in order to obtain a normalization rate.

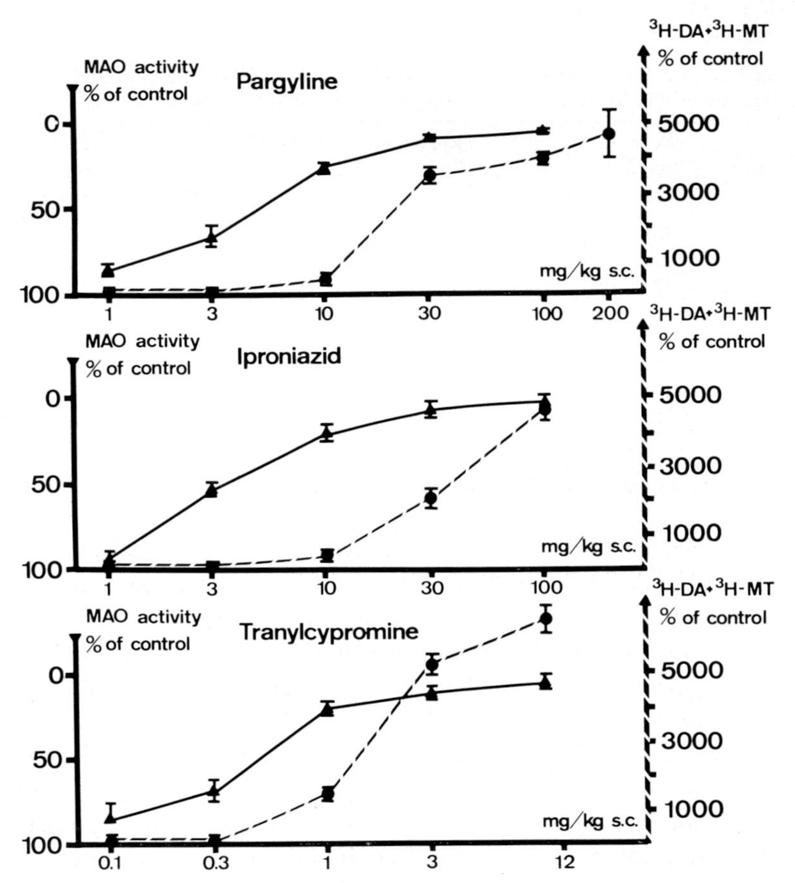

FIG. 6. Comparison of the dose–response curves of the effects of monoamine oxidase inhibitors on MAO activity and [^{3}H]dopamine + [^{3}H]methoxytyramine (MT) accumulation in rat brain. MAO activity was determined with dopamine as substrate. Both accumulation of ^{3}H-labelled amines and MAO activity were measured four hours after pargyline, 16 hours after iproniazid, and two hours after tranylcypromine. Each point represents the mean of four (MAO activity) or five ([^{3}H]dopamine + [^{3}H]MT accumulation) determinations.

iproniazid and tranylcypromine needed to inhibit MAO were definitely smaller than those required to increase the accumulation of the ^{3}H-labelled amines (Fig. 6). For example, the threshold dose of pargyline for MAO inhibition was about 1 mg/kg, s.c., and for the increase in [^{3}H]dopamine + [^{3}H]MT accumulation, 10 mg/kg, s.c. The corresponding figures for iproniazid were 1 mg/kg, s.c. and 10 mg/kg, s.c., and for tranylcypromine 0.1 mg/kg, s.c. and about 0.5 mg/kg, s.c.

The [^{3}H]dopa test system has provided information on the effect of MAO inhibitors on the accumulation of newly formed dopamine and methoxytyramine. It was important to know what deamination products were formed from [^{3}H]dopamine and [^{3}H]MT under these conditions. Neutral and acidic deaminated metabolites present in the brain 30 minutes after the injection of [^{3}H]dopa were analysed by solvent extraction and subsequent combined thin layer chromatography and thin layer electrophoresis. In controls and pargyline-pretreated animals the only detectable product was [^{3}H]HVA. As would be expected, pargyline pretreatment (50 mg/kg, s.c.) decreased the amount of [^{3}H]HVA. However, the duration of action of pargyline on [^{3}H]HVA production was short; that is to say, [^{3}H]HVA had reverted to normal after 24 hours (P. C. Waldmeier & L. Maître, in preparation). This is in marked contrast to the persistent effect of the same dose of pargyline on pool HVA (half-life 9.5 days; see Fig. 2).

It is well established that MAO exists in multiple forms in mammalian tissues (for recent reviews see Eiduson 1972 and Sandler & Youdim 1972). One of the isoenzymes studied by Collins and his colleagues (1970) showed a preferential action on dopamine. The rate of oxidation of dopamine was many times greater than that of other substrates, in contrast to other isoenzymes. The K_m for dopamine was about 100 times lower than that of the other isoenzymes (Youdim 1972*a*). Moreover, this 'dopamine-monoamine oxidase' was particularly resistant to the action of the hydrazine-type MAO inhibitor and less so to tranylcypromine (Youdim 1972*b*).

When Youdim's description of dopamine-monoamine oxidase is compared with our *in vivo* results, it appears conceivable that the large increase in [^{3}H]dopamine + [^{3}H]MT could be a consequence of the inhibition of this particular isoenzyme.

The fact that an MAO possessing a high degree of affinity for dopamine occurs in the rat brain and is active *in vivo* thus seems clearly established. Whether this MAO activity is associated with neuronal systems, and what its functional significance might be, still remain to be clarified. It is tempting to speculate that it corresponds, at least in part, to the striatal intraneuronal 'dopamine MAO' described recently by Agid *et al.* (1973).

Neither of these questions can be answered at present. However, there are some arguments in favour of the idea that dopamine-MAO activity is associated with neuronal systems. It has been shown that two hours after treatment with MAO inhibitors the endogenous methoxytyramine in the whole brain of the rat is increased by a factor ranging from 15 to 60, depending on the inhibitor used. In these experiments, tranylcypromine evoked the greatest and iproniazid the least effect (Kehr 1974). In 1963, Carlsson & Lindqvist already reported that neuroleptic drugs caused an increase in MT and normetanephrine in the brains of nialamide-pretreated mice. These results indicate that the MT

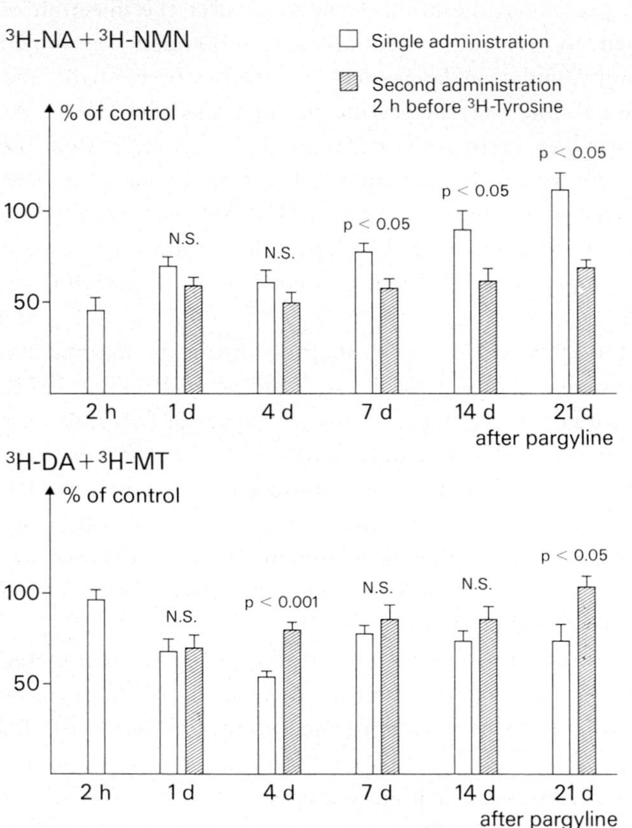

FIG. 7. Time-course of the effect of pargyline on the accumulation of ^{3}H-labelled amines 40 minutes after [^{3}H]tyrosine. The blank columns show the time-course of the effect of a single administration of pargyline (50 mg/kg, s.c.) on the accumulation of [^{3}H]noradrenaline (NA) + [^{3}H]normetanephrine (NMN) (*upper part*) and [^{3}H]dopamine (DA) + [^{3}H]methoxytyramine (MT) (*lower part*). The shaded columns illustrate the effect of a second injection of pargyline (50 mg/kg, s.c.) 80 minutes before [^{3}H]tyrosine ($n = 5$).
Control values were: [^{3}H]NA + [^{3}H]NMN: 410 ± 24 c.p.m./g. [^{3}H]DA + [^{3}H]MT: 661 ± 40 c.p.m./g.

formed after MAO inhibition is related to dopamine released from the nerve endings.

Further evidence has emerged from experiments analogous to the [^{3}H]dopa test system described above, but in which [^{3}H]dopa was replaced by [^{3}H]-tyrosine. In these experiments, pargyline caused a decrease in the accumulation of [^{3}H]noradrenaline and [^{3}H]dopamine and of their *O*-methylated metabolites formed from [^{3}H]tyrosine. This is probably due to feedback inhibition of tyrosine hydroxylation by elevated catecholamine levels (Spector *et al.* 1967). This effect of pargyline was long-lasting, its half-life being similar to that of the effect on overall MAO activity (Fig. 7).

In an experiment continued over a period of three weeks, a second dose of pargyline was injected each time two hours before [^{3}H]tyrosine. The accumulation of [^{3}H]noradrenaline + [^{3}H]normetanephrine during the initial phase following the second dose of pargyline (Fig. 7, shaded columns) was decreased to about the same value as after a single dose, irrespective of the time of pretreatment with the first dose of the inhibitor. This could mean that the feedback inhibitory effect on [^{3}H]noradrenaline accumulation was enhanced (compared with the effects of single doses of pargyline). By contrast, the accumulation of [^{3}H]dopamine was not diminished, but regularly increased (Fig. 7). This has been interpreted as being a consequence of the potent but transient inhibitory effect of pargyline on the deamination of newly formed dopamine (P. C. Waldmeier & L. Maître, in preparation).

In view of the observed effects of MAO inhibitors on [^{3}H]dopa metabolism, it was of interest to study their action on [^{3}H]5HTP metabolites as well. An experiment strictly analogous to those illustrated in Fig. 5 was performed, in which [^{3}H]dopa was replaced by [^{3}H]5HTP. The radioactivity in the fractions containing [^{3}H]5HT and [^{3}H]5HIAA was determined 30 minutes after the injection of the labelled precursor. The influence of the interval following pargyline pretreatment on the formation of these metabolites is shown in Fig. 8. [^{3}H]5HT increased initially by a factor of 15–20, then declined between six and 48 hours with a half-life of 18 hours (Fig. 8). As was mentioned above, [^{3}H]5HT accumulation from [^{3}H]5HTP in pargyline-pretreated rats was not as great as the accumulation of [^{3}H]dopamine + [^{3}H]MT from [^{3}H]dopa.

It is interesting to note that the effect of pargyline on the accumulation of [^{3}H]5HIAA was biphasic (Fig. 9). The maximum initial decrease was observed after an interval of two hours. Later, the effect was reversed, normal levels being regained at about 10 hours, followed by a longer-lasting overshoot. Control levels were reached again after 48 hours. It is striking that during the early phase, when the overall MAO activity is largely blocked, the accumulation of [^{3}H]5HIAA is greater than that seen in controls. Whether this overshoot in 5HIAA accumulation is due to the onset of a compensatory mechanism or to a shift of the metabolism of 5HTP or 5HT to other routes is unknown.

In conclusion, evidence is presented showing that MAO inhibitors exert at least two clearly distinguishable effects on brain monoamines. One of them is the well-known, long-lasting blockade of overall MAO, with the concomitant

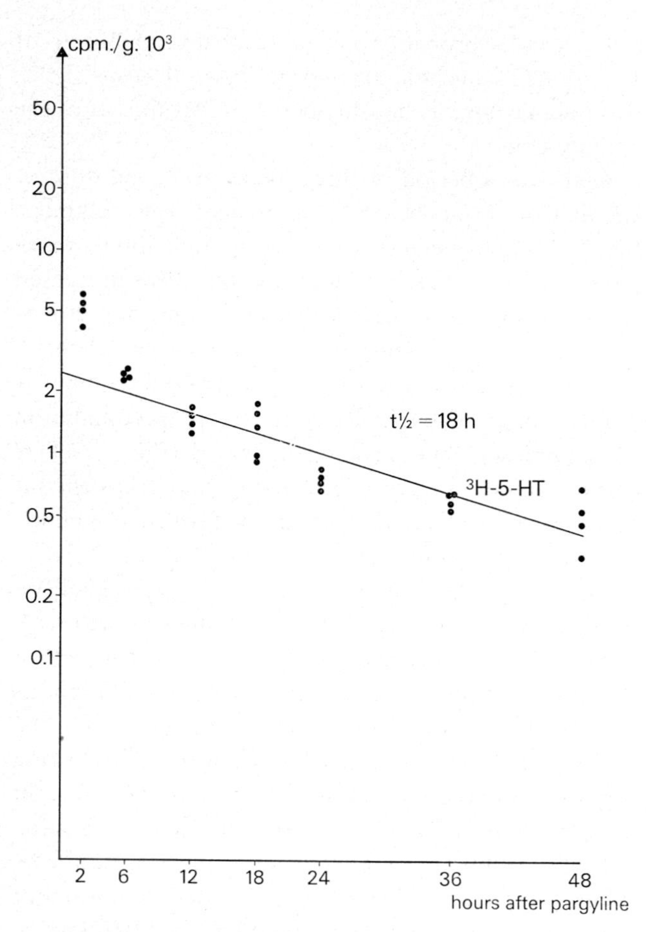

FIG. 8. Time-course of the effect of pargyline (50 mg/kg, s.c.) on the accumulation of [^{3}H]5-hydroxytryptamine (5HT) 30 minutes after the injection of [^{3}H]5-hydroxytryptophan. Points represent values obtained by subtraction of the average of the control values from the values of individual treated animals.

prolonged decreases in the endogenous contents of HVA and DOPAC and in tyrosine hydroxylation. The second effect is of much shorter duration. Although it occurs in response to doses 3–20 times higher than those required for overall MAO inhibition, it also seems to be attributable to the MAO-inhibitory properties. This suggests the existence of at least two separate MAO pools in the rat brain. The main pool has a half-life of several days. The second has a rapid turnover rate (half-life: 12–18 hours); it is much smaller than the main pool and probably deaminates dopamine preferentially.

The duration of several pharmacological effects of MAO inhibitors in the

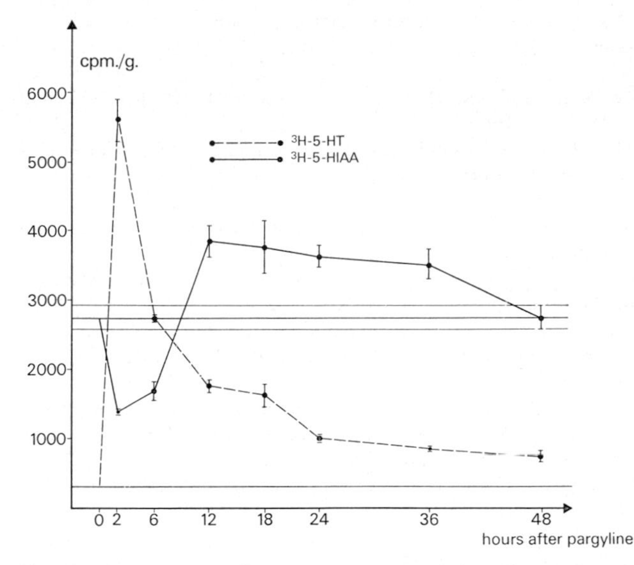

FIG. 9. Comparison of the time-courses of the effects of pargyline (50 mg/kg, s.c.) on the accumulation of [^{3}H]5-hydroxytryptamine (5HT) and [^{3}H]5-hydroxyindoleacetic acid (5-HIAA) 30 minutes after the injection of [^{3}H]5-hydroxytryptophan. Each point represents the mean (± S.E.M.) of five determinations.

rat seems to be more closely related to their interaction with the small and rapidly turning over MAO pool than with the main MAO pool.

ACKNOWLEDGEMENTS

For the generous supply of iproniazid and tranylcypromine we wish to thank Hoffmann-La Roche Ltd. and Smith, Kline and French Laboratories. Our thanks are also due to Dr M. Wilhelm and Dr A. Storni for synthesizing pargyline and deprenyl. We are indebted to Mrs E. Braun, Miss A. M. Buchle, Miss O. Gereben, Miss E. Paul and Mr K. Stöcklin for their excellent technical assistance.

References

AGID, Y., JAVOY, F. & YOUDIM, M. B. H. (1973) Monoamine oxidase and aldehyde dehydrogenase activity in the striatum of rats after 6-hydroxydopamine lesion of the nigrostriatal pathway. *Br. J. Pharmacol. 48*, 175-178

ANDÉN, N.-E., ROOS, B.-E. & WERDINIUS, B. (1963) On the occurrence of homovanillic acid in brain and cerebrospinal fluid and its determination by a fluorometric method. *Life Sci. 7*, 448-458

ATACK, C. V. & MAGNUSSON, T. (1970) Individual elution of noradrenaline (together with adrenaline), dopamine, 5-hydroxytryptamine and histamine from a single, strong cation exchange column, by means of mineral acid-organic solvent mixtures. *J. Pharm. Pharmacol. 22*, 625-627

BLASCHKO, H. & CHRUSCIEL, T. L. (1960) The decarboxylation of amino acids related to tyrosine and their awakening action in reserpine-treated mice. *J. Physiol. (Lond.) 151*, 272-284

BLASCHKO, H., RICHTER, D. & SCHLOSSMANN, H. (1937) The oxidation of adrenaline and other amines. *Biochem. J. 31*, 2187-2196

CARLSSON, A. & LINDQVIST, M. (1963) Effect of chlorpromazine or haloperidol on formation of 3-methoxy-tyramine and normetanephrine in mouse brain. *Acta Pharmacol. Toxicol. 20*, 140-144

CARLSSON, A., LINDQVIST, M. & MAGNUSSON, T. (1957) 3,4-dihydroxyphenyl-alanine and 5-hydroxytryptophan as reserpine antagonists. *Nature (Lond.) 180*, 1200

COLLINS, G. G. S., SANDLER, M., WILLIAMS, E. D. & YOUDIM, M. B. H. (1970) Multiple forms of human brain mitochondrial monoamine oxidase. *Nature (Lond.) 225*, 817-820

EIDUSON, S. (1972) Ontogenic development of monoamine oxidase. In *Monoamine Oxidases—New Vistas* (Costa, E. & Sandler, M., eds.) *(Adv. Biochem. Psychopharmacol. 5)*, pp. 271-287, Raven Press, New York

EVERETT, G. M. (1961) Some electrophysiological and biochemical correlates of motor activity and aggressive behaviour. In *Neuro-psychopharmacology*, vol. 2 (Rothlin, E. ed.), pp. 479-484, Elsevier, Amsterdam, London, New York & Princeton

EVERETT, G. M., DAVIN, J. C. & TOMAN, J. E. P. (1959) Pharmacological studies of monoamine oxidase inhibitors. *Fed. Proc. 18*, 388

JOHNSTON, J. P. (1968) Some observations upon a new inhibitor of monoamine oxidase in brain tissue. *Biochem. Pharmacol. 17*, 1285-1297

KEHR, W. (1974) A method for the isolation and determination of 3-methoxytyramine in brain tissue. *N.S. Arch. Pharmacol. 284*, 149-158

MURPHY, G. F., ROBINSON, D. & SHARMAN, D. F. (1969) The effect of tropolone on the formation of 3,4-dihydroxyphenylacetic acid and 4-hydroxy-3-methoxyphenylacetic acid in the brain of the mouse. *Br. J. Pharmacol. 36*, 107-115

NEFF, N. H. & GORIDIS, C. (1972) Neuronal monoamine oxidase: specific enzyme types and their rates of formation. In *Monoamine Oxidases—New Vistas* (Costa, E. & Sandler, M., eds.) *(Adv. Biochem. Psychopharmacol. 5)*, pp. 307-323, Raven Press, New York

PLANZ, G., QUIRING, K. & PALM, D. (1972) Rates of recovery of irreversibly inhibited monoamine oxidases: a measure of enzyme protein turnover. *N.S. Arch. Pharmacol. 273*, 27-42

SANDLER, M. & YOUDIM, M. B. H. (1972) Multiple forms of monoamine oxidase: functional significance. *Pharmacol. Rev. 24*, 331-348

SHARMAN, D. F. (1971) Methods of determination of catecholamines and their metabolites. In *Methods of Neurochemistry*, Vol. 1 (Fried, R, ed.), pp. 83-127, Dekker, New York

SPECTOR, S., GORDON, R., SJOERDSMA, A. & UDENFRIEND, S. (1967) End product inhibition of tyrosine hydroxylase as a possible mechanism for regulation of norepinephrine synthesis. *Mol. Pharmacol. 3*, 549-555

SPECTOR, S., HIRSCH, C. W. & BRODIE, B. B. (1963) Association of behavioural effects of pargyline, a non-hydrazide MAO inhibitor, with increase in brain norepinephrine. *Int. J. Neuropharmacol. 2*, 81-93

UDENFRIEND, S., WEISSBACH, H. & BOGDANSKI, D. F. (1957) Effect of iproniazid on serotonin metabolism *in vivo*. *J. Pharmacol. Exp. Ther. 120*, 255-260

WALDMEIER, P. C. & MAÎTRE, L. (1975) Lack of significance of MAO-B for *in vivo* deamination of dopamine. *N.S. Arch. Pharmacol.* Suppl. to vol. *287*, R2

WIEGAND, R. G. & PERRY, J. E. (1961) Effect of L-DOPA and *N*-methyl-*N*-benzyl-2-propynylamine-HCl on L-dopa, dopamine, norepinephrine, epinephrine and serotonin levels in mouse brain. *Biochem. Pharmacol. 7*, 181-186

WIRTH, W., GRÖSSWALD, R., HÖRLEIN, U., RISSE, K. H. & KREISKOTT, H. (1958) Zur Pharmakologie acylierter Phenothiazin-Derivate. *Arch. Int. Pharmacodyn. Ther. 115*, 1-31

WURTMAN, R. J. & AXELROD, J. (1963) A sensitive and specific assay for the estimation of monoamine oxidase. *Biochem. Pharmacol. 12*, 1439-1441

YOUDIM, M. B. H. (1972*a*) Multiple forms of monoamine oxidase and their properties. In *Monoamine Oxidases—New Vistas* (Costa, E. & Sandler, M., eds.) *(Adv. Biochem. Psychopharmacol. 5)*, pp. 67-77, Raven Press, New York
YOUDIM, M. B. H. (1972*b*) Multiple forms of mitochondrial monoamine oxidase. *Br. Med. Bull. 29*, 120-122

Discussion

Oreland: Dr Maître, have you considered the possibility that the postulated small enzyme pool with a rapid turnover might be localized in a different compartment with, say, a lower concentration of inhibitor entering it and therefore causing only the reversible phase of the otherwise irreversible inhibition? If so, you would not get a true value of MAO turnover in that compartment.

Maître: It is possible that there is a difference in compartmentation but it seems unlikely, because MAO inhibitors of various chemical classes act in exactly the same manner and with the same rapidity.

Oreland: In that case, how do you explain the relatively large difference between different inhibitors in the slowly turning over pool of MAO?

Maître: We find practically the same turnover values for iproniazid, pargyline and deprenyl. Only tranylcypromine acts more quickly. This effect has already been described in other test systems with tranylcypromine by Planz *et al.* (1972). They suggested that tranylcypromine is only partially an irreversible inhibitor. This could explain why, in many experiments, tranylcypromine acts more rapidly, when the duration of the action is shorter.

Trendelenburg: Could one not suppose that the so-called short-lasting effects are apparent only when MAO is inhibited by more than 70% or so? Your data seem to indicate that the time courses for recovery from inhibition (from 100 to about 70%) and for the short-lasting effect were roughly the same.

Maître: This is possible. But when we gave a second injection of pargyline, one day after the first, we found an increase in the accumulation of dopamine plus methoxytyramine which was roughly three times that seen after one day, and at the same time an inhibition of overall MAO which was practically the same at the two times.

Trendelenburg: We usually think in terms of a 95–98% inhibition of MAO activity, but we might really have to think in terms of 5% or 2% activity *remaining* after inhibition. The difference between 5% and 2% may mean a lot, biologically, whereas in the actual determination of the activity, 95% inhibition looks very much like 98% inhibition. The second dose of pargyline may make all the difference (that is, between 95 and 98% inhibition).

Maître: That is a possibility, and we have seen from Dr Green's paper that the few per cent remaining might have considerable functional activity. We have to consider, however, that the very small effect of deprenyl also disappears with the same half-life (see Fig. 5, p. 258). After treatment with this dose of deprenyl (10 mg/kg, subcutaneously), the inhibition of MAO A is only about 60%.

Neff: I don't see why you need to postulate multiple pools of MAO, Dr Maître; you are simply following the return of enzyme activity that will deaminate the substrate you inject. You will have enough enzyme to do that shortly after giving the MAO inhibitor. The results with tranylcypromine can be explained on the grounds that it is a slowly reversible inhibitor.

Maître: In an experiment in which labelled HVA was measured after the injection of labelled dopa, it was not possible to relate the rapid normalization of HVA production to the overall MAO inhibition, because normalization is very fast and occurs at a time when MAO activity is surely inhibited completely, a few hours after giving 50 mg/kg pargyline (P. C. Waldmeier & L. Maître, unpublished 1975).

van Praag: You observed an increase in 5HIAA some hours after giving the MAO inhibitor and 5HTP—an overshoot, as it were. There may be an alternative explanation to the one you gave, namely inhibition of the efflux of 5HIAA. Ashcroft *et al.* (1968) demonstrated that some MAO inhibitors were able to block the efflux of 5HIAA, somewhat as probenecid does. This recalls an experiment we did in man. We couldn't explain it at the time, but it could be that iproniazid is also able to block the transport of 5HIAA (van Praag & Leijnse 1962). We loaded normal controls and depressed patients under treatment with iproniazid first with 5HT, and observed a decrease in urinary 5HIAA. Then we loaded them with 5HIAA and saw a comparable decrease of urinary 5HIAA. We couldn't explain this. Perhaps this is clinical support for the idea that MAO inhibitors can block the efflux of 5HIAA from several biological systems.

Maître: This interpretation would mean that the efflux of 5HIAA would be blocked between 6 and 24 hours after giving pargyline, and not before, since the level of labelled 5HIAA in brain was decreased for more than six hours, and was practically normal between 6 and 12 hours (Fig. 9, p. 265), meaning that it became obvious just at the change-over point from the decrease to the overshoot. So I do not think this is a likely explanation; if this effect occurred it would have been operating before and the initial decrease should be a combination of this effect and a much greater decrease.

Sharman: The MAO inhibitor that caused all the trouble with transport was phenelzine (Moir 1972). I do not think that pargyline has the same effect.

Green: What happens to endogenous 5HT and 5HIAA? If there are effects on transport it will probably also be evident in the endogenous compounds.

Maître: With pargyline, 50 mg/kg, injected subcutaneously, the endogenous content of 5HIAA decreases to 40–50% after six hours and returns slowly to normal values by the end of about one week. The same dose of pargyline increases the 5HT content in the same animal to reach a plateau at about 190% between 12 and 24 hours, then decreasing slowly and becoming normal between one and two weeks. The elevation of 5HT concentration lasts a little bit longer than the normalization of DOPAC.

Sharman: Your animals received labelled dopa intravenously, Dr Maître. Did you ever inject it into the ventricular system?

Maître: No.

Sharman: This is something that worries me. If you are giving L-dopa intravenously, the first MAO that the dopamine formed from L-dopa will come up against is surely in the blood vessel walls, because the L-dopa will be decarboxylated there. I wonder if your short-term effect is something that happens in the blood vessels whereas your endogenous HVA, of course, is produced near the dopaminergic neurons. Did you look for other radioactive metabolites in these rats, when the short-term effect had worn off?

Maître: Yes. In the experiments in which we measured HVA, we looked at all deaminated acidic and neutral metabolites. HVA was the only metabolite detectable.

Sharman: After your short-term effect on the accumulation of the amines, was there a corresponding decrease in the acid metabolite?

Maître: Yes, in the period between 2 and 48 hours after giving pargyline, there was a decrease and a normalization of tritiated HVA, but no overshoot could be detected, as in the case of tritiated 5HIAA.

Sharman: I am still inclined to think that this short-term effect may have something to do with MAO in the blood vessels.

Maître: That is a very good suggestion.

Blaschko: As a general comment, I think biochemists should be very humble in approaching the situation being discussed here, for reasons that Professor Singer mentioned earlier. The other problem that I am then confronted with is suggested by what Dr Iversen told us about the amount of MAO in the amine-containing neurons. This will to a certain extent come into this discussion. If you measure MAO activity, you determine the overall inhibitory effect. This will depend, at a particular time, on the distance that the newly formed mitochondria have travelled and what contribution to the total pool of mitochondria they represent. Even in the amine-containing neurons, the newly formed mitochondria haven't yet reached the endings; they couldn't have much effect on the accumulation of amines and amine metabolites, but the percentage inhibition in a homogenate will depend on the overall activity of all mitochondria.

Knoll: Dr Maître, you observed that MAO activity inhibited by tranylcypromine recovers rapidly whereas iproniazid and pargyline have long-lasting effects. In iproniazid and pargyline the suicidal group is attached to the nitrogen and in tranylcypromine it is located between the aromatic ring and the nitrogen. If a three-point attachment of the inhibitor to the enzyme is considered, it is evident that the enzyme-killing group fits into the same area in the case of iproniazid and pargyline and to another site of the enzyme in the case of tranylcypromine.

On the other hand, only tranylcypromine contains a cyclopropyl group for killing the enzyme. The possibility, therefore, that it is the chemical nature and not the location of the suicidal group which leads to differences, cannot be ruled out. Lilly 51641 might be a useful tool for further research, because the cyclopropyl group in this inhibitor is attached to the nitrogen.

Youdim: Phenylethylhydrazine, which is a hydrazine inhibitor, does not attack FAD at the same site as deprenyl or clorgyline (Youdim 1976).

References

Ashcroft, G. W., Dow, R. C. & Moir, A. T. B. (1968) The active transport of 5-hydroxyindol-3-ylacetic acid and 3-methoxy-4-hydroxyphenylacetic acid from a recirculatory perfusion system of the cerebral ventricles of the unanaesthetized dog. *J. Physiol. (Lond.) 199*, 397-424

Moir, A. T. B. (1972) Interaction in the cerebral metabolism of the biogenic amines. Effects of phenelzine on the cerebral metabolism of the 5-hydroxyindoles in dog brain. *Br. J. Pharmacol. 45*, 249-264

Planz, G., Quiring, K. & Palm, D. (1972) Rates of recovery of irreversibly inhibited monoamine oxidases: a measure of enzyme protein turnover. *Naunyn-Schmiedeberg's Arch. Pharmacol. 273*, 27-42

van Praag, H. M. & Leijnse, B. (1962) The influence of so-called monoamineoxidase-inhibiting hydrazines on oral loading-test with serotonin and 5-hydroxyindoleacetic acid. *Psychopharmacologia 3*, 202-203

Youdim, M. B. H. (1976) Rat liver mitochondrial monoamine oxidase; an iron requiring flavoprotein, in *Flavins and Flavoproteins* (Singer, T. P., ed.), Elsevier Scientific Publishing Company, Amsterdam, in press

Introduction to clinical aspects of monoamine oxidase inhibitors in the treatment of depression

C. M. B. PARE

Department of Psychological Medicine, St Bartholomew's Hospital, London

Abstract The Medical Research Council's 1965 trial of phenelzine, imipramine and electroconvulsive therapy showed that phenelzine in a dose of 45 mg per day was no use for a certain type of depressive illness. It was falsely inferred that monoamine oxidase inhibitors were useless for treating depression of any sort. As a result of positive effects of trials in phobic states, a similarly illogical inference is now being made that monoamine oxidase inhibitors are only useful for treating anxiety.

A review of the literature shows that there is abundant evidence that monoamine oxidase inhibitors benefit certain types of depression, and that this is likely to be a specific treatment rather than a weaker kind of tricyclic antidepressant. After a discussion of clinical and pharmacogenetic factors, future problems are raised, including dosage, the combination of monoamine oxidase inhibitors with tricyclic antidepressants, lithium and L-tryptophan, and new drugs of the monoamine oxidase inhibitor type.

In 1965 the Medical Research Council conducted a trial of imipramine, electroconvulsive therapy (ECT), phenelzine and placebo in depressed patients, as a result of which a great many psychiatrists in England assumed that whereas imipramine and ECT were effective treatments for depression, phenelzine was no more effective than placebo (Medical Research Council 1965). In fact, what the trial showed was that phenelzine in a dose of 45–60 mg/day for four weeks was no use for patients between the ages of 40 and 69 years who were inpatients in mental hospitals and whose type of depression was judged by the consultants in charge to be suitable for ECT.

Recently, several controlled trials have shown that monoamine oxidase (MAO) inhibitors are effective in the treatment of phobic states (Tyrer *et al.* 1973; Lipsedge *et al.* 1973; Solyom *et al.* 1973). Stemming from this, Tyrer (1974) has drawn what I believe to be a conclusion which is equally invalid:

that the MAO inhibitors benefit patients with depression because of their anxiolytic properties and have no specific antidepressant effect.

To my mind, there is abundant evidence from controlled trials to support the view that MAO inhibitors are effective and specific antidepressants when given in adequate doses, for an adequate period of time, to the correct type of patient.

TYPE OF PATIENT

The broad distinction between endogenous, reactive and neurotic depressions has been the subject of considerable investigation. Less attention has been paid to the sub-classification of 'neurotic' depressions, which probably comprise the majority of cases. Sargant and his colleagues (West & Dally 1959) coined the term 'atypical' depression, stating that such patients did not respond well to ECT but responded well to MAO inhibitors. However, many of the clinical features of 'atypical' depression, such as anxiety, initial insomnia, absence of guilt and morning worsening, are typical of neurotic depressions as a whole (Kendell 1968; Carney *et al.* 1965). In my experience, evidence of an 'illness' factor is equally important; for instance, a reasonable premorbid personality, with a degree of symptoms out of proportion to what one might expect from environmental happenings.

Using a cluster analytic technique, Paykel (1971) defined four groups in a mixed sample of depressed patients. Of particular interest to this discussion was the finding that psychotic (endogenous) depressives were found predominantly as inpatients and responded well to amitriptyline. The anxious depressives, on the other hand, responded poorly to amitriptyline and were found predominantly among the outpatients. These patients had many characteristics of the so-called 'atypical' depressions. They were moderately depressed and had the highest scores on depressed feelings and suicidal tendencies; they had the greatest number of previous episodes of depression, and had the highest neuroticism scores (Maudsley Personality Inventory, MPI).

If we now consider the controlled trials of MAO inhibitors, Bennett (1967) reviewed 70 controlled trials of antidepressants involving 10 000 patients. He concluded that whereas the tricyclic antidepressants were definitely superior in inpatient trials, the MAO inhibitors were at least as effective in those controlled trials conducted on outpatients. Such a finding could hardly be reached if the MAO inhibitors were a similar type of antidepressant to the tricyclics, but weaker in action. They must have different effects, and/or be acting on a different group of patients.

TABLE 1

Effect of monoamine oxidase inhibitors on the concentration of 5-hydroxytryptamine (5HT) in human brainstem, expressed as a percentage of control values

Drug	*Daily dose (mg)*	*No. of patients*	*5HT in brainstem*	*No of cases < 150%*
Isocarboxazid	30	12	190	3/12
Isocarboxazid	30	11	207	3/11
Nialamide	300	8	180	2/8
Clorgyline	20	13	216	2/13
Tranylcypromine	30	9	256	1/9
Nialamide	500 mg, i.m. 3 × per week	8	253	1/8
Isocarboxazid	60	7	246	0/7

CHOICE OF DRUG AND DOSAGE

Clinical experience and a review of the controlled trials show that tranylcypromine is the most potent MAO inhibitor, followed perhaps by iproniazid, and nialamide is perhaps the least effective, though with fewer side-effects. In the course of other investigations we have studied the effect of various MAO inhibitors on the concentrations of 5-hydroxytryptamine (5HT) in the brainstem of patients dying from various terminal illnesses (Maclean *et al.* 1965; Bevan Jones *et al.* 1972). Table 1 is culled from those patients, who had received an MAO inhibitor for at least 10 days before death. As one might expect from clinical experience, tranylcypromine in a dose of 30 mg/day is roughly equivalent to 60 mg of isocarboxazid and results in higher 5HT concentrations than 30 mg of isocarboxazid or 300 mg of nialamide. More important, perhaps, is the marked individual variability and the finding that a quarter of the patients on the less potent drug/dosage regime had increases in 5HT concentrations of less than 50% (Fig. 1). Such patients might be expected to respond poorly from a therapeutic point of view.

The effect of dosage can best be studied by considering a single MAO inhibitor. For the summary of trials of phenelzine shown in Table 2 I am indebted to Dr E. S. Paykel. Overall this table suggests that either a high dose or prolonged treatment are necessary for response. It would seem that a dose of 45 mg daily and a treatment period of three weeks are too low when combined, but one is adequate provided the other is greater. The most important

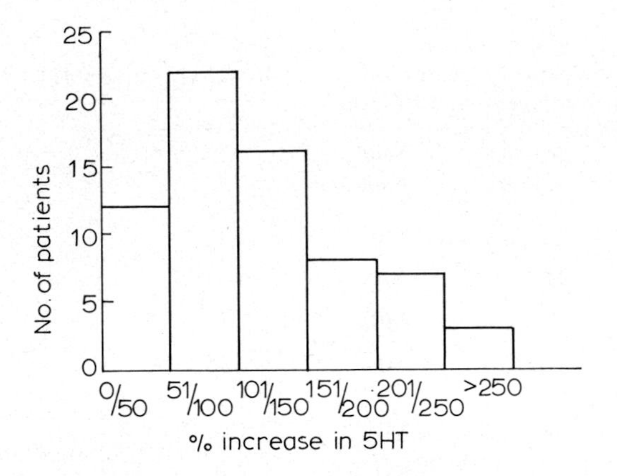

FIG. 1. The variation in concentrations of 5-hydroxytryptamine (5HT) in the brainstem of patients who had been receiving monoamine oxidase inhibitors, expressed as percentages of control values.

evidence that dose is important comes from the study by Nies *et al.* (1975) showing that after six weeks of treatment, 60 mg/day of phenelzine is superior to placebo but 30 mg is not. The most direct evidence that length of treatment is important derives from the finding of Tyrer *et al.* (1973), using a dose of 45 mg of phenelzine daily in phobic patients, that phenelzine was superior to placebo at eight weeks but not at four weeks.

SPECIFICITY OF MAO INHIBITORS IN DEPRESSION

Controlled trials are essential in evaluating drugs in psychiatry but clinical experience is equally important. A clinician who sees a severely depressed patient unresponsive to tricyclic antidepressants and ECT who responds dramatically to an MAO inhibitor, relapses when the drug is withdrawn and responds again to the reintroduction of the drug, experiences something which is hard to convey in a paper. I personally am convinced that although there is a considerable overlap, some patients respond only to tricyclics and some only to MAO inhibitors. This response is usually dramatic and not simply a general euphoriant effect. It works or it doesn't work. Hamilton (personal communication 1974) has produced preliminary data supporting such a drug specificity (Fig. 2). His figures suggest an all-or-none response and the fact that the patients who did not respond to the MAO inhibitor benefited from subsequent ECT, in contrast to the patients who failed to respond to the tricyclic, suggests to me that these drugs are benefiting two different groups of patients (Hamilton

TABLE 2

Controlled trials of phenelzine against placebo

	Authors	*Patient type*	*Source of sample*	*Daily dose (mg)*	*Length of treatment (weeks)*
Depression					
Positive studies	Rees & Davies 1961	Predominantly endogenous	Inpatient	90	3
	Lascelles 1961	Atypical	Outpatient	45	4
	Johnstone & Marsh 1973	Neurotic	Outpatient	90	3
	Robinson *et al.* 1973	Neurotic/atypical	Outpatient	60	6
	Nies *et al.* 1975	Neurotic/atypical	Outpatient	60 (no effect with 30 mg)	6
Doubtfully positive	Hutchinson & Smedberg 1960	Endogenous	Inpatient	45	2
	Hare *et al.* 1962	Mixed	Day patients	30	2
Negative studies	Greenblatt *et al.* 1964	Mixed	Inpatient	60	8
	Medical Research Council 1965	Mixed	Inpatient	60	4
	Bellak & Rosenberg 1966	Mixed	Inpatient	Not given	3–5
	Raskin *et al.* 1974	Mixed	Inpatient	45	4
	Mountjoy & Roth 1975	Neurotic depressives, anxiety states, phobics (negative in all groups)	Mixed	75	4
Anxiety					
Positive studies	Tyrer *et al.* 1973	Phobic	Outpatient	45 (no effect at 4 wk)	8
	Solyom *et al.* 1973	Phobic	Outpatient	45	12

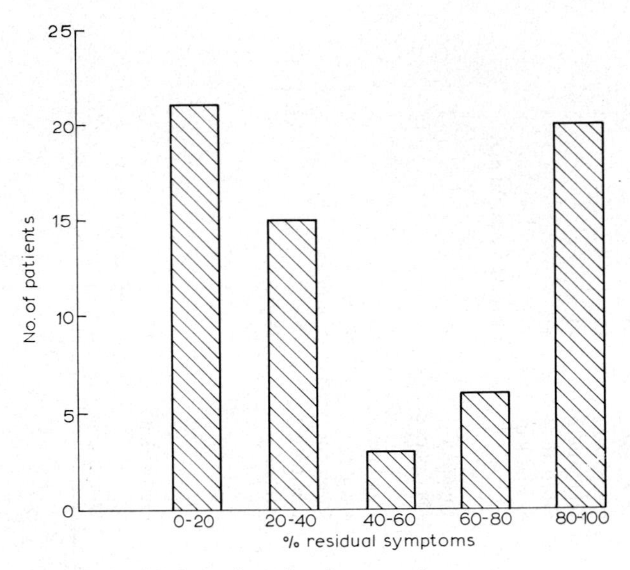

(*a*) Effects of imipramine after four months.

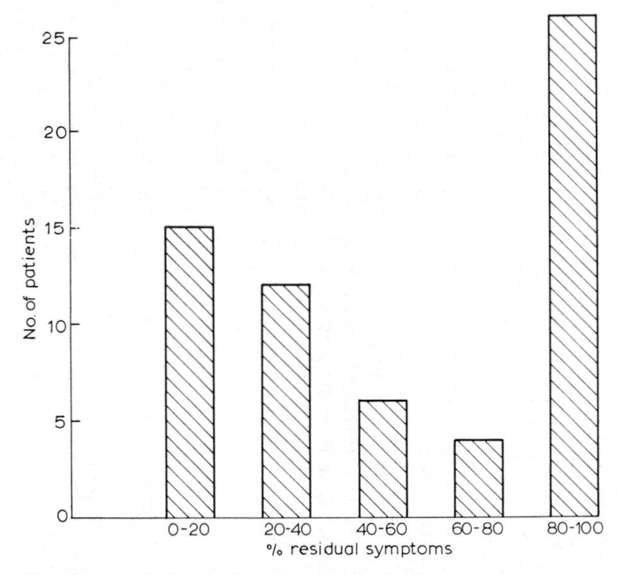

(*b*) Effects of phenelzine after four months.

FIG. 2. The effects of treatment with imipramine or phenelzine on primary depressive illness. Patients were randomly allocated to treatment with imipramine in a minimum dose of 150 mg/day or phenelzine in a minimum dose of 45 mg/day. (M. Hamilton, personal communication 1974).

1974). Furthermore, although I agree that the MAO inhibitors are more likely to benefit the atypical type of depression, I am impressed by the wide variety of patients, including the most typically endogenous type of patient, who may only respond, and quite dramatically, to the MAO inhibitors. It is because of this that we have suggested a genetically distinct type of MAO inhibitor-responsive depression and have published evidence in support of this hypothesis (Pare *et al.* 1962; Pare & Mack 1971).

COMBINATIONS OF MAO INHIBITORS WITH OTHER DRUGS

Tryptophan and MAO inhibitors

All trials of which I am aware show that tryptophan markedly potentiates the antidepressant effect of MAO inhibitors (Coppen *et al.* 1963; Pare 1963; Glassman & Platman 1969; Ayuso Gutierrez & López-Ibor Aliño 1971). Tryptophan has been said to have an antidepressant effect in its own right (Broadhurst 1970; Coppen *et al.* 1972) but the simultaneous appearance of side-effects similar to those found with large doses of MAO inhibitors suggests to me that the beneficial effect of tryptophan when given with MAO inhibitors is a true potentiation. In practice, a combination of tryptophan and MAO inhibitors is less important therapeutically than for its academic interest.

Lithium and MAO inhibitors

Grahame-Smith & Green (1974) suggested on pharmacological evidence that lithium might potentiate the antidepressant effect of MAO inhibitors and, using tranylcypromine in a controlled trial with lithium and placebo, Jenner and his colleagues showed that this was indeed so (M. Dimitrakovdi & F. A. Jenner, in preparation). In this trial, however, the potentiation was not dramatic and to my mind any such benefit is outweighed by the inference that lithium would not prevent a swing into hypomania in bipolar depressives on long-term treatment with MAO inhibitors. This indeed happened to a patient of mine, who swung into hypomania after I added lithium to his iproniazid.

Tricyclic antidepressants and MAO inhibitors

A clinician's job day in, day out, is to judge whether the possible side-effects of any particular treatment are justified, and in doing so he takes into account the degree of disability, resistance to previous treatment, and the possibility of improvement or deterioration in the absence of the treatment under consideration. A proportion of depressed patients are resistant to treatment and it is noteworthy that in the Medical Research Council's trial (1965) 20% of the patients were still hospitalized six months later, having received imipramine, ECT and in many cases phenelzine as well. There is no doubt in my mind that the longer the depression continues not only is it more resistant to treatment but personality, domestic and social factors all tend to deteriorate. It is in such treatment-resistant patients that a combination of MAO inhibitors and tricyclic antidepressants may be justified (Gander 1965; Sethna 1974).

With experience it is becoming clear that combining the antidepressants is much safer than had been anticipated and that adverse effects are due to mishandling of the drugs (Shuckit *et al.* 1971). Giving a tricyclic antidepressant to a patient already on MAO inhibitors is liable to produce sympathetic over-activity; headache may be experienced, hyper-reflexia with muscle rigidity and twitching may occur, while the blood pressure may be high, low or normal. This reaction is especially likely to occur with clomipramine (Beaumont 1973) and in my opinion this latter tricyclic should never be used in combination with an MAO inhibitor. If, however, an MAO inhibitor and tricyclic antidepressant are started together, and the dose is gradually increased, side-effects are only slightly more marked than with either drug given separately and even then are confined to increased weight, postural hypotension and impotence. Clinical experience confirms the findings of Loveless & Maxwell (1965) that trimipramine and amitriptyline are the safest drugs to use in combination and because of their sedative properties are best given at night and the MAO inhibitors in the day. I have never seen a 'cheese' reaction with the combined drugs and I just wonder whether in fact the combination, properly used, might even be safer than MAO inhibitors by themselves. Most of the tricyclic drugs will inhibit the uptake of tyramine into noradrenaline storage sites and might in this way afford some protection from a dietary indiscretion. It would be nice if a pharmacologist could give the clinicians an answer to this.

NEW TYPES OF MAO INHIBITOR

In this brief introduction I have necessarily had to skate over many important aspects and I can only touch on the implications which the finding of different isoenzymes of MAO have to the clinician. These findings have made us look again at the MAO inhibitors already in use and the potency of tranylcypromine might possibly be related to its being the one MAO inhibitor which to date has been found to inhibit isoenzyme 4 and to increase dopamine concentrations in human brain (Bevan Jones *et al.* 1972; Youdim *et al.* 1972).

Deprenyl, a type B inhibitor of MAO, is said to have antidepressant properties (Varga & Tringer 1967) in spite of its lack of effect on 5HT, and is said not to produce a hypertensive reaction with tyramine. The clinical reports are particularly interesting, as it appears to have a more stimulating effect than the MAO inhibitors in current use, and of course one would like to know whether it benefits the same or a different type of patient.

References

AYUSO GUTIERREZ, J. L. & LÓPEZ-IBOR ALIÑO, J. J. (1971) Tryptophan and an MAOI (Nialamide) in the treatment of depression. A double-blind study. *Int. Pharmacopsychiatry 6*, 92-97

BEAUMONT, G. (1973) Clomipramine (Anafranil) in the treatment of pain, enuresis and anorexia nervosa. *J. Int. Med. Res. 1*, 435-437

BELLAK, L. & ROSENBERG, S. (1966) Effects of antidepressant drugs on psychodynamics. *Psychosomatics 7*, 106-114

BENNETT, I. F. (1967) Is there a superior antidepressant? *Proc. 5th Int. Congr. of the C.I.N.P.* (Washington, D. C., March, 1966) *Antidepressant Drugs*, pp. 375-393

BEVAN JONES, A. B., PARE, C. M. B., NICHOLSON, W. J., PRICE, K. & STACEY, R. S. (1972) Brain amine concentrations after monoamine oxidase inhibitor administration. *Br. Med. J. 1*, 17-19

BROADHURST, A. D. (1970) L-Tryptophan versus ECT. *Lancet 1*, 1392-1393

CARNEY, M. W. P., ROTH, M. & GARSIDE, R. F. (1965) The diagnosis of depressive syndromes and the prediction of ECT response. *Br. J. Psychiatr. 111*, 659-674

COPPEN, A., SHAW, D. M. & FARRELL, J. P. (1963) Potentiation of the antidepressive effect of a monoamine oxidase inhibitor by tryptophan. *Lancet, 1*, 79-81

COPPEN, A., SHAW, D. M., HERZBERG, B., MAGGS, R. & PRANGE, A. J. (1972) The comparative antidepressant value of L-tryptophan and imipramine with and without attempted potentiation by liothyronine. *Arch. Gen. Psychiatr. 26*, 234-241

GANDER, D. R. (1965) Treatment of depressive illnesses with combined antidepressants. *Lancet, 2*, 107-109

GLASSMAN, A. H. & PLATMAN, S. R. (1969) Potentiation of a monoamine oxidase inhibitor by tryptophan. *J. Psychiatr. Res. 7*, 83-88

GRAHAME-SMITH, D. G. & GREEN, A. R. (1974) The role of brain 5-hydroxytryptamine in the hyperactivity produced in rats by lithium and monoamine oxidase inhibition. *Br. J. Pharmacol. 52*, 19-26

GREENBLATT, M., GROSSER, G. H. & WECHSLER, H. (1964) Differential response of hospitalized depressed patients to somatic therapy. *Am. J. Psychiatr. 120*, 935-943

HAMILTON, M. (1974) Drug resistant depressions: response to E.C.T. *Pharmakopsychiatr. Neuro-Psychopharmakol. 7*, 205-206

HARE, E. H., DOMINIAN, J. & SHARPE, L. (1962) Phenelzine and dexamphetamine in depressive illness. *Br. Med. J. 1*, 9-12

HUTCHINSON, J. T. & SMEDBERG, D. (1960) Phenelzine (Nardil) in treatment of endogenous depression. *J. Ment. Sci. 106*, 704-710

JOHNSTONE, E. C. & MARSH, W. (1973) Acetylator status and response to phenelzine in depressed patients. *Lancet 1*, 567-570

LASCELLES, R. G. (1966) Atypical facial pain and depression. *Br. J. Psychiatr. 112*, 651-659

KENDELL, R. E. (1968) *The Classification of Depressive Illnesses*, Maudsley Monograph No. 18, Oxford University Press, London

LIPSEDGE, M. S., HAJIOFF, J., HUGGINS, P., NAPIER, L., PEARCE, J., PIKE, D. J. & RICH. M. (1973) The management of severe agoraphobia: a comparison of iproniazid and systematic desensitization. *Psychopharmacologia 32*, 67-80

LOVELESS, A. H. & MAXWELL, D. R. (1965) A comparison of the effects of imipramine, trimipramine, and some other drugs in rabbits treated with a monoamine oxidase inhibitor. *Br. J. Pharmacol. 25*, 158-170

MACLEAN, R., NICHOLSON, W. J., PARE, C. M. B. & STACEY, R. S. (1965) Effect of monoamine-oxidase inhibitors on the concentrations of 5-hydroxytryptamine in the human brain. *Lancet 2*, 205-208

MEDICAL RESEARCH COUNCIL (1965) Clinical trial of the treatment of depressive illness. *Br. Med. J. 1*, 881-886

MOUNTJOY, C. Q. & ROTH, M. (1975) A controlled trial of phenelzine in anxiety, depressive and phobic neuroses, in *Neuropsychopharmacology* (Boissier, J. R., Hippius, H. & Pichot, P., eds.), International Congress Series No. 359, Excerpta Medica, Amsterdam

NIES, A., ROBINSON, D. S., RAVARIS, C. L. & IVES, J. O. (1975) The efficacy of the monoamine oxidase inhibitor phenelzine: dose effects and prediction of response, in *Neuropsychopharmacology* (Boissier, J. R., Hippius, H. & Pichot, P., eds.), International Congress Series No. 359, Excerpta Medica, Amsterdam

PARE, C. M. B. (1963) Potentiation of monoamine oxidase inhibitors by tryptophan. *Lancet 2*, 527-528

PARE, C. M. B. & MACK, J. W. (1971) Differentiation of two genetically specific types of depression by the response to antidepressants. *J. Med. Genet. 8*, 306-309

PARE, C. M. B., REES, L. & SAINSBURY, M. J. (1962) Differentiation of two genetically specific types of depression by the response to anti-depressants. *Lancet 2*, 1340-1343

PAYKEL, E. S. (1971) Classification of depressed patients: a cluster analysis derived grouping. *Br. J. Psychiatr. 118*, 275-288

RASKIN, A., SCHULTERBRANDT, J., REATIG, N., CROOK, T. H. & ODLE, D. (1974) Depression subtypes and response to phenelzine, diazepam and a placebo. *Arch. Gen. Psychiatry 30*, 66-75

REES, L. & DAVIES, B. (1961) Controlled trial of phenelzine (Nardil) in treatment of severe depressive illness. *J. Ment. Sci. 107*, 560-566

ROBINSON, D. S., NIES, A., RAVARIS, C. L. & LAMBORN, K. R. (1973) The monoamine oxidase inhibitor, phenelzine, in the treatment of depressive anxiety states. A controlled clinical trial. *Arch. Gen. Psychiatry 29*, 407-413

SETHNA, E. R. (1974) A study of refractory cases of depressive illnesses and their response to combined antidepressant treatment. *Br. J. Psychiatr. 124*, 265-272

SHUCKIT, M., ROBINS, E. & FEIGHNER, J. (1971) Tricyclic antidepressants and monoamine oxidase inhibitors. Combined therapy in the treatment of depression. *Arch. Gen. Psychiatry 24*, 509

SOLYOM, L., HESELTINE, G. F. D., MCCLURE, D. J., SOLYOM, C., LEDWIDGE, B. & STEINBERG, G. (1973) Behaviour therapy versus drug therapy in the treatment of phobic neuroses. *Can. Psychiatr. Assoc. J. 18*, 25-31

TYRER, P. (1974) Indications for combined antidepressant therapy. *Br. J. Psychiatr. 124*, 620

TYRER, P., CANDY, J. & KELLY, D. (1973) Phenelzine in phobic anxiety: a controlled trial. *Psychol. Med. 3*, 120-124

VARGA, E. & TRINGER, L. (1967) Clinical trial of a new type of promptly acting psycho-energetic agent (phenyl-isopropylmethyl-propynylamine-HCl, E-250). *Acta Med. Acad. Sci. Hung. 23*, 289-295

WEST, E. D. & DALLY, P. J. (1959) Effects of iproniazid on depressive syndromes. *Br. Med. J. 1*, 1491-1494

YOUDIM, M. B. H., COLLINS, G. G. S., SANDLER, M., BEVAN JONES, A. B., PARE, C. M. B. & NICHOLSON, W. J. (1972) Human brain monoamine oxidase: multiple forms and selective inhibitors. *Nature (Lond.) 236*, 225-228

Discussion

Green: We have studied the effect of a combination of lithium and MAO inhibitors in rats (Grahame-Smith & Green 1974). We injected 3 mequiv./kg lithium chloride subcutaneously (using sodium chloride as control) twice a day for three or more days. This dose produces a plasma concentration of around 1 mequiv./l, which is about the therapeutic level in man. If one then gives tranylcypromine or pargyline, after about 50 minutes the rats start showing all the signs of hyperactivity seen with tranylcypromine and tryptophan (Fig. 1). In animals treated with lithium chloride, 5-hydroxytryptamine (5HT) synthesis was increased by 70%. A relationship exists between the degree of hyperactivity and the increased rate of synthesis of 5HT, and the animals appeared to be more active than one would have expected from the 70% increase in synthesis. We wondered whether lithium also causes a change in the release of 5HT. Fig. 2 (p. 282) shows the effect of one injection of lithium chloride followed four hours later by tranylcypromine and tryptophan. The animals treated with lithium chloride developed hyperactivity more rapidly. There was no indication that this dose of lithium produced any change in the rate of accumulation of 5HT. We therefore suggested that lithium also alters the amount of 5HT available for release at the nerve ending.

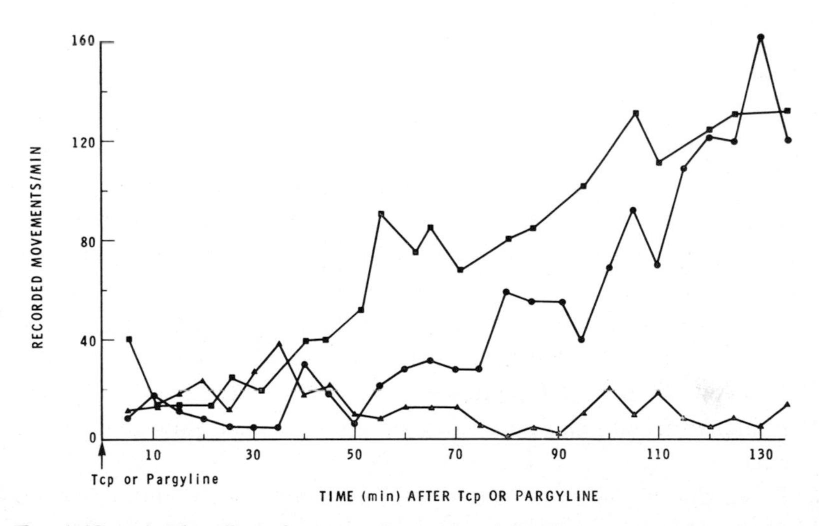

FIG. 1 (Green). The effect of monoamine oxidase inhibition on the activity of rats pretreated with NaCl (3 mequiv./kg) or LiCl (3 mequiv./kg) twice daily for three days. Tranylcypromine (Tcp; 20 mg/kg) or pargyline (75 mg/kg) was injected on the fourth day. (▲), NaCl + pargyline; (●), LiCl + pargyline; (■), LiCl + Tcp. For clarity NaCl + Tcp has not been shown but the activity was essentially the same as with NaCl + pargyline.

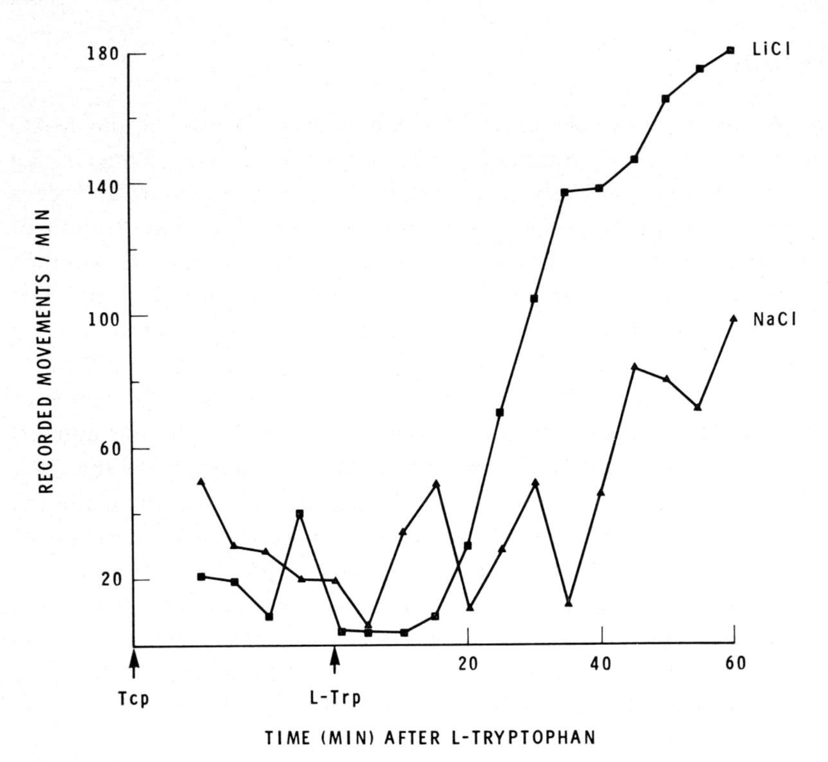

FIG. 2 (Green). Activity of rats given NaCl (3 mequiv./kg) or LiCl (3 mequiv./kg) followed four hours later with tranylcypromine (Tcp, 20 mg/kg) and L-tryptophan (L-Trp, 50 mg/kg) 30 min later. (■), LiCl + Tcp + L-Trp; (▲), NaCl + Tcp + L-Trp.

While we are discussing drug combinations, I would like to mention an isolated case in which a woman with intractable depression was treated with a combination of an MAO inhibitor, tryptophan and reserpine. If you give an animal reserpine to deplete the stores of 5HT, hyperactivity commences sooner after the injection of tranylcypromine and tryptophan because the increased amine formed has to 'spill over' immediately, as it cannot be taken up into storage vesicles (Grahame-Smith 1971). Grahame-Smith (1973) has tried, in a woman with retarded depression resistant to other therapies, giving an MAO inhibitor, tryptophan and reserpine. She showed quite remarkable improvement.

We suggested that lithium and tranylcypromine might be an antidepressant merely because it seemed more than coincidence that both tranylcypromine and tryptophan and tranylcypromine and lithium produce hyperactivity, and Dr Coppen had shown the antidepressant action of the former combination.

Sandler: Dr Pare pinpointed a lot of the problems in this area very well in his paper. Depression is an endpoint, probably for a lot of different metabolic disturbances. To untangle them is a terrible problem; often one can feel one's way by trial and error only and with a variety of different drugs. It is puzzling why some MAO inhibitors should be more effective than others; it is probably a manifestation of the 'dirtiness' of the drug, an effect of the MAO inhibitor on the other enzyme systems or, perhaps, of subtleties in its action on different MAO subcomponents, although these still have to be proved to exist *in vivo*. But, if one can postulate that some amines increase in particular sites of the brain in preference to other amines under the action of particular drugs, one may be some way towards explaining why some patients react in a particular way and others react differently. The latest puzzle in this situation is the claim of Fischer and his group (Spatz *et al.* 1975) that phenylalanine is a useful treatment for depression, and they claim, in fact, that D-phenylalanine is more effective than L-phenylalanine! L-Phenylalanine leads us to phenylethylamine, which forms the basis of a recent hypothesis of affective behaviour (Sabelli & Mosnaim 1974), but we know only one way that D-phenylalanine can be metabolized, namely by D-amino acid oxidase to a keto acid, perhaps before racemization back to the L-form. We tend to focus on the monoamines because we think we understand them better, but perhaps we ought to cast a wider net and take such compounds as the keto acids, which are chemically highly reactive, into our deliberations (Sandler 1973).

Pare: Professor Kety, of course, was the originator of the practice of feeding amino acids; he started the use of tryptophan. As far as I can remember, Dr Kety, you gave phenylalanine and didn't find any elevation of mood?

Kety: It was more complicated than that. We thought we would reduce the number of trials and since we wanted to test a number of amino acids, we gave them in combination with the idea that if we found positive effects we would then try to separate them out. In the first trial we gave phenylalanine and methionine. There was an exacerbation of psychosis in schizophrenic patients. When we separated them, that effect was produced by methionine, while phenylalanine did nothing. The first trial, however, with the combination gave us the most severe reactions, but that may have been because it was the first trial (Pollin *et al.* 1961).

Pare: We have given phenylalanine with MAO inhibitors but we didn't find any mood-elevating effect.

van Praag: With regard to your attempts to differentiate the clinical action of tricyclics and MAO inhibitors, Dr Pare, there is good evidence now that phenelzine is of some value in neurotic depression (anxious depression or hostile depression), and less active in what is called endogenous depression. I think, however, that there is no good evidence that one can generalize about the indication of MAO inhibitors. We have investigated iproniazid and isocarboxazid extensively and our data indicate that these drugs are active in endogenous depression (van Praag 1962). So it is not justified simply to claim that MAO inhibitors are more prone to react favourably in patients with neurotic depression and that tricyclics are better for endogenous depression.

Secondly, with regard to reserpine and MAO inhibitors, we tried to evaluate the hypothesis that the clinical action of MAO inhibitors was related to an increase of monoamines in the central nervous system in man (van Praag 1962). In this context we studied the effect of a combination of reserpine and MAO inhibitors, in particular iproniazid. This combination is supposed to increase the concentration of free—that is, physiologically active—monoamines in the brain. We formed three groups: one group started directly with iproniazid and reserpine, 3 mg per day; the second group of patients had not reacted favourably to MAO inhibitors after 4–6 weeks; we then gave them a combination of the MAO inhibitor with reserpine; the third group was also a group of 'failures', who hadn't responded to MAO inhibitors after 4–6 weeks. We then infused 5 mg reserpine under strict control of blood pressure and other parameters while they were still on MAO inhibitors. We couldn't find any effect, neither a potentiation of the antidepressant effect of the MAO inhibitor nor an earlier occurrence. But we didn't combine it with tryptophan, as Dr Grahame-Smith did; it was just the MAO inhibitor plus reserpine.

Coppen: I agree with Professor Sandler and Dr Pare that in many ways clinical psychopharmacology is in a terrible mess at present and it is hard to make any rational deductions from what few data we have. There are considerable deficiencies in our methodology, beginning with the whole question of classification—the description of the patients we are investigating—which we shall be discussing after Sir Martin Roth's paper. I want to mention one vital area, the question of dosage. Most investigations have used a fixed dosage of drugs, but it is clear that in giving a fixed dose of, say, a tricyclic, which has been studied rather more than the MAO inhibitors, you are doing a very unstandardized thing to different people. Their plasma levels, for example, vary on a fixed dosage of, say, amitriptyline from 20 to 200 ng/ml. The important question, of course, is whether this matters. This is an area that is just beginning to be investigated. Clearly, a dose of 150 mg is doing different things to different people, for various reasons, including firstly, genetic differences in the rate at which individuals metabolize these drugs; secondly, the interaction with other drugs; and thirdly, the fact that half the patients don't take the drugs that are given to them anyway. All this makes the interpretation of trials very difficult unless you have some biological measure: perhaps the plasma level of the drug or, in the case of MAO inhibitors, the degree of MAO inhibition in platelets, for example. Without some measure of what you are doing to the patient it is hard to interpret the results of the trials.

A very nice example of a trial in which this strategy was used was reported by Robinson *et al.* (1973). He looked at an MAO inhibitor in depression, and did two vital things. He had some measure of MAO inhibition in blood platelets, and an adequate clinical description, based mainly on the Newcastle scale, of the type of patients he was looking at. He found that to obtain a therapeutic effect it was necessary to have 85% inhibition of MAO. Not all patients had 85% inhibition on the dosage given. This means that until a trial is done in which the dosage is adjusted to give 85% inhibition, one can't make any statements about a patient being resistant or non-resistant. Secondly, he had a clinical description of the patients, and it turned out that he had more of the reactive type of patient than the endogenous type.

I feel that not until we move into this area of seeing what we are doing to individual patients, working out the adequate dosage in terms of plasma level or, preferably, some biological effect, will we make sense of investigations such as these. Moreover, if you take 100 depressed patients, we know that in a month 30% will recover spontaneously, in whatever way they are treated, and 20% won't recover, whatever you do to them; so you have 50% potentially treatable patients. Of that 50%, half are probably not taking the drug and the other half aren't getting it in adequate dose. So the fact that we have some positive results is a tribute to the potential value of these drugs, when we learn how to use them.

Murphy: We also should not forget that these MAO inhibitors have adverse effects. Dosage, time and factors like this have been very much our concern. We have used a rating scale which picks up multiple kinds of behaviour. We found all kinds of things happening in different individuals at different times during treatment with phenelzine, with a number of adverse effects at about the two-week point, before some of the antidepressant effects appear—which often did not appear at all! We saw everything, including some increase in psychosis and anxiety ratings, the apparent precipitation of some manic episodes, and quite definite antidepressant effects in a few patients. There was no doubt that we are giving a very active agent, but it produced very diverse behavioural effects which only in a small number of patients could be classed as 'antidepressant'.

Youdim: I agree with Dr Coppen about the need for biochemical standardization. Some years ago Dr Pare, Dr Sandler and I looked at brains obtained at autopsy from patients who had been treated with different MAO inhibitors (Youdim *et al.* 1972). Our purpose was to see whether these drugs inhibited MAO activity in the same way in eleven specific brain areas. The three inhibitors tranylcypromine, isocarboxazid and clorgyline had different effects when we used four substrates (dopamine, tyramine, kynuramine and tryptamine) to measure MAO activity. Clorgyline gave uninterpretable results because it showed differences in the different areas of brain, but isocarboxazid showed the same pattern of inhibition in all areas tested, greater inhibition being observed with tryptamine than with the other substrates. Tranylcypromine inhibited dopamine deamination by about 100% and the least inhibition was observed with tryptamine as substrate. It is important to be able to get clinical and biochemical data for these drugs and to show whether, if they are effective, they can be correlated with inhibition of a particular form of the enzyme, or the deamination of a particular substrate.

Crow: We have evidence to support Dr Coppen's point that the metabolism of the drug does matter, for MAO inhibitors as well as for tricyclic antidepressants. This is work by Dr Eve Johnstone (Johnstone & Marsh 1973). She was studying people who were fast or slow acetylators, acetylation being the way that phenelzine has been said to be metabolized in the body. It was found that slow acetylators differ from controls in their response to phenelzine after three weeks, whereas fast acetylators showed no difference from controls. These were all neurotically depressed outpatients. More recently, Dr Johnstone (unpublished work) has studied the clinical effects of phenelzine in fast and slow acetylators in relation to the degree of monoamine oxidase inhibition, and excretion of phenelzine. Fig. 1 shows the clinical effects assessed by the Goldberg rating scale. There were 30 patients. The difference is large and persists into the follow-up period (Fig. 2). The differences were also found on visual analogue scales, which the patient does himself. The therapeutic response can be correlated with the urinary excretion of tryptamine (as an index of monoamine oxidase inhibition); the fast acetylators show less tryptamine excretion, but catch up after a period of time (Fig. 3). The levels of unchanged phenelzine in urine also are much lower in the fast acetylators than in the slow acetylators (Fig. 4), which presumably shows that the active drug persists longer in patients who are slow acetylators. These results of Dr Johnstone's therefore show a very nice relationship between the way in which the drug is metabolized, its biochemical effects and the therapeutic response.

van Praag: Another crucial point in evaluating the effect of drugs, besides psychopathology, dosage and pharmacokinetics, is the biochemical 'typology' of the patient. There is accumulating evidence that certain parameters in central monoamine metabolism, as measured in c.s.f., affect the response one can expect from giving tricyclic drugs or from monoamine precursors (van Praag 1976). We did a preliminary study of the significance of disturbances in 5HT metabolism in depression. There is evidence that in certain patients with

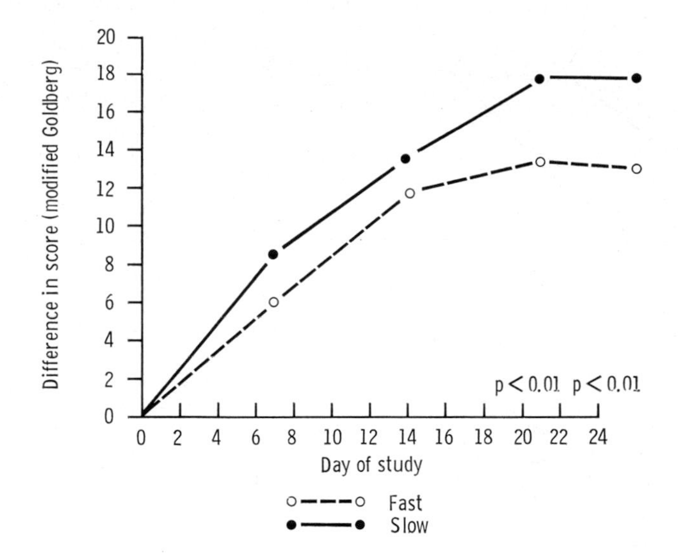

FIG. 1 (Crow). The clinical effects of phenelzine in fast and slow acetylators, assessed as the difference between initial and subsequent scores on the Goldberg rating scale (Eve C. Johnstone, unpublished).

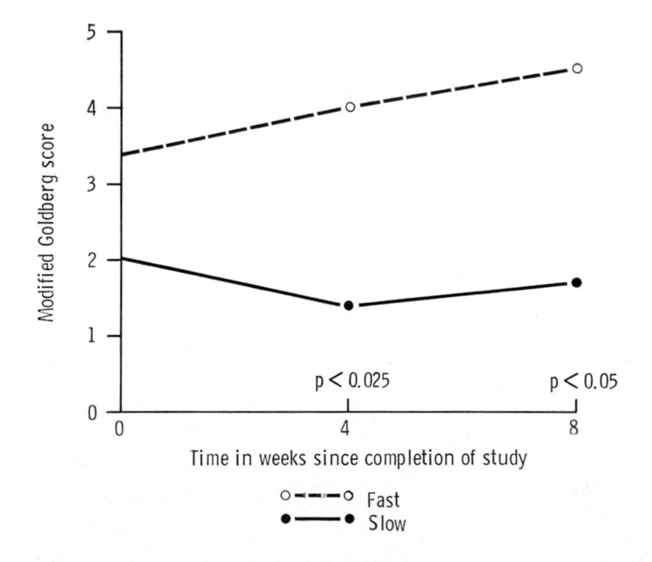

FIG. 2 (Crow). Finishing and follow-up scores (Goldberg rating scale) of fast and slow acetylators after treatment with phenelzine (Eve C. Johnstone).

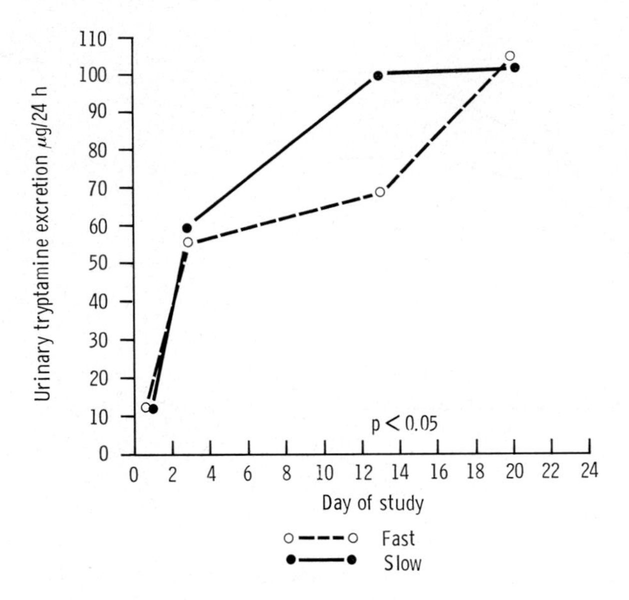

FIG. 3 (Crow). Urinary tryptamine excretion in fast and slow acetylators during treatment with phenelzine (Eve C. Johnstone).

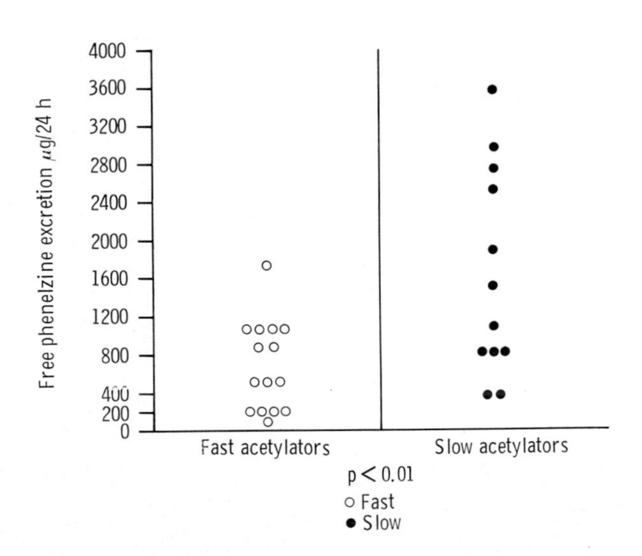

FIG. 4 (Crow). Excretion of free phenelzine in fast and slow acetylators on day 13 of study (Eve C. Johnstone).

TABLE 1 (van Praag)

Prophylactic action of L-5-hydroxytryptophan in unipolar and bipolar depression

Patients	five (female 3, male 2)
Age	35–55 years
Syndrome	vital (= endogenous) depression
Aetiology	hereditary and psychogenic factors
Course	bipolar 3, unipolar 2
Pathogenesis	subnormal 5-hydroxytryptamine turnover
Relapse rate	three or more depressive phases in the past 10–14 months

what we call vital depression—that is, the syndrome of endogenous depression —there is a decreased 5HIAA response in the lumbar c.s.f. after probenecid (van Praag *et al.* 1973; van Praag & Korf 1971). This phenomenon is indicative of a lowered turnover of 5HT in the CNS. One peculiar feature is that in certain patients it is not syndrome-dependent; that is, this disturbance of 5HT metabolism remains after the depressive syndrome has disappeared. There are two possible explanations: that there is no relation between 5HT and depression, or that the lowered 5HT turnover is not a causal factor but a predisposing factor which makes the patient more liable, or vulnerable, to depressive symptomatology. If the latter explanation were true, it would make sense to try to abolish the disturbance in 5HT metabolism by administering 5HT precursors, which are supposed to increase the rate of synthesis of 5HT in man. We tried, therefore, to measure whether 5-hydroxytryptophan (5HTP) has any prophylactic value in these patients with a disturbed 5HIAA response to probenecid (Table 1). So far we have treated five patients, all with the syndrome of endogenous depression, but different from the point of view of aetiology. The course was bipolar in three patients and unipolar in two. They all had subnormal central 5HT turnover. They were chosen because their relapse rate was high. So far the results have been rather remarkable. They were successfully treated with tricyclics in hospital for two months and the treatment was continued for about four months after discharge. This medication was then discontinued and one month later they started a one-year treatment with 5HTP in combination with a peripheral decarboxylase inhibitor.

The 5HIAA response after probenecid was subnormal before treatment. After tricyclics, as we know, there is a decreased response of 5HIAA, which may be a reflection of the decreased turnover of 5HT which one can find in test animals. After the discontinuation of tricyclic drugs there was an increase in the 5HIAA response, but it remained subnormal. One month after we started the 5HTP treatment, the 5HIAA response had risen substantially, which indicates that these patients are capable of transforming 5HTP into 5HT

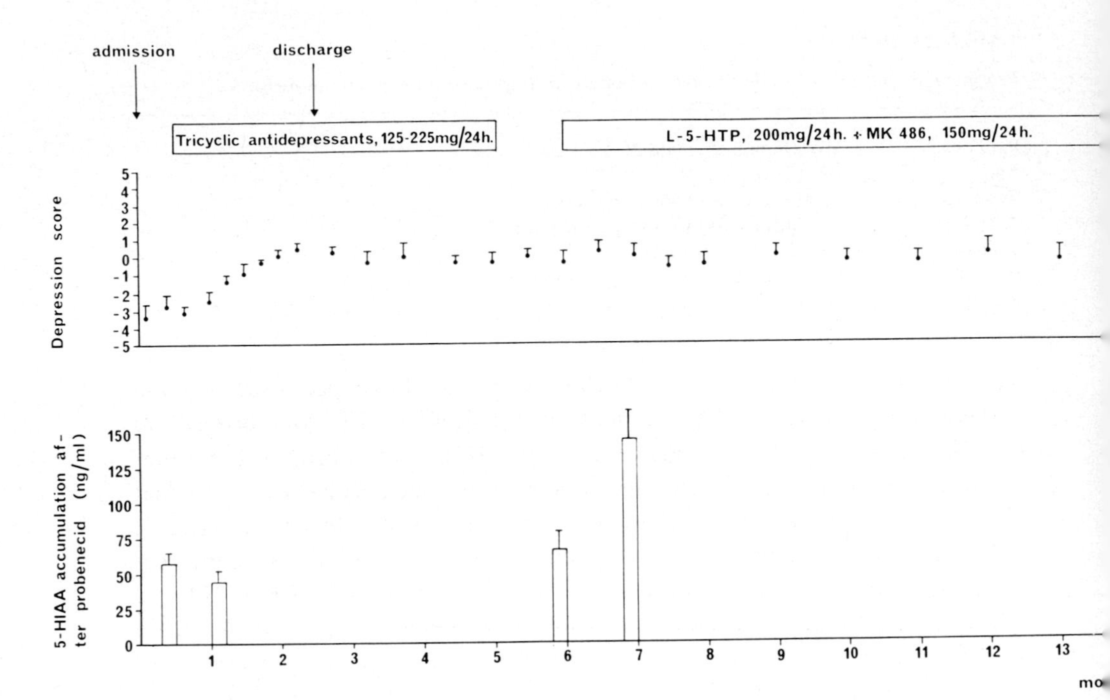

FIG. 1 (van Praag). Biochemical and clinical effect of chronic administration of L-5-hydroxytryptophan in combination with a peripheral decarboxylase inhibitor in unipolar and bipolar depression.

in the central nervous system. During the year on 5HTP there has been no relapse so far (Fig. 1). We are now planning to confirm this result and to replace 5HTP with placebo. This preliminary study was only to demonstrate the possible prophylactic value of 5HTP, and also the value of biochemical parameters in choosing the right treatment strategy in depression.

Sourkes: I would like a little practical advice here. I have two laboratories, and one of them is supposed to be translating research data into the clinic. Can the biochemical psychiatrists here tell me whether, in a patient who is on an MAO inhibitor, to take a simple case, we should measure platelet MAO levels and on the basis of this result we would be able to tell the psychiatrist if his treatment is working well? In another case, if the patient is on desipramine, what should we measure? The blood level of the drug? What should we be doing to tell the clinician whether he is giving enough of the drug, for example? I would like advice on this.

Pare: Ideally one wants a pharmacological measure, like platelet amine oxidase inhibition, rather than blood levels of MAO inhibitors. But for the tricyclic antidepressants this isn't so easy, and most people measure the plasma levels of the drug. However, we are starting to measure the uptake of amines into platelets, using different amines, as a pharmacological measure of the effect of the tricyclic antidepressants, and correlating this with plasma levels of drug and clinical response to see if this provides useful correlations.

Coppen: I think we are in the unenviable position of not knowing how valuable such investigations are. It's somewhat of a reproach to us that after 15–20 years of these drugs we still don't know how to use them or what the most useful measure is. There are very few investigations of the relationship between the plasma level of nortriptyline and therapeutic response; we have one study of amitriptyline, as far as I know, on this subject, and one or two studies on imipramine. This more or less constitutes the world literature on the subject (see Angst 1973). So it is really still a question of research rather than giving a service, and it is the same with the inhibition of uptake of amines and so on. It's a research problem but one that will be very rewarding as we learn how to use the drugs, with a subsequent reduction of morbidity.

Dr Kety mentioned (p. 2) the accidental finding that the anti-tuberculous drugs had a tremendously euphoric effect on patients. I have never seen a euphoric effect of an MAO inhibitor, and I wonder how this report came about and whether it was a false report that luckily has led to interesting further observations. It occurs to me that alternatively there may have been one or two patients with bipolar depression who were also having tuberculosis. What are your views, Dr Pare, on the treatment of bipolar depression by MAO inhibitors or by any other drugs: that is, patients with a history of mania but who are depressed?

Pare: In my experience, some such patients have *only* responded to MAO inhibitors.

Coppen: We have a number of patients with bipolar depression who seem to do quite well on the combination of lithium and MAO inhibitor, but on the whole one would be rather careful.

Pare: The problem, as raised by Dr Green and in my paper, is whether the primary clinical benefit of lithium in preventing swings into mania will work when it is combined with an MAO inhibitor.

Blaschko: I just wonder whether the question that Dr Coppen asked would require us to know what the response of normal subjects to MAO inhibitors is? I presume that the TB patients who were initially observed were people who would be considered to be normal subjects?

Coppen: I understand that MAO inhibitors have little effect on normal subjects, but I have no direct experience of this. But there is evidence that depressed people and normal subjects are different in their reactions to tricyclics.

Normal subjects are more vulnerable to the autonomic side-effects of tricyclics than depressives.

van Praag: My wife and I have taken iproniazid and isocarboxazid for several weeks, for scientific purposes, and we felt no effects, but perhaps our constitution is such that we can resist pharmacological mood elevation.

Kety: The doses given to tuberculosis patients may have been very large. In animals you can produce activation with an MAO inhibitor, surely?

Green: Only when combined with tryptophan, at least in our model. Even with large doses of MAO inhibitor alone, the hyperactivity syndrome does not appear.

Kety: Perhaps in the sanatorium the patients were on high tryptophan diets!

Murphy: I have recently examined much of the earlier literature on the treatment of tuberculosis and psychiatric disorders with MAO-inhibiting drugs such as iproniazid and looked for histories of bipolar depression and so forth in the case-reports of manic and psychotic episodes during the administration of MAO inhibitors (Murphy 1975). In about 20–30% of the cases a past history of hypomania or mania was noted, but there were also many instances where it was stated that the individuals never previously had hypomanic or manic behaviour.

Coppen: Do you know of any systematic data on normal subjects given MAO inhibitors, apart from Dr van Praag and his wife? In recovered depressives, one doesn't see this picture of hypomania at all; the MAO inhibitors are not euphorics by any means.

Murphy: No, not in all patients, but a whole variety of behavioural changes can be observed if one uses sensitive daily measures of behaviour. Most patients receiving currently available MAO inhibitors do not show euphoric activity like that described in the early literature on iproniazid. Drug dosage is another important issue, as iproniazid was used in some of the studies in doses which are disproportionately large in comparison with doses of the drugs now being used. In addition, not only tryptophan, but also dopa and perhaps other substances like dihydroxyphenylserine (dops), will interact with MAO inhibitors in clinical situations. Lower doses of these amine precursors will produce behavioural activation, once MAO inhibition is present.

Youdim: We received a patient who had tried to commit suicide by taking thirty tablets of 30 mg phenelzine; she was in a coma and we decided to follow her platelet MAO activity to see whether we could correlate it with any behavioural changes. Her platelet MAO was inhibited up to six weeks. Unfortunately we were unable to obtain a report of her behaviour during that period.

Fuller: I want to comment on the use of amine uptake inhibitors in combination with MAO inhibitors and emphasize that there is a high degree of selectivity among monoamine uptake inhibitors just as there is among MAO inhibitors.

Two new compounds being studied extensively in our laboratories are amine uptake inhibitors with the following structures:

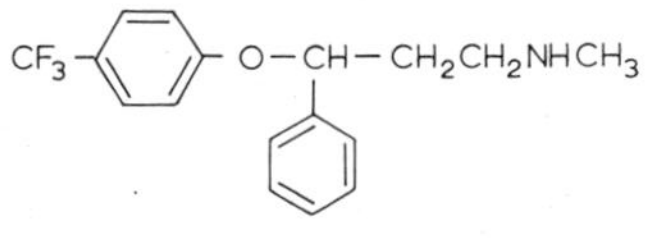

CF_3–C_6H_4–O–CH(–C_6H_5)–$CH_2CH_2NHCH_3$

Lilly 110140

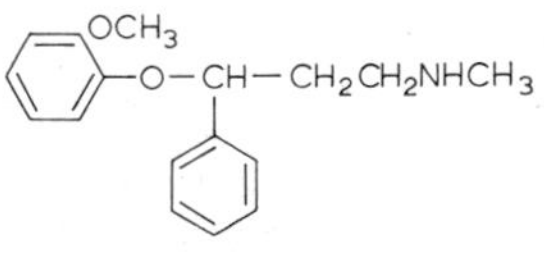

OCH_3–C_6H_4–O–CH(–C_6H_5)–$CH_2CH_2NHCH_3$

Lilly 94939

These compounds differ only in the substituent on the phenoxy group, but they have very different effects on monoamine uptake both *in vitro* and *in vivo*. To evaluate uptake inhibition *in vivo* we have studied the depletion of monoamines by agents that require active uptake, by the amine pump, for entry into the neuron. For 5HT neurons, the depleting agent we use is *p*-chloroamphetamine, and for noradrenaline neurons the depleting agent is 6-hydroxydopamine. Blockade of the amine-depleting effects of these drugs is a measure of the inhibition of uptake into 5HT and noradrenaline neurons. We tested a series of uptake inhibitors at doses up to 32 mg/kg, and a few results are shown in Table 1. Thus it is possible to manipulate very selectively the uptake of noradrenaline and of 5HT, and if one wants to combine an uptake inhibitor with an MAO inhibitor, one can achieve selectivity with both drugs.

TABLE 1 (Fuller)

ED_{50} values of various agents for protection against depletion of monoamines (mg/kg)

Uptake inhibitor	*5HT depletion*	*Noradrenaline depletion*
110140	0.4	No effect
94939	No effect	1.2
Clomipramine	10.0	6.0
Desipramine	No effect	0.25

Knoll: Did you analyse the 5HT receptor blocking effects of your new compounds?

Fuller: They do not block 5HT receptors in isolated peripheral smooth muscle, but those receptors may not be the same as brain receptors and it's difficult to study brain receptors.

Knoll: The guinea pig ileum contains neuronal 5HT receptors which might serve as models.

Fuller: We have tested that. It is not blocked by these uptake inhibitors.

Rafaelsen: We are particularly interested in membrane transport in the elucidation of manic-depressive psychosis, or manic-melancholic disorder. We were therefore intrigued by Cotzias's report that MAO inhibitors might play a specific regulatory role in the transport of substances such as dopamine (see Cotzias *et al.* 1974). Mice pre-treated with nialamide or iproniazid and then given dopamine showed brisk motor responses characteristic of treatment with its precursor, L-dopa. They suggested that binding of nialamide and iproniazid might lead to inactivation of monoamine oxidase and also to transport and subsequent release. We were intrigued by this. Although the blood platelet or leucocyte might be the best available model for mitochondrial or synaptosomal membranes in the brain, a better model for the blood–brain barrier may be the red blood cell, even if it is far from ideal. We have therefore done some preliminary experiments. We took a 25% suspension of normal human red cells and incubated it *in vitro*. The uptake of L-tryptophan and L-leucine was reduced with iproniazid and nialamide. The inhibitor and amino acids were both at a concentration of 2mM. Although these are very preliminary results, perhaps we should bear in mind the possibility that MAO inhibitors might have an action related directly or indirectly to membrane transport in psychotic disorders.

We plan to continue this work with red cells from depressed patients in parallel experiments.

Pare: We have studied the uptake of 5HT and dopamine into platelet-rich plasma. We find that in patients with endogenous depression the uptake of 5HT and dopamine is much lower than in control patients, individually matched for age and sex.

Rafaelsen: It would be interesting to compare platelets and erythrocytes, for these reasons.

Green: I am interested in these results with tryptophan because Boullin and I published data on tryptophan uptake into the platelet, which looked reasonably like the synaptosome in terms of uptake kinetics (Boullin & Green 1972).

However, I wonder if you are looking at active transport, since at a concentration of 2 mM-tryptophan you will mainly observe diffusion. At 10^{-4}M-tryptophan we only found a 3:1 concentration of tryptophan in the platelet, whereas we could reach 16:1 at 10^{-7}M and uptake was blocked by metabolic inhibitors.

Rafaelsen: The kinetic analysis with this concentration seems to indicate, however, that we were interfering with facilitated transport.

Green: I am also interested in Dr Pare's finding of differences in 5HT uptake in platelets from depressed patients and normals. I recall that previous studies have not shown this difference.

Murphy: Our studies of 5HT uptake in platelets from depressed patients compared to normals have been negative, but it is a technically difficult procedure because of problems in preserving the platelets in the same state after isolation. As techniques improve, I think the question should be re-examined.

References

ANGST, J. (ed.) 1973) *Classification and Prediction of Outcome of Depression*, F. K. Schattauer Verlag, Stuttgart & New York

BOULLIN, D. J. & GREEN, A. R. (1972) Mechanisms by which human blood platelets accumulate glycine, GABA and amino acid precursors of putative neurotransmitters. *Br. J. Pharmacol. 45*, 83-94

COTZIAS, G. C., TANG, L. C. & GINOS, J. Z. (1974) Monoamine oxidase and cerebral uptake of dopaminergic drugs. *Proc. Natl. Acad. Sci. U.S.A. 71*, 2715-2729

GRAHAME-SMITH, D. G. (1971) Studies *in vivo* on the relationship between brain tryptophan, brain 5-HT synthesis and hyperactivity in rats treated with a monoamine oxidase inhibitor and L-tryptophan. *J. Neurochem. 18*, 1053-1066

GRAHAME-SMITH, D. G. (1973) in discussion of *Serotonin and Behavior* (Barchas, J. & Usdin, E., eds.), pp. 563-564, Academic Press, New York

GRAHAME-SMITH, D. G. & GREEN, A. R. (1974) The role of brain 5-hydroxytryptamine in the hyperactivity produced in rats by lithium and monoamine oxidase inhibition. *Br. J. Pharmacol. 52*, 19-26

JOHNSTONE, E. C. & MARSH, W. (1973) Acetylator status and response to phenelzine in depressed patients. *Lancet 1*, 567-570

MURPHY, D. L. (1975) The behavioral toxicity of MAO-inhibiting drugs. *Adv. Pharmacol. Chemother*, in press

POLLIN, W., CARDON, P. V., JR & KETY, S. S. (1961) Effects of amino acid feedings in schizophrenic patients treated with iproniazid. *Science (Wash. D.C.) 133*, 104-105

ROBINSON, D. S., NIES, A., RAVARIS, C. L., IVES, J. O. & LAMBORN, K. R. (1973) Treatment response to MAO inhibitors: relation to depressive typology and blood platelet MAO inhibition, in *Classification and Prediction of Outcome of Depression* (Angst, J., ed.), F. K. Schattauer Verlag, Stuttgart & New York

SABELLI, H. C. & MOSNAIM, A. D. (1974) Phenylethylamine hypothesis of affective behaviour. *Am. J. Psychiatry 131*, 695

SANDLER, M. (1973) Closing remarks, in *Frontiers in Catecholamine Research* (Usdin, E. & Snyder, S. H., eds.), pp. 1187-1188, Pergamon, New York

SPATZ, H., HELLER, B., NACHON, M. & FISCHER, E. (1975) Effects of D-phenylalanine on clinical picture and phenethylaminuria in depression. *Biol. Psychiatr. 10*, 235-239

VAN PRAAG, H. M. (1962) Een kritisch onderzoek naar de betekenis van monoamine-oxidaseremming als therapeutisch principe bij de behandeling van depressies [A critical study of the significance of MAO inhibition as a therapeutic principle in the treatment of depressions], Dissertation, Utrecht

VAN PRAAG, H. M. (1976) Monoamines and affective disorders, in *Current Developments in Psychopharmacology*, vol. 3, Spectrum, New York, in press

VAN PRAAG, H. M. & KORF, J. (1971) Endogenous depressions with and without disturbances in the 5-hydroxytryptamine metabolism: a biochemical classification? *Psychopharmacologia, 19*, 148-152

VAN PRAAG, H. M., KORF, J. & SCHUT, T. (1973) Cerebral monoamines and depression. An investigation with the probenecid technique. *Arch. Gen. Psychiatry 28*, 827-831

YOUDIM, M. B. H., COLLINS, G. G. S., SANDLER, M., BEVAN JONES, A. B., PARE, C. M. B. & NICHOLSON, W. J. (1972) Human brain monoamine oxidase; multiple forms and selective inhibitors. *Nature (Lond.) 236*, 226-228

The relationship between classification and response to drugs in affective disorders—problems posed by drug response in affective disorders

M. ROTH, C. GURNEY, C. Q. MOUNTJOY, T. A. KERR and K. SCHAPIRA

Department of Psychological Medicine, The Royal Victoria Infirmary, Newcastle upon Tyne

Abstract The literature of affective disorder offers certain suggestions about the relationship between the clinical picture and the therapeutic response to various antidepressive drugs. Thus in controlled trials undertaken in certain groups of depressed patients, monoamine oxidase inhibiting drugs proved inactive, in contrast to tricyclic compounds which elicited a significant therapeutic response. However, the recorded observations are contradictory and no firm conclusions are possible.

Therapeutic response was one of the problems investigated in the course of the present enquiry into the classification of affective disorders. In this investigation of 145 cases, the predominantly anxious patients were found to differ from those with a predominantly depressed mood in respect of a wide range of variables. These included indices of stability and adaptation in the early developmental years and measures of premorbid personality. There was a substantial overlap in respect of affective disturbance, but certain clinical features were relatively specific for the predominantly depressed and anxious patient groups respectively.

With the aid of principal components and discriminant function analyses, a diagnostic index was developed. This made it possible to achieve clear separation between patients with anxiety states on the one hand and those with depressive illness on the other. This distinction between the groups was validated by a number of findings in the course of a follow-up study by independent observers. The depressed patients had achieved a significantly better recovery than those with anxiety states throughout the period of observation. The groups differed significantly in response to the treatments prescribed for them by independent observers. Very little cross-over of clinical type occurred in the follow-up period. Finally, the clinical features which best predicted outcome in the anxiety states were different from the best predictors in the depressions. The findings accord better with a dichotomous than a continuum view of depressive and anxiety states.

The two groups were thus separated by difference in course and outcome, personality measures, and prognostic indices, as well as certain clinical features. The evidence suggests that they are also differentiated by pattern of therapeutic response. It is in the depressive groups that favourable results with tricyclic compounds have been most consistently achieved. On the other hand, a number of recent studies suggest that monoamine oxidase inhibitors alleviate certain features of anxiety and phobic states. These features await more clear definition.

The conflict of evidence regarding the effects of monoamine oxidase (MAO) inhibitors poses clear and challenging questions about the classification of affective disorders. The area of enquiry defined by these effects can be best delineated by contrasting the situation with that which prevails in relation to the action of tricyclic compounds. There is now a large body of evidence that they are effective in the treatment of depressive illness (Kiloh *et al.* 1962; Medical Research Council 1965; Kay *et al.* 1973). Moreover, four studies of maintenance therapy with tricyclic compounds are now on record, three undertaken in the United States and one under the aegis of the Medical Research Council of Great Britain (Prien *et al.* 1974; Covi *et al.* 1974; Klerman *et al.* 1974 and Mindham *et al.* 1972 and 1973), to testify to the efficacy of these substances in reducing the rate of relapse. In fact, the claims made on behalf of these compounds are far-ranging. A number of enquiries have reported them as being effective in the treatment of neurotic as well as endogenous depression (Hordern *et al.* 1965; Covi *et al.* 1974) and some workers have described beneficial effects in a wider range of neuroses (Mendel & Klein 1969).

The situation in relation to monoamine oxidase inhibitors is very different. Perhaps the most powerful body of evidence to call in question their efficacy came from the authoritative Medical Research Council trial of 1965. This not only found phenelzine to be ineffective in a group of severely depressed inpatients but suggested it might be inferior to placebo, although the results were not, in this respect, significant statistically.

The range of disorders treated in this trial was restricted but Raskin *et al.* (1974), comparing phenelzine, diazepam and placebo in different syndromes, concluded that phenelzine 'did not appear to be an effective treatment for any of the depressive subtypes under consideration'. Kay *et al.* (1973) reported amitriptyline as superior to phenelzine in the treatment of depressed outpatients and after reviewing the subject had difficulty in specifying indications for its use.

However, claims have been advanced for a number of years for the value of phenelzine in 'atypical depression' (Sargant & Dally 1962) and an early report claimed striking success in agoraphobic states (King 1962). More recently, confirmatory evidence from controlled trials has come from Tyrer *et al.* (1973) who reported overall improvement and relief of secondary phobias in agoraphobia at six weeks and Robinson *et al.* (1973) who described mitigation of anxiety, fatigue, phobia or other somatic complaints in depressed patients.

The range of effectiveness appears to include anxious, depressive and phobic states.

Two types of interpretation have been advanced to resolve this paradoxical evidence. According to the first, tricyclic compounds are the only drugs effective in the treatment of affective disorders. As the evidence that they are more effective than placebo or non-specific substances such as benzodiazepines, is slender, the benefits ascribed to MAO-inhibiting substances are judged illusory. This view has been stated first because it represents the working policy of a high proportion of experienced, practising clinical psychiatrists who use only tricyclic substances in the treatment of depression and have wholly abandoned MAO-inhibiting drugs. However, it is liable to be overlooked that the therapeutic benefits of tricyclic compounds are confined to a limited domain of depressions. In the remaining cases—and even in the best-designed trials whose verdict has proved favourable to their effects these often add up to a third of the total group—effective agents have yet to be discovered.

According to the second interpretation, the discrepancies in the evidence have stemmed from the fact that tricyclic compounds and MAO inhibitors exert therapeutic benefits in different groups of affective disorders, those in which the latter are effective not yet having been clearly defined. This implies the existence of qualitatively distinct syndromes which would be incompatible with the unitarian view. Failure to achieve clear and consistent results with MAO inhibitors could, according to this reading, have been due to the employment, by different investigators, of heterogeneous populations that varied in composition.

There is evidence from enquiries into classification that have made use of factor analytic and related techniques that this might indeed be partly responsible. Several studies on record ostensibly confined to depressive disorders have isolated a cluster of anxiety features and of patients in whom such features dominated the clinical picture. A clear correlation with an unfavourable response to treatment with tricyclic compounds is a consistent finding. Studies undertaken in general practice offer similar suggestions (Fahy *et al.* 1969). Many clinical psychiatrists appear to regard anxiety states of every kind as depressive equivalents (Sargant & Dally 1962) and MAO inhibitors were introduced to psychiatry as antidepressive agents (Klein 1964). It would seem, therefore, that affective disorders with predominant depression and those with predominant anxiety are in some manner correlated with one another. However, the precise magnitude and character of this correlation and the manner of association of the underlying syndromes are unclear and need elucidation through appropriate enquiries. Merely to assert that they constitute a unity is to beg the most important questions.

DARWIN ON DEPRESSION AND ANXIETY

However, it is possible that to study the clinical features for an orderly arrangement of the phenomena is to look in the wrong place. Increasing attention has been directed in recent years to the role of life events in the causation of psychiatric disorders. As the founder of modern biology, the most discerning of observers, regarded extreme emotions as reactive, his comments on anxiety and depression merit close attention. The following passage is taken from Darwin's *The Expression of the Emotions in Man and Animals:* 'After the mind suffers from a paroxysm of grief and the cause still continues, we fall into a state of low spirits or are utterly dejected. Prolonged pain generally leads to the same state of mind. If we expect to suffer, we are anxious, if we have no hope of relief, we despair'. It will be noted that Darwin has related depression to events in the past and anxiety to happenings apprehended in future. For many clinical observers, this contrast between the closure, finality and the despairing retrospective rumination of the depressive, and the doom-laden anticipation of future events, inherent in anxiety, will have a quality of authenticity.

Further, it is implicit in this formulation that the two affects are closely related, for those who continue in despair would be liable to grow anxious. Suicide notes often contain delusional anticipations of defeat and disaster. Equally, those who continually expect to suffer over long periods are bound to become despondent. In short, it seems probable that the problems of classifying anxiety states and depressions are inextricably bound up with one another. In such enquiries it would, therefore, appear sound strategy to arrange, by explicit definition, for the inclusion of representative groups both of predominantly depressed and predominantly anxious patients.

DEPRESSIVE AND ANXIETY STATES AS REACTIONS TO LIFE EVENTS

Darwin evidently regarded anxiety as a reaction to the prospect of ordeals to come and depression as a response to grief and loss suffered. Is it possible that the affective disturbances, observed by the psychiatrist, can be more logically and clearly classified on the basis of the stressful circumstances in the social environments that precede them rather than the clinical features he daily evaluates? Recent investigations have studied the problems thus posed with the aid of a method for separating consequences from possible causes and the mathematical analysis of observations recorded from patients and controls. Several groups of workers have, in recent years, studied the role of life events (Rahe 1972; Paykel 1972) but reference is made here specifically to the investigations of Brown and his colleagues (Brown *et al.* 1973*a*; Brown *et al.* 1973*b*; Brown 1974; Brown *et al.* 1975).

Three of their conclusions are relevant in this context. The first is that in contrast to schizophrenia, where they appear to act as mere precipitants, life events play a significant part in the causation of depressive illness. They arrive at this view from calculation of the period of time by which the depression has been anticipated or 'brought forward' as a result of the life event. This estimated 'brought forward time' has proved so long in depressions that, in the absence of onerous life circumstances, illness would have been long postponed or might have never developed. The second conclusion relates to the effect of background social factors in increasing vulnerability to breakdown. Three of these—loss of mother before the age of 11, having three or more children under 15 at home, and the absence of a close, confiding relationship with husband or man-friend—were far commoner in working-class than middle-class women and accounted for most of the increase in prevalence of depression among those in the former group, who had been exposed to adverse life circumstances. Finally, so great were the divergences in the prevalence of event-related depression among women of different classes, that for these authors (Brown *et al.* 1975) the problems presented by depression are intimately bound up with the issue of social injustices.

The main objective of studies of classification is ultimately to advance knowledge of causation and improve the treatment of mental disorders. If most of the variation, in respect of severe depression, is attributed to social factors such studies are pre-empted and rendered purposeless. But any such verdict would be premature. In the first place, that the contribution of social adversity is unlikely to decline in the foreseeable future is not, from the scientific point of view, the most important consideration. More cogent is the fact that those predisposed by personality traits to the commonest forms of affective disorder are liable to generate life events and circumstances of their own, so that cause and consequence are difficult to disentangle. The contribution of such factors may be made manifest in such ways as assortative mating, choice of occupation, impulsiveness and accident-proneness, and patterns of sexual and social relationships and drinking habits, which are difficult to measure. To the extent that such factors are not controlled in comparative studies of life-events, the contribution made by such events will be overestimated. The range and variety of physical and psychiatric disorders that can be related to broadly similar happenings (Rahe 1972) gives force to this point. For it detracts from the importance of life events as causes and correspondingly enhances that of the inherent factors. These must be mainly instrumental in deciding that where in one individual a bereavement is followed by a myocardial infarction or an attack of ulcerative colitis, others respond with severe depression, an agoraphobic state or some other neurotic illness. We have thus come full circle to the affective disturbances which are the focus of interest in this paper and the need to classify them in the most valid and heuristically useful manner.

METHODOLOGICAL CONSIDERATIONS

The clinical psychiatrist approaching the problems of classifying affective disorders will usually have some prior conceptual scheme that he uses in everyday diagnosis. This may take the form of a multidimensional continuum or one entailing distinct categories of disorder. In the light of this there is much to be said for making a tentative clinical diagnosis in explicit terms. For any judgement, whether it places the patient in a category or along a continuum, is bound to exert bias and it is preferable that this should be made known.

However, there are advantages to be gained from paring down the diagnostic criteria to a group of essential, nuclear features. This makes it possible to examine the question of whether the groups of patients selected on the strength of such criteria also differ in respect of variables not included in and independent of the defining features. The hypotheses always implicit in diagnosis can, in this manner, be exposed to tests that could serve to call in question, modify or invalidate the initial criteria.

For this reason, tightly defined domains, such as those proposed by Murphy *et al.* (1974) in the case of affective disorders and Feighner *et al.* (1972) in the form of working definitions for a wide range of psychiatric conditions, may be undesirable for scientific purposes. The possibilities of refining discrimination between groups of conditions are sacrificed in advance. Moreover, enquiries into classification which make such tight clinical criteria their starting point beg the question of whether the original symptom clusters do indeed cohere and have the robustness of clinical syndromes.

Perhaps an even more important argument against definitions that are widely inclusive goes back to the whole purpose of studies of classification. Ultimately, such enquiries should bring to light new knowledge about aetiology. But if they are to do this, concepts and definitions which presuppose aetiological associations have to be avoided. For these reasons no distinctions have been made between 'primary' and 'secondary' depressions (or anxiety states) along the lines suggested by the St. Louis group (Robins *et al.* 1972; Murphy *et al.* 1974; Feighner *et al.* 1972) except to exclude affective disorders associated with schizophrenia or other specific psychiatric disorder. On the other hand, affective disturbances associated with physical disease have not been classed in a separate group. There is indeed a body of evidence in favour of a significant relation between affective disorders and physical illness. Many clinicians would also be prepared to agree that there is a considerable overlap, uncertain in origin and extent, between depression on the one hand and chronic alcoholism on the other. But to designate such cases as 'secondary depression' is to assume in advance that which requires proof. The strength of the association is not known. And if one sets a tightly defined 'secondary depression' aside as a separate category it cannot be determined. What may be a useful clinical conjecture about aetiology is given premature and spurious finality and cannot be refined into more precise factual knowledge.

In the light of such complexities, compounded by preconceptions and their halo effects—the circular element inherent in all reasoning that starts from hypothetical clinical syndromes—one might have supposed that, scientifically speaking, exercises of this nature are foredoomed to failure. This would be an erroneous judgement, for good clinical descriptions and classifications have stood the test of time remarkably well. They have often led to and then been substantiated by advances in therapy. And having regard to the statistical crudity of the operation in which an observer selects one cluster of features from an innumerable variety of possible combinations and designates them as a syndrome, the results have never ceased to astonish the statisticians.

THE NEWCASTLE STUDY OF DEPRESSIVE AND ANXIETY STATES

We should like now to summarize the results of an enquiry in which the main hypothesis tested was that anxiety states and depressive disorders were distinct entities or, in other words, that the continuum view of these disorders could be refined. The starting point of the enquiry was a cohort of 145 individuals judged on clinical grounds to be suffering from a primary mood disorder of anxiety, depression or both combined, which was not secondary to some other indubitably distinct condition, such as schizophrenia. The results have been fully published (Roth *et al.* 1972; Gurney *et al.* 1972; Kerr *et al.* 1972; Schapira *et al.* 1972) and comments will be confined to certain findings. On the basis of clinical criteria relating to the initial complaints, the course of illness before interview and the predominant mood change observed during the examination, each patient was allocated to one of three groups: anxiety state (68 patients); depressive illnesses (62 patients); and 'doubtful cases' (15 patients) in whom diagnosis was indeterminate. The initial allocation to groups was undertaken on the basis of an assessment of the current clinical picture alone.

The practical and theoretical importance of the subject has already been

TABLE 1

Prevalence of depressive symptoms within two main groups of affective disorder

	Anxiety state (68) %	*Depressive illness (62)* %
Depression worst a.m.	11	48***
Depression reactive to change	74***	41
Early waking	15	65***
Suicide: acts	17	37**
ideas	34	35
Retardation	5	48***
Delusions	5	17*
Episodes of hypomania	0	6
Pessimistic outlook	87	92
Ideas of guilt	37	52
Agitation	21	28

*, ** and ***: differences significant at 0.05, 0.01 and 0.001 levels respectively. (With permission of the *British Journal of Psychiatry*.)

discussed. The literature on the subject of discriminating between anxiety and depressive syndromes reflects considerable disagreement. The majority of factor analytic and related forms of analysis of rating scales have failed to identify distinct anxiety and depressive factors (Zubin & Fleiss 1971; Spitzer *et al.* 1967). More recently, Prusoff & Klerman (1974), using a 58-item patient self-report inventory, were able to achieve separation of anxious and depressive states. But as many as 35% of the patients could not be correctly assigned.

Some results from the comparison of presenting features of the current illness in the two groups are shown in Tables 1 and 2. There was marked overlap in respect of certain items and seeming specificity of other features which had initially been expected to prove common to the two groups to some extent. For example, transient or variable anxiety or depression did not distinguish between the groups, being common in both. But depression that was severe and unremitting was significantly more common in depressed patients and the reverse pattern held in respect of severe and persistent anxiety. Certain classical features, widely associated with depression, such as diurnal variation of mood, early-morning waking and retardation were significantly more common in the depressed group. But others, including pessimistic outlook, ideas of guilt and agitation, failed to differentiate between them. This was also true for suicidal ideas. The finding that suicidal acts had been recorded in almost half as many anxious as in depressed patients was unexpected.

A whole range of other features proved to be significantly more prevalent

TABLE 2

Prevalence of anxiety symptoms within two main groups of affective disorder

	Anxiety state *(68)* %	*Depressive illness* *(62)* %
Panic attacks	86***	17
Loss of confidence	90*	77
Increased vasomotor responses	74***	23
Emotional lability	86***	52
Intolerance to noise	68	51
Irritability	90	81
Initial insomnia	72	57
Restless sleep	50	37
Poor concentration	80	91
Dizzy attacks	68**	40
Attacks of unconsciousness	22**	3
Agoraphobias: marked	41***	2
mild	36	26
Depersonalization: marked	36***	4
mild	11	17
Derealization	36***	0
Perceptual distortions	42***	8
Déjà vu and/or entendu	22**	5
Idea of a presence	30*	11
Hypnagogic phenomena	43**	17

*, ** and ***: differences significant at 0.05, 0.01 and 0.001 levels respectively. (With permission of the *British Journal of Psychiatry*.)

among the anxiety states. To the extent that any differences were independent of halo effects, they were not just due to greater severity of all clinical features in any one group. A few selected observations must suffice here (Table 2). Panic attacks showed marked specificity; cases with severe and frequent panics, feelings of imminent collapse or death, were almost entirely confined to the anxious group. Situational anxiety experienced in buses, shops or streets but easily overcome with some effort is not specific; but more intractable phobias which were severely restrictive and confined the patient to the home and feelings of unreality related to the perception of the outside world, as distinct from the individual's own body and its movement, were entirely confined to the anxious group.

The groups also differed significantly in respect of features unrelated to the initial clinical criteria by which patients were allocated to the different groups. Those with anxiety states had a higher prevalence of psychiatric disorder among their first-degree relatives (Table 3) and had more often exhibited

TABLE 3

Psychiatric illness in first-degree relatives

		Anxiety state (68) %	*Depressive illness (62)* %
Parents:	Neurotic illness	20*	5
	Personality disorder	68***	28
Siblings:	Neurotic illness	17	9
	Personality disorder	32*	14

* and ***: differences significant at 0.005 and 0.001 levels respectively. (With permission of the *British Journal of Psychiatry*.)

neurotic traits and maladjustment in childhood. Among anxious women, frigidity had been significantly more commonly present before illness ($P < 0.01$) than among depressed women. Other forms of maladjustment showed similar trends but without significant differences.

Personality measures provided some independent validation for the higher prevalence of maladjustment among the anxious group recorded on the basis of clinical interview. The neuroticism scores of the anxious patients on the Maudsley Personality Inventory (MPI) were significantly higher than those of the depressives ($P < 0.02$). It is of interest that when patients were re-examined by independent observers 42 months later, these differences were still in evidence although a proportion in each group had recovered or improved. Also, the difference in the degree of extroversion which had been in evidence in the initial examination was also in evidence on follow-up (Kerr *et al.* 1970). The personality ratings made on a clinical basis were consistent with these psychometric test results. The anxious patients were recorded as having been significantly more often shy and self-conscious, poor mixers, hypersensitive, given to anticipatory anxiety and somatic discomfort. They were more markedly dependent on relatives and familiar environments, and more often immature emotionally, than the depressed group. All these differences were highly significant statistically.

The clustering of features and patients

In the light of these findings it appeared desirable to examine, with the aid of a more precise method than a simple comparison, the clustering and dispersion of features and of patients. A principal components analysis was therefore undertaken. More than 300 variables had been scored for each patient; they

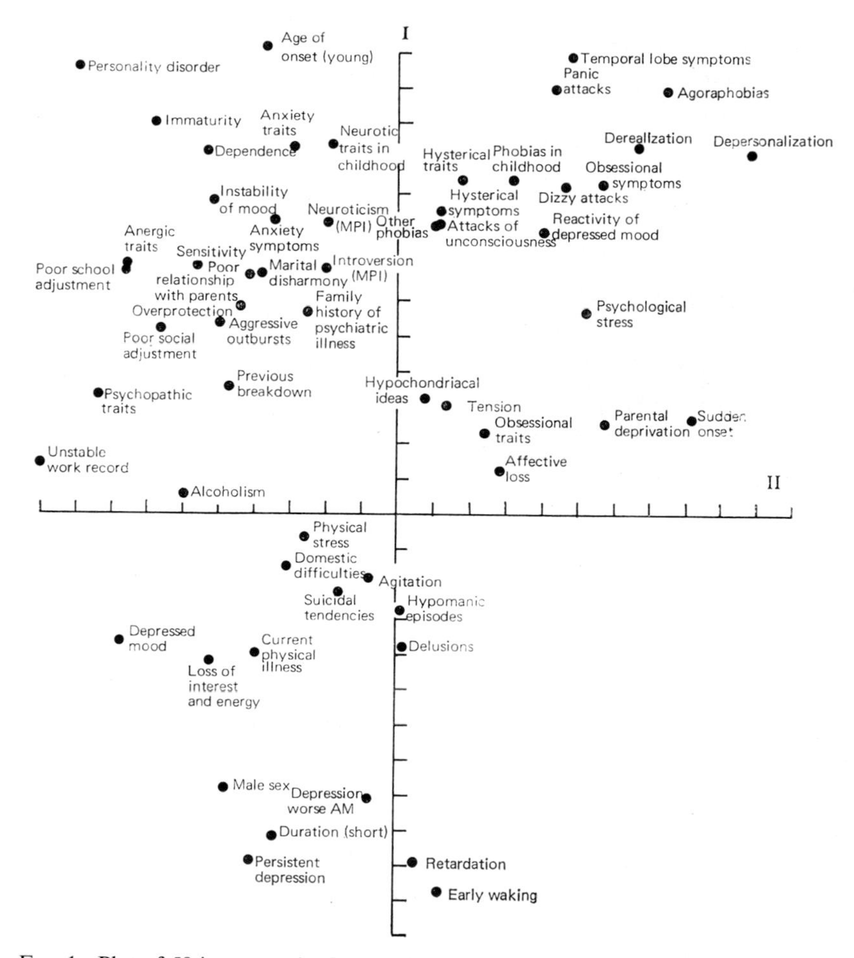

FIG. 1. Plot of 58 items on the first two principal components.

were compressed into 58 items for the purposes of the analysis. In this technique the components representing the main independent group of co-varying features are extracted in order of their importance. Fig. 1 represent loadings of these 58 items. The first component is bipolar and accounts for 14% of the total variance. Its character becomes clear when the features found above and below the horizontal line are examined. Definitions of the various features will be found in the original publications (Roth *et al.* 1972). Taking symptoms first, those related to depression are mostly well below the line, having high negative loadings on the component. They include early waking, persistent depression, retardation, short duration of illness, depression worse in the morning

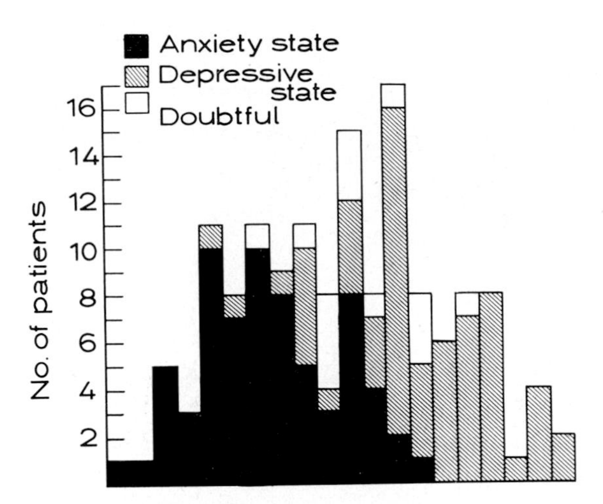

FIG. 2. The distribution of patients' summated scores on the first component extracted from the principal component analysis, the initial clinical diagnoses having been re-applied.

and delusions. Clustered near the opposite pole are panic attacks, agoraphobia, unreality feelings, 'temporal lobe' features, dizzy attacks, obsessional and hysterical symptoms, phobias in childhood, attacks of unconsciousness and reactivity or variability of depression, all with large positive loadings. The component, therefore, depicts the polarization of the syndromes of anxiety and depression.

Personality features did not figure among the diagnostic criteria. However, most of the maladaptive and neurotic premorbid personality traits—'immaturity', 'dependence', 'neuroticism on MPI', 'instability of mood'—are found closer in space to the anxiety than to the depressive clinical features, indicating that they are more highly correlated with the former.

When patients' summated scores on this first component were calculated from the symptoms they had exhibited and plotted in the form of a distribution, most of the patients who had been diagnosed as anxiety states were found to occupy a different part of the distribution from that occupied by the depressive and 'doubtful' cases, who mainly fell in the middle (Fig. 2). The difference between the mean component scores of the two groups was highly significant statistically. But there was considerable overlap and the distribution did not depart significantly from a normal one. At this stage the findings were not inconsistent with a continuum theory entailing gradual transition from states of neurotic anxiety at one extreme end to those of psychotic depression at the other.

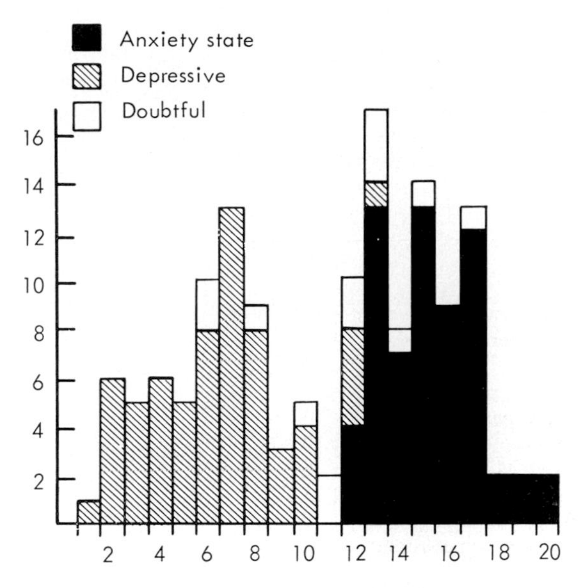

FIG. 3. The distribution of patients' summated scores on an Anxiety/Depression index derived from discriminant function analysis.

Discrimination between anxiety states and depressive disorders

The possibility of a more clear and unequivocal separation between the two groups remained open, for the scores of two distinct populations on a given set of variables may overlap markedly for a whole variety of reasons. In such a situation, the technique of discriminant function analysis may serve to separate the distribution clearly into its constituent parts.

This was applied in the following manner. Cases in which diagnosis could be confidently made were used in the first stage to extract quantitative scores for each of the 58 compressed items that had been utilized in the principal components analysis. This made it possible to select the 13 items with the highest discriminating value. The weights or scores of these items were then used to calculate the summated symptom scores of the total group of 145 patients, including the 15 doubtful cases. The plotted distribution of patients' scores was clearly bimodal (Fig. 3), departing to a highly significant degree from a normal distribution. Moreover, most of the doubtful cases who had previously concentrated near the middle were now allocated to one or the other side of the central dip. On the basis of weights derived from the discriminant function analysis, a simple diagnostic index was constructed (Table 4).

The degree of separation is, therefore, substantially greater than that achieved

TABLE 4

Items in anxiety/depression index

Item	*Score*	
Neurotic traits in childhood	3 or more	+10
	2 or less	+ 5
	none	0
Dependence	present	+ 6
	absent	0
Physical stress	severe	+16
	mild/moderate	+ 8
	none	0
Panic attacks	3 or more per week	+20
	2 or less per week	+10
	none	0
Situational phobias	marked	+ 6
	mild/moderate	+ 3
	none	0
Derealization	marked &/or persistent	+ 2
	mild	+ 1
	none	0
Anxiety symptoms	6 or more	+12
	3 to 5	+ 6
	less than 3	0
Depressed mood	severe	−18
	mild/moderate	− 9
	none	0
Early waking	present	− 4
	absent	0
Suicidal tendencies	attempt	−12
	ideas	− 6
	none	0
Retardation	present	− 6
	absent	0
Obsessional symptoms	marked	+ 4
	mild	+ 2
	none	0
Neurotic (MPI)	0– 8	−15
	9–16	−12
	17–24	− 9
	25–32	− 6
	33–40	− 3
	41–48	0

Ranges

Anxiety range	+11	+78
Doubtful range	− 3	+10
Depressive range	−55	− 2

(With permission of the *British Journal of Psychiatry*.)

in previous studies. That of Prusoff & Klerman (1974) in which 30–40% overlap was found provides an instructive contrast. The difference may have been partly due to the fact that theirs was a population of outpatients, in contrast to the inpatient status of the present sample. However, a comparison of the Boston index with the Newcastle scale group offers some other explanations.

A number of specific anxiety features, to which the analysis had assigned high scores, are absent from the Boston scale. They include panic attacks, a distinctive symptom which in its severe forms was almost entirely confined to the anxiety group in the present study. They have a close relationship to agoraphobic symptoms and disabilities. This is also true of syncopal attacks and less easily characterized attacks of unconsciousness. They do not figure as items in the full Boston scale but they differentiate significantly between the depressive and anxiety groups. They are not among the final 13 items of the present index, partly because their high correlation with panic attacks limits their independent discriminating power. Another discrepancy is the presence, in the Newcastle scales, of certain items related to the antecedents of illness and the premorbid setting. Physical illness that bore an impressive association with onset and personality features such as psychological dependence and the history of many neurotic traits in childhood also achieve substantial scores. In contrast, the Boston scale is confined to the clinical features of the current illness.

The possibility that the separation achieved merely reflects a differentiation between more or less intensely ill patients has to be considered. This could have been the case if the depressive group, which included all the psychotic and endogenous cases, had scored higher on all or a majority of items. But this was not the case. A whole range of clinical features already specified had been scored more highly in or were wholly confined to the anxiety group. Different clinical configurations characterized the groups and not just differences in severity.

It is the frequency of a variable depression in both groups that has probably caused these patients to be described, in much of the literature of affective disorders, as 'depressive' and 'anxious–depressive' cases. But it is *syndromes* that have been differentiated; isolated depressive and anxiety *symptoms* as such proved, in fact, to have a small positive correlation. And as the multivariate statistical methods used are highly sensitive and liable to yield varying results with minor changes in the composition of patient groups, the differentiation into distinct anxiety and depressive syndromes has to be considered tentative until replicated in other samples and validated by independent methods of observation.

A certain amount of independent evidence from further samples of patients with affective disorders has recently become available. The subjects were 117 patients with affective disorders, admitted to a therapeutic trial, from which endogenous depressions and cases of affective disorder judged from their character and severity to require immediate treatment with tricyclic compounds, were excluded. The hypothesis that separation into depressive and anxiety syndromes was possible was, therefore, exposed to a stringent test. The scores of the 117 patients, on the Newcastle anxiety–depression scale, were plotted. Although the distribution did not fall into two entirely distinct modes, only 20% of the population, initially given a diagnosis of depressive illness or anxiety state, were mis-classified by the diagnostic index scores (C. Q. Mountjoy & M. Roth, unpublished observations 1975). This result provides confirmation from a second sample for the separation of the anxiety and depressed states already described. The results of independent validation from follow-up studies will be discussed later (p. 314).

Subdivisions within the main groups

In the present context, it is the distinction between the anxiety and depressive states that is at the focus of interest. However, as certain subdivisions have been shown to have some relationship with the efficacy of specific pharmacological agents, brief reference will be made to them. Within the category of anxiety states there are non-specific anxiety neuroses, with a blend of somatic and psychic tension on the one hand and phobic states on the other. Among the phobic anxiety states, the two main groups are agoraphobic and social phobic states. In the former, feelings of tension are engendered by leaving the shelter of the home environment and are exacerbated to the point of panic in situations such as crowded shops or public transport. In social phobic states anxiety is evoked by exposure to the scrutiny of others when eating, drinking or speaking, so that social gatherings are liable to provoke feelings of tension and flight from the situation. While benzodiazepines often confer much relief in the first, they make little impression on the crippling disabilities of the latter group.

As far as depressions are concerned, there has been a long-lasting controversy about the unity or diversity of this group of disorders. The subject has been reviewed in publications by different authors in recent years and comment will be confined here to a few salient points. There is perhaps less scope for disagreement about the existence of more than one form of depressive illness, than about the possibility of defining distinct categorical entities within the depressions (Eysenck 1970). As far as the present data are concerned, the depressed patients have been separated, with the aid of a discriminant function analysis, into two distinct groups, one corresponding to endogenous and the other a broad group of 'neurotic' depressions, a result similar to that recorded by Carney *et al.* (1965). A growing body of evidence has accumulated to establish at least one of these, endogenous depressions, as a categorical entity (Fahy *et al.* 1969).

The work on unipolar states, which have been differentiated as a result of the studies of Perris (1966) in Sweden, Angst (1966) in Switzerland and Winokur & Clayton (1967) in the USA, lends force to this, in that there can be no doubt that manic-depressive illness is sharply distinct and that at least a proportion of other endogenous depressions fail to qualify as bipolar only because they are not far enough advanced in their course to have suffered an unequivocal attack of mania.

The situation is more uncertain in relation to neurotic depression. It is possible, as Kiloh *et al.* (1972) have suggested, that this group comprises a continuum of different variants of the norm that merge with one another. Paykel (1971) reached a similar conclusion, describing neurotic depression as a heterogeneous entity made up of anxious depressives, hostile depressives and young depressives with personality disorder. However, in view of the findings on the new frontiers of depression with anxiety states, the concept of neurotic depression needs reappraisal.

As both the depressive and anxiety states are divisible, differentiation entails more than a separation of the two groups that have attracted most attention in studies of the classification of depression. The problem is of great complexity, but a first approach utilizing the observations accumulated about the present material of 145 patients seemed worth while. A variant of discriminant function analysis appropriate for more than two groups—canonical variate analysis—was used in an attempt to obtain optimal separation between the four hypothetical groups: endogenous depression, neurotic depression, agoraphobia and simple anxiety states, into which patients were tentatively allocated. The few bipolar cases were included with endogenous depression and social phobia with agoraphobia. Once again the technique involves using cases in which the diagnosis could be made with reasonable confidence to extract quantitative weights for each feature and so plotting patients' summated scores on each variate. The representation on a plain surface is a poor substitute for the multidimensional model required to show the spatial relationship of several patient clusters, but Fig. 4 attempts so to depict the results achieved when patients scores on variates I and II were plotted.

A very satisfactory degree of separation was achieved with the aid of the items of highest discriminating value, but the result has to be regarded as

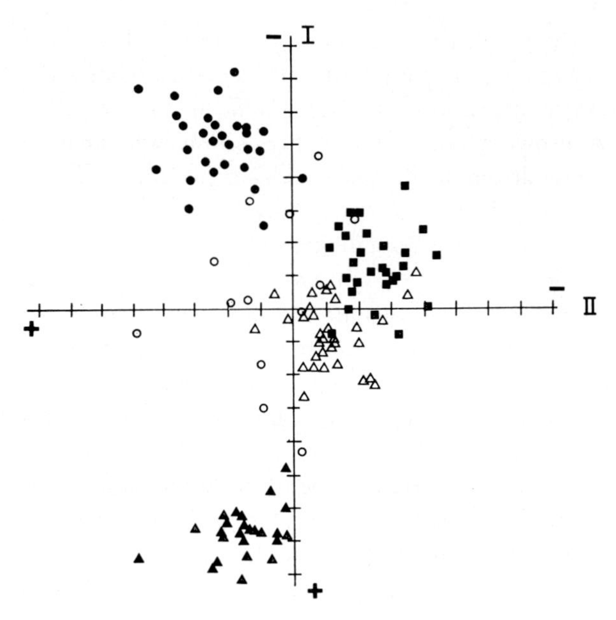

FIG. 4. Canonical variate analysis (58 variables) on variates I and II. Anxiety state, ■; agoraphobia, ●; doubtful, ○; reactive depression, △; and endogenous depression, ▲.

tentative until replicated in further samples and by independent investigators.

Validation from follow-up studies

All taxonomic investigations that have clinical observation as their starting point are unavoidably informed by hypotheses, explicit or implicit, and are bound, therefore, to be circular. But this problem did not begin with the emergence of factor analysis. It is an inherent limitation of this form of enquiry. Hence escape from being logically imprisoned within a circular argument has to be sought. This can be achieved with the aid of some form of independent validation. The distinctions postulated between hypothetical groups should stand up to the objective evaluation of course, outcome or the results of treatment, among other variables.

A follow-up study of the anxiety and depressive groups was, therefore, undertaken by observers who were unaware of the diagnoses originally made. Several lines of evidence upheld the validity of the distinction between these two main groups. The depressed patients had responded significantly better than the anxious patients both to ECT and tricyclic antidepressants (Gurney *et al.* 1970). As regards course and outcome, there was a higher proportion of

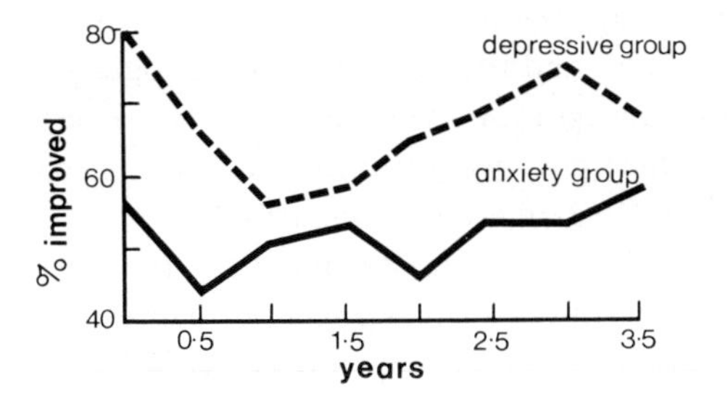

FIG. 5. Course of patients with anxiety state and depressive illness during the follow-up period.

improved patients among the depressed than among the anxious patients, throughout the follow-up period (Fig. 5), and the differences were statistically significant (Kerr *et al.* 1972; Kerr *et al.* 1974). Very little cross-over was found in this longitudinal study between the two main groups. In other words, anxiety states tended on relapse to break down into anxiety neuroses and depressives developed depressive illnesses (Kerr *et al.* 1972). Finally, the clinical features present during illness which best predicted outcome in the anxiety states were entirely different from those which predicted outcome in the depressions. Moreover, there was only limited overlap between the items of predictive value in the affective group as a whole and

(*i*) items that predicted outcome in anxiety states (3 out of 9) and

(*ii*) items that did so in depressions (5 out of 12) (Kerr *et al.* 1974).

The findings accord better with a dichotomous than a continuum view of depressive and anxiety states.

CONCLUDING REMARKS

The polarity of syndromes dominated by anxiety, as against those dominated by depression, proves meaningful in ways that had not been anticipated. At the right-hand extreme of the distribution (Fig. 3, p. 309) are the agoraphobic subjects; at the opposite extreme, the endogenous depressives. In general these are, during breakdown, the most severely ill and incapacitated patients in their respective groups. The former registered the highest neuroticism scores on the MPI and were also judged to have the most neurotic personalities on the strength of clinical criteria; the latter obtained the lowest scores. The groups inbetween occupy an intermediate position in respect of these measures. The anxious groups are less stable and adjusted than the depressives in terms of biographical criteria, and again the most extreme trends are displayed at the limits of the distribution. During follow-up, the anxious patients exhibited a less favourable outcome than the depressives and the best and worst prognoses were those of endogenous depressives and agoraphobics, respectively. These were also found, at follow-up, to be the most extroverted and introverted groups of subjects as judged from the MPI score (Kerr *et al.* 1970).

Finally, evidence from many controlled trials suggests that tricyclic compounds are effective agents in the treatment of classical endogenous depression (Kiloh *et al.* 1962). The situation is more complex in relation to affective disorders in which anxiety is a prominent feature. Raskin *et al.* (1970) report a favourable response to imipramine in predominantly anxious patients. However, in a number of enquiries in which the correlation of response to individual clinical items has been studied, prominent anxiety has been found to portend an unfavourable outcome (Kiloh *et al.* 1962; Hordern *et al.* 1965). Other evidence also suggests that the prognostic significance of severe anxiety is in an opposite sense to that of severe and retarded depression (Carney *et al.* 1965; Gurney *et al.* 1970; Kerr *et al.* 1972). In the studies of Hollister and his group (Hollister *et al.* 1971) imipramine proved most effective in retarded depressions, while tranquillizers proved most beneficial in the anxious depressives, benzodiazepines proving superior to major tranquillizers in this group.

On the other hand, evidence has begun to accumulate from controlled as well as earlier uncontrolled studies that MAO inhibitors exert beneficial effects on certain symptoms at the anxiety end of the distribution. The clinical profiles that are reported to respond in each case include anxiety phobias and depression (King 1962; Sargant & Dally 1962; Robinson *et al.* 1973), and mitigation of agoraphobia has been reported (Tyrer *et al.* 1973). In a recent Newcastle study, a first analysis of a patient population comprising neurotic depressives and cases of anxiety neurosis, using a global rating scale, suggested that phenelzine and diazepam had no advantage over diazepam alone. However, when the results in respect of other rating scales were analysed and drop-outs from the trial were omitted, a significant result in favour of phenelzine was recorded in phobic neuroses, anxiety neurosis and neurotic depression with four of the rating scales used, three of them (one in each diagnostic group) being measures of anxiety (Mountjoy & Roth 1975; Mountjoy & Roth, unpublished observations).*

More enquiries are needed to define with greater precision the indications for the use of MAO-inhibiting drugs. But the available evidence already suggests that the polarity of depression as against anxiety syndromes, discussed

* This further analysis was made after presentation of this paper at the symposium.

here, may be closely related to the differential effects of drugs of proven therapeutic value in the affective disorders. If this is the case, there must be corresponding differences in the underlying biochemical mechanisms concerned in the genesis or perpetuation of the different syndromes.

But developments in classification may prove significant not only for pharmacological research. Refinement of psychopathology is just as dependent upon such advances. In the present context the finding that anomalies of personality, maladjustment, and historical and psychodynamic factors appear to loom larger among anxiety states than among depressive disorders may be pertinent. The difference is, of course, quantitative not qualitative, and in both groups disorders often commence in an abrupt step-like manner. But the question posed by this contrast is whether the greater success achieved to date in the treatment of depressive illness than of anxiety states is a related phenomenon. For antidepressive treatments tend to have a global effect upon depressive illness, if they act at all, whereas the therapies used in anxiety states generally influence only a limited range of symptoms. These contrasts offer a challenge to pharmacological and biochemical research. But the problems that arise, for those interested in psychopathology and in classification, may well be related. A simultaneous approach on all these fronts may be needed to achieve a significant advance in knowledge of affective disorders.

References

ANGST, J. (1966) *Zur Ätiologie und Nosologie endogener depressiver Psychosen*, Springer, Berlin

BROWN, G. W. (1974) Meaning measurement and stress of life events, in *Stressful Life Events: Their Nature and Effects* (Dohrenwend, B. S. & B. P., eds.), Wiley, New York

BROWN, G. W., BHROLCHAIN, M. N. & HARRIS, R. (1975) Social class and psychiatric disturbance among women in an urban population. *Sociology, 9*, 225

BROWN, G. W., SKLAIR, F., HARRIS, T. O. & BIRLEY, J. L. T. (1973*a*) Life events and psychiatric disorders. *Psychol. Med. 3*, 74

BROWN, G. W., HARRIS, T. O. & PETO, J. (1973*b*) Life events and psychiatric disorders. Part 2: nature of causal link. *Psychol. Med. 3*, 159

CARNEY, M. W. P., ROTH, M. & GARSIDE, R. F. (1965) The diagnosis of depressive syndromes and the prediction of E.C.T. response. *Br. J. Psychiatr. 111*, 659-674

COVI, L., LIPMAN, R. S., DEROGATIS, L. R., SMITH, J. E., III & PATTISON, J. H. (1974) Drugs and group psychotherapy in neurotic depression. *Am. J. Psychiatry 131*, 191

EYSENCK, H. J. (1970) The classification of depressive illness. *Br. J. Psychiatr. 117*, 241-250

FAHY, J. T., BRANDON, S. & GARSIDE, R. F. (1969) Clinical syndromes in a sample of depressed patients: a general practice material. *Proc. R. Soc. Med. 62*, 331-335

FEIGHNER, J. P., ROBINS, E., GUZE, S. B., WOODRUFF, R. A., WINOKUR, G. & MUNOZ, R. (1972) Diagnostic criteria for use in psychiatric research. *Arch. Gen. Psychiatry 26*, 57-63

GURNEY, C., ROTH, M., KERR, T. A. & SCHAPIRA, K. (1970) The bearing of treatment on the classification of the affective disorders. *Br. J. Psychiatr. 117*, 251-266

GURNEY, C., ROTH, M., GARSIDE, R. F., KERR, T. A. & SCHAPIRA, K. (1972) Studies in the classification of affective disorders. The relationship between anxiety states and depressive illnesses. II. *Br. J. Psychiatr. 121*, 162-166

HOLLISTER, L. E., OVERALL, J. E., POKORNY, A. D. & SHELTON, J. (1971) Acetophenazine and diazepam in anxious depressions. *Arch. Gen. Psychiatr. 24*, 273

HORDERN, A., BURT, C. G., HOLT, N. F. & CADE, J. F. J. (1965) *Depressive States: A Pharmacotherapeutic Study*, Thomas, Springfield, Ill.

KAY, D. W. K., GARSIDE, R. F. & FAHY, T. J. (1973) A double blind trial of phenelzine and amitriptyline in depressed out-patients. A possible differential effect of the drugs on symptoms. *Br. J. Psychiatr. 123*, 63-67

KERR, T. A., SCHAPIRA, K., ROTH, M. & GARSIDE, R. F. (1970) The relationship between the Maudsley Personality Inventory and the course of affective disorders. *Br. J. Psychiatr. 116*, 11-19

KERR, T. A., ROTH, M., SCHAPIRA, K. & GURNEY, C. (1972) The assessment and prediction of outcome in affective disorders. *Br. J. Psychiatr. 121*, 167-174

KERR, T. A., ROTH, M. & SCHAPIRA, K. (1974) Prediction of outcome in anxiety states and depressive illnesses. *Br. J. Psychiatr. 124*, 125-133

KILOH, L. G., BALL, J. R. B. & GARSIDE, R. F. (1962) Prognostic factors in the treatment of depressive states with imipramine. *Br. Med. J. 1*, 1225-1227

KILOH, L. G., ANDREWS, G., NEILSON, M. & BIANCHI, G. N. (1972) The relationship of the syndromes called endogenous and neurotic depression. *Br. J. Psychiatr. 121*, 183-196

KING, A. (1962) Phenelzine treatment of Roth's calamity syndrome. *Med. J. Aust. 1*, 879

KLEIN, D. F. (1964) Delineation of two drug-responsive anxiety syndromes. *Psychopharmacologia 5*, 397-403

KLERMAN, G. L., DIMASCIO, A., WEISSMAN, M., PRUSOFF, B. & PAYKEL, E. S. (1974) Treatment of depression by drugs and psychotherapy. *Am. J. Psychiatry 131*, 186

MEDICAL RESEARCH COUNCIL (1965) Clinical trial of the treatment of depressive illness. *Br. Med. J. 1*, 881-886

MENDEL, J. G. C. & KLEIN, D. F. (1969) Anxiety attacks with subsequent agoraphobia. *Comp. Psychiatr. 10*, 190-195

MINDHAM, R. H. S., HOWLAND, C. & SHEPHERD, M. (1972) Continuation therapy with tricyclic antidepressants in depressive illness. *Lancet 2*, 854

MINDHAM, R. H. S., HOWLAND, C. & SHEPHERD, M. (1973) An evaluation of continuation therapy with tricyclic antidepressants in depressive illness. *Psychol. Med. 3*, 5

MOUNTJOY, C. Q. & ROTH, M. (1975) A controlled trial of phenelzine in anxiety, depressive and phobic neuroses, in *Neuropsychopharmacology* (Boissier, J. R., Hippius, H. & Pichot, P., eds.), International Congress Series No. 359, Excerpta Medica, Amsterdam

MURPHY, G. E., WOODRUFF, R. A. & HERJANIC, M. (1974) Primary affective disorder. *Arch. Gen. Psychiatr. 31*, 181-184

PAYKEL, E. S. (1971) Classification of depressed patients; a cluster analysis derived grouping. *Br. J. Psychiatr. 118*, 275

PAYKEL, E. S. (1972) Environmental precipitants of depression. Paper to the Royal College of Psychiatrists, London

PERRIS, C. (1966) A study of bipolar (manic depressive) and unipolar recurrent depressive psychoses. *Acta Psychiatr. Scand. 42*, Suppl. 194

PRIEN, R. F., KLETT, C. J. & CAFFEY, E. M. JR (1974) Lithium prophylaxis in recurrent affective illness. *Am. J. Psychiatry 131*, 198-203

PRUSOFF, B. & KLERMAN, G. L. (1974) Differentiating depressed from anxious neurotic outpatients. *Arch. Gen. Psychiatry 30*, 302-309

RAHE, R. H. (1972) Subjects' recent life changes and their near future illness reports. *Ann. Clin. Res. 4*, 250

RASKIN, A., SCHULTERBRANDT, J. G., REATIG, N. & MCKEON, J. J. (1970) Differential response to chlorpromazine, imipramine and placebo. *Arch. Gen. Psychiatry 23*, 164

RASKIN, A., SCHULTERBRANDT, J. G., REATIG, N., CROOK, T. H. & ODLE, D. (1974) Depression subtypes and response to phenelzine, diazepam and a placebo. *Arch. Gen. Psychiatry 30*, 66-75

ROBINS, E., MUNOZ, R. A., MARTIN, S. & GENTRY, K. A. (1972) Primary and secondary affective disorders, in *Disorders of Mood* (Zubin, J. & Freyhan, F. A., eds.), Johns Hopkins Press, Baltimore & London

ROBINSON, D. S., NIES, A., RAVARIS, C. L. & LAMBORN, K. R. (1973) The monoamine oxidase inhibitor, phenelzine, in the treatment of depressive anxiety states. A controlled clinical trial. *Arch. Gen. Psychiatry 29*, 407-413

ROTH, M., GURNEY, C., GARSIDE, R. F. & KERR, T. A. (1972) Studies in the classification of affective disorders. The relationship between anxiety states and depressive illness. I. *Br. J. Psychiatr. 121*, 147-161

SARGANT, W. & DALLY, P. (1962) Treatment of anxiety states by antidepressant drugs. *Br. Med. J. 1*, 7-9

SCHAPIRA, K., ROTH, M., KERR, T. A. & GURNEY, C. (1972) The prognosis of affective disorders: the differentiation of anxiety states from depressive illness. *Br. J. Psychiatr. 121*, 175-181

SPITZER, R. L., FLEISS, J. L., ENDICOTT, J. & COHEN, J. (1967) Mental status: properties of factor analytically derived scales. *Arch. Gen. Psychiatry 16*, 479-493

TYRER, P., CANDY, J. & KELLY, D. (1973) Phenelzine in phobic anxiety: a controlled trial. *Psychol. Med. 3*, 120-124

WINOKUR, G. & CLAYTON, P. (1967) Family history studies. I. Two types of affective disorders separated according to genetic and clinical factors. *Recent Adv. Biol. Psychiatr. 9*, 35

ZUBIN, J. & FLEISS, J. (1971) Current biometric approaches to depression, in *Depression in the 1970's (Modern Theory and Research)* (Fieve, R. R., ed.), pp. 7-10, Excerpta Medica, New York

Discussion

van Praag: I am fascinated by this beautiful piece of work, which is a great contribution to psychopathology. I have two questions, one of a nosological character. You tend to differentiate between neurotic and endogenous depression. Within the concept of endogenous depression, one can differentiate between a retarded and an agitated form. What, in your system, is the relation of agitated ('vital' or endogenous) depression to anxiety states, anxiety neuroses, neurotic depression?

Roth: The whole question of 'agitated' depression is thrown into the melting pot by the results of many enquiries that have examined the phenomenology of depressions systematically. I am uncertain where typical 'agitated' depressions belong, because 'agitation' did not figure among the defining criteria of depression in our study. To qualify for the depressive group patients had to manifest such features as depressive affect and thought content, feelings of inadequacy or guilt, and diminished drive.

For what it is worth, what emerged was a marked preponderance of retardation. There was no significant difference between anxiety states and depressions in the prevalence of 'agitation'.

It is of interest that Perris (1973) finds in the symptomatology of unipolar depressions (in contrast to bipolar ones) more agitation, higher anxiety scores and more somatic complaints. These would be regarded by many psychiatrists as more characteristic of depressive and anxiety neuroses than psychoses. Yet Leonhard (1957), who introduced the concepts of bipolar and unipolar depression, was referring exclusively to endogenous psychoses. One wonders how far the concept of depression may have become diluted under the influence of such treatments as ECT, which resolves a variety of disorders for a few days. 'Depressive Illness' may in this way have widened in meaning to include forms of neurosis in which anxiety is predominant. This doesn't mean that I do not believe agitated depressions exist. True agitated depressions, with guilty, nihilistic and other delusions, do exist. But they appear to be rare as compared with retarded depressions.

van Praag: My second question is more general. You mentioned Brown's work and the importance of life events in the occurrence of depressive illnesses and you suggested that this could be of importance to workers interested in biological factors in psychiatry. I felt that you hinted at some contradiction between concepts of abnormal behaviour provoked mainly by serious and disturbing life events, and that provoked by biological factors, such as disturbances in brain function. Personally, I see no contradiction (van Praag 1971, 1972). I see no abnormal behaviour without some disturbance in brain function, irrespective of whether or not life events have been of importance. I think that in either case biological factors are very important. But perhaps I am wrong and you didn't mean to suggest any contradiction between sociological and biological points of view?

Roth: If Brown's evidence and arguments are accepted (Brown *et al.* 1975), they demand a radical revision of the concepts of psychiatric illness. The prevalence of depressive disorders was very much higher among working-class than middle-class women. And this difference proved to be closely related to the frequency of adverse life events and of social difficulties of longer duration, both types of adversity occurring more commonly among working-class women. For example, the 'underprivileged' suffered more often from lack of a close confiding relationship with a spouse, from having more than three children at home, lack of job opportunities, and separation from or death of the mother before the age of 11 years. In fact the greater parts of the variation in respect of depressive illness in the population of women studied appeared to be accounted for by social factors.

However, some of the assumptions made in estimating the role of life events

(Brown *et al.* 1973*a*,*b*) are questionable. A sample of the general population was used in comparison. But such individuals do not provide appropriate controls. For a valid comparison, subjects similar in all other respects to the experimental group should have been used. They should have been comparable in terms of heredity, developmental history and previous adjustment. As such constitutional and personality factors were not controlled, the study may have exaggerated the importance of life events. Of course, I agree that there is no inherent contradiction between causation by life events and by biological factors. But Brown's results, if taken at face value, allow little scope for the latter.

van Praag: One could say the same for tuberculosis, which was also a disease very prevalent in the lower socioeconomic classes. Nevertheless, it is undoubtedly a 'biological' disease, among other things.

Roth: Brown, I suspect, would say that you cannot make a close comparison between two such conditions.

Rafaelsen: I am also somewhat surprised by the way you seem to accept Brown's views, because what we knew previously was that there are *no* class differences in breakdown. The Helgason (1964) study in Iceland indicated an over-representation in social class 1 of manic-depressive disorders. Also in our catchment area in Greater Copenhagen there seems to be *no* class difference in the number of depressions coming into hospital. From a psychodynamic point of view, the finding of Kielholz's group in Switzerland was that although they had presumed to be able to explain most depressions by life events, they were only able to do so in 15% of first cases, declining to 5% in the third, fourth and fifth attack of depression (Kielholz 1959).

I should also mention a spin-off of Baastrup's lithium discontinuation study in Denmark (Baastrup *et al.* 1970). Of 21 patients who were taken off lithium treatment, in a double-blind study, every one when they got a depression could explain it by some life event. Their argument went even further, that because they were being treated with lithium, it could not be an endogenous depression, but had to be a reactive depression. The moment the patients heard that they had been taken off lithium, they knew that it was again an endogenous depression. In other words, people very eagerly seek an explanation and causation in life. Therefore if the life events are not objectively established in a study, but are seen through the patient's eyes, they are simply not trustworthy as evidence.

Roth: Brown and his colleagues have, in fact, tried to deal with the possible distortion of such 'search for meaning' effects. However, as I said, there may be other flaws in the argument. Brown and his colleagues report no differences in the role of life events among different forms of depressive illness. But when adverse circumstances preceded a bipolar or psychotic illness we are very likely dealing with the model described by Edwards (1969) as 'that of breaking the backs of camels by adding straws, the last straw being only fortuitously more traumatic than the first'. The non-specificity of life events is a further point. If they are indeed associated with myocardial infarction in some (Theorell 1970), shortened life expectation in others (Parkes *et al.* 1969), depression in others still, and also a wide variety of diseases of high prevalence (Gunderson & Rahe 1974), one has to ask what weight should be attached to them in relation to other factors.

Coppen: I want to emphasize how important Sir Martin Roth's work is. What we lack is an approach to any agreement on how we describe the patients we are looking at, whether we are treating them or investigating their biochemistry. We have no internationally or even nationally agreed scales that we can use. The Hamilton rating scale for severity, although it has deficiencies, is at least something we can all refer to. But for the actual type of patient we have a complete lack of universally acceptable dimensions, and I think Sir Martin Roth's work may be leading to something that we can all start using.

We have been doing a study, mainly on plasma amitriptyline levels and therapeutic response. In our second study we thought it was important to try to state more precisely the type of patient we were looking at. We decided to adopt the Newcastle criteria, so we have excluded from our patients those at the anxiety end of the Newcastle scale and have concentrated on patients suffering from depressions classified as either reactive or endogenous. We have been following them very carefully. The only data yet are on therapeutic outcome at six weeks. At that point there is absolutely no difference between the therapeutic response to amitriptyline in the reactive and endogenous groups, which is of some interest.

Finally, I am intrigued by the patients who show a dramatic improvement when taken off a tricyclic drug. Are there any systematic observations on this? I don't see many anxiety patients being given tricyclics. I have the impression that they respond rather badly to them.

Roth: To take the last point, of those patients with anxiety states in our study who had been treated by their consultants, those treated with tricyclic compounds responded very poorly; those treated with MAO inhibitors and/or benzodiazepines fared significantly better. The findings of Hollister (1972) with anxious patients on imipramine are similar.

Dr Coppen also raises an important methodological point. The scale I described is a discriminating scale which can be used to assign individuals to one of two categories. It is not a descriptive scale. If we had wished to develop a descriptive scale we would have carried out a different kind of analysis on the data. To depict the severity of anxiety or depression an additional measure would be needed.

Coppen: Classification of the type of patient and severity of their symptoms are different, then?

Roth: Yes; although each item was described on a graded scale, the separation into groups is unrelated to severity of illness.

Sourkes: As a non-clinician, I learnt a great deal from Dr Pare's paper and from Sir Martin's about psychiatry and classification in relation to drugs, but one thing that I have become confused about, and where I can't make the leap from my area—biochemistry—is this: if we follow Dr Pare's prescription of measuring certain things, do the results depend upon the *chance factors* that go into the psychiatrist's viewpoint about psychiatry and classification? Is there any point in going ahead and making biochemical measurements that seem important, and seem to be related to things going on in the brain? Will the results of these studies be vitiated because the psychiatrist with whom one works happens to be more dynamically oriented and another more biologically? I would like some clarification on this, because those of us who are biochemists and are going to work with psychiatrists cannot vouch for our colleagues ideologically, except that we have some mutual understanding of what we are trying to achieve. Would the same biochemical results be analysed differently, in other words, in two different places?

Roth: I believe that so long as patients are depicted with no more than a label such as 'depressive illness', comparison with others' results has little value. The criteria employed in selection should be stated, and some objective method of characterizing the clinical state is also needed to make replication possible.

Coppen: What method of classification should be used?

Roth: You need in the first place a diagnosis based on stated operational criteria. One requires also a quantitative score derived from a differentiating scale, in this context one that indicates where the patient falls along the anxiety–depression distribution. Finally, a severity scale appropriate for the specific clinical disorder, such as one of the Hamilton scales. One therefore needs a number of indices: diagnosis together with quantitative scores derived from scales shown to have a respectable degree of reliability and validity.

Murphy: Your results were presented in terms of diagnostic categories only; did you mean to imply that there was no relationship between any of the variables you studied independently and the response to phenelzine? Further, in the Medical Research Council's trial in 1965 the dose was quoted earlier as being 45–60 mg of phenelzine per day. There is some feeling now that this may be an inadequate dose and I would think there are still grounds for looking at MAO inhibitors further, even in patients with endogenous depression.

Roth: We administered 15 mg three times a day in the first fortnight and this dose was increased to 75 mg daily in the second fortnight in those exhibiting little or no response. We had to limit ourselves to a trial of four weeks because we wanted to secure the cooperation of as many clinicians as possible. It appeared reasonable to expect some therapeutic benefit at the end of four weeks.* But this may be one reason for the discrepancy between our findings and the results of some other trials which suggest that in affective disorders with prominent anxiety, MAO inhibitors are beneficial. Tyrer *et al.* (1973), for example, observed no response within the first four weeks of their trial of agoraphobic patients.

Another reason may be the high response rate among those treated with diazepam alone. Half the depressed and 57% of the anxious patients responded, and a drug requires a very high success rate to improve on such results. These may in part have reflected placebo responses, which are often ill-sustained. A longer observation period might therefore have served to reveal any more specific effects produced by phenelzine. I agree with Dr Murphy that more studies are needed. But I doubt whether MAO inhibitors will prove beneficial in endogenous depressions.

Pare: To be deliberately controversial, and for the psychiatrically unsophisticated, an argument can be made that what Sir Martin has done is to separate people who have previously been reasonable personalities, but have become ill, from people who have never been adequate at any time in their life. The people in the anxious group have from childhood been neurotic, dependent, and inadequate, with hysterical traits. I don't believe that any pill can make them better! This would be like giving a pill to someone with mental deficiency, hoping to give him a normal I.Q. A doctor treats illness: he is *not* trying to 'cure' a person who has never been competent. I am being deliberately controversial here, of course. Sir Martin Roth is comparing the effects of drugs on people with endogenous depressions who by all criteria were effective but had become ill, and on people who never have been any good. This is an important point on the question of classification, and on what is a fair test of MAO inhibitors, quite apart from Dr Murphy's point about dosage and duration.

Roth: We were not necessarily dealing with 'inadequate' people in any of the groups. Many of those with agoraphobia certainly had some well concealed and circumscribed difficulties, but were none-the-less competent housewives and mothers and workers who often functioned well. You appear to be con-

* In fact, a more detailed analysis of the data subsequently revealed significant improvement in respect of four of the measures used to evaluate the clinical state. The results of the trial are to be published in full.

verting quantitative gradations into all-or-none differences. You mention hysterical features, but they are merely placed at a certain point along the scale. The fact that they figure above the horizontal line doesn't mean that there were none with hysterical traits among the endogenous depressions. It merely reflects their relationship to the total population traits and patients. Many of the agoraphobic patients, before the onset of an abrupt change—which can follow a trivial circumstance like fainting in a lavatory, after which they become housebound—had coped quite well.

References

BAASTRUP, P. C., POULSEN, J. C., SCHOU, M., THOMSEN, K. & AMDISEN, A. (1970) Prophylactic lithium: double blind discontinuation in manic-depressive and recurrent-depressive disorders. *Lancet 2*, 326-330

BROWN, G. W., BHROLCHAIN, M. N. & HARRIS, R. (1975) Social class and psychiatric disturbance among women in an urban population. *Sociology 9*, 225

BROWN, G. W., SKLAIR, F., HARRIS, T. O. & BIRLEY, J. L. T. (1973*a*) Life events and psychiatric disorders. *Psychol. Med. 3*, 74

BROWN, G. W., HARRIS, T. O. & PETO, J. (1973*b*) Life events and psychiatric disorders. Part 2: nature of causal link. *Psychol. Med. 3*, 159

EDWARDS, J. H. (1969) Familial predisposition in man. *Br. Med. Bull. 25*, 58-64

GUNDERSON, E. K. E. & RAHE, R. H. (1974) *Life Stress and Illness*, Thomas, Springfield, Ill.

HELGASON, T. (1964) Epidemiology of mental disorders in Iceland. *Acta Psychiatr. Scand. 40*, Suppl. 173, 1-258

HOLLISTER, L. E. (1972) Clinical use of psychotherapeutic drugs. 2: Antidepressive and antianxiety drugs and special problems in the use of psychotherapeutic drugs. *Drugs 4*, 361-410

KIELHOLZ, P. (1959) Diagnosis and therapy of the depressive states. *Acta Psychosomatica, Documenta Geigy 1*, 1-63

LEONHARD, K. (1957) *Aufteilung der endogenen Psychosen*, Akademie-Verlag, Berlin

PARKES, C. M., BENJAMIN, B. & FITZGERALD, R. G. (1969) Broken heart: a statistical survey of increased mortality among widowers. *Br. Med. J. 1*, 740

PERRIS, C. (1973) The heuristic value of a distinction between bipolar and unipolar affective disorders, in *Classification and Prediction of Outcome of Depression* (Angst, J., ed.), F. K. Schattauer Verlag, Stuttgart & New York

THEORELL, T. (1970) Psychosocial factors in relation to the onset of myocardial infarction and to some metabolic variables, Karolinska Institutet, Stockholm

TYRER, P., CANDY, J. & KELLY, D. (1973) Phenelzine in phobic anxiety: a controlled trial. *Psychol. Med. 3*, 120-124

VAN PRAAG, H. M. (1971) The position of biological psychiatry among the psychiatric disciplines. *Compr. Psychiatry 12*, 1-7

VAN PRAAG, H. M. (1972) Biologic psychiatry in perspective: the dangers of sectarianism in psychiatry. *Compr. Psychiatry 13*, 401-410

Variations in monoamine oxidase activity in some human disease states

M. SANDLER

Bernhard Baron Memorial Research Laboratories and Institute of Obstetrics and Gynaecology, Queen Charlotte's Maternity Hospital, London

Abstract Monitoring *in vivo* monoamine oxidase activity and its possible fluctuations in human disease has been hindered by a dearth of suitable methods of investigation. The shortcomings of early indirect approaches led to the present widespread practice of measuring enzyme activity directly in individual tissues, particularly blood platelets. Whilst the platelet monoamine oxidase assay, employing multiple substrates, has been helpful, for example, in demonstrating a deficit of the B form of the enzyme in migraine, caution must be advised in interpreting such data, for the platelet enzyme is atypical. What is needed is some approach to the *functional* activity of the enzyme in its whole-body environment. Evidence has recently been obtained to indicate that an observed deficit of tyramine conjugation in depressive illness after oral tyramine challenge may best be interpreted in terms of increased *in vivo* MAO activity, a greater proportion of orally administered amine being diverted from the smaller degradation pathway (conjugation) to the larger (oxidative deamination).

Certain physiological and pathological states appear to be associated with variations in monoamine oxidase (amine: oxygen oxidoreductase [deaminating] [flavin-containing]; EC 1.4.3.4) (MAO) activity. Thus, fluctuations have been observed in the different phases of the menstrual cycle (Southgate *et al.* 1968), and may be associated with changes in diet (Truitt *et al.* 1963) or gut flora (Phillips *et al.* 1962), whilst an evergrowing list of disease states, including toxaemia of pregnancy (Sandler & Coveney 1962), thyrotoxicosis (Levine *et al.* 1962), megaloblastic (Latt *et al.* 1968) and iron-deficiency (Youdim *et al.* 1975) anaemia and coeliac disease (Challacombe *et al.* 1971) may manifest with abnormal enzyme activity.

Profound decreases in activity are commonly induced in clinical practice by the administration of MAO-inhibiting drugs (Pletscher *et al.* 1966), yet relatively few attempts to quantify their action have been reported in the literature. The reason is simple; although many different *in vivo* assay procedures have been devised, none is completely satisfactory. Indirect procedures, such as estimating urinary tryptamine excretion (Sjoerdsma *et al.* 1959) or 5-hydroxyindoleacetic acid output after an oral 5-hydroxytryptamine (5HT) load (Sjoerdsma *et al.* 1958) may give rise to conflicting data (Sandler & Baldock 1963; Sandler *et al.* 1967). Direct assay may also present problems, however; measuring enzyme activity in buccal scrapings (Wurtman & Axelrod 1963) has given variable results in this laboratory (unpublished), whilst the operation to obtain jejunal biopsy specimens for this purpose (Levine & Sjoerdsma 1962; Challacombe *et al.* 1971) is not completely without risk to the patient. Even though its substrate specificity pattern differs profoundly from that of most other tissues (Collins & Sandler 1971), the blood platelet enzyme (Paasonen *et al.* 1964) which, as in other tissues, is mitochondrial (Solatunturi & Paasonen 1966), has increasingly come under scrutiny as an index of *in vivo* MAO activity in man (e.g. Robinson *et al.* 1968; Zeller *et al.* 1969; Sandler *et al.* 1974). Despite its drawbacks, if one employs a battery of different substrates, a profile is obtained which may be used to characterize a particular disease state (Meltzer & Stahl 1974; Sandler *et al.* 1974; Youdim *et al.* 1975). In this way, a relative deficit of platelet phenylethylamine-oxidizing ability, an index of MAO B activity, was recently demonstrated in migraine (Sandler *et al.* 1974).

Even so, it may be argued that a single tissue represents only itself and that what is needed for monitoring MAO inhibitor therapy, and perhaps for diagnosis, is an integrated *in vivo* index of whole-body MAO activity and, preferably, of MAO A separately from MAO B, if these *in vitro* manifestations do indeed represent a true *in vivo* phenomenon.

IN VIVO EXISTENCE OF MULTIPLE FORMS?

The classification of MAO into A and B forms (see Houslay & Tipton 1974) is based on the differential inhibitory ability of certain MAO inhibitors towards a variety of substrates (Johnston 1968). Type A is sensitive to clorgyline (Johnston 1968) and oxidatively deaminates noradrenaline and 5-hydroxytryptamine, but not phenylethylamine. Type B is preferentially inhibited by deprenyl (Knoll & Magyar 1972) and oxidatively deaminates phenylethylamine, but not noradrenaline and 5HT (Yang & Neff 1973). Tyramine is a substrate of both. Very localized concentrations of a clorgyline-insensitive 5HT metabolizing variant, termed MAO C, have recently been detected histochemically in rat brain (Gascoigne *et al.* 1975). Both A and B have been demonstrated in tissue homogenates, whether the differential inhibitors were added *in vitro*

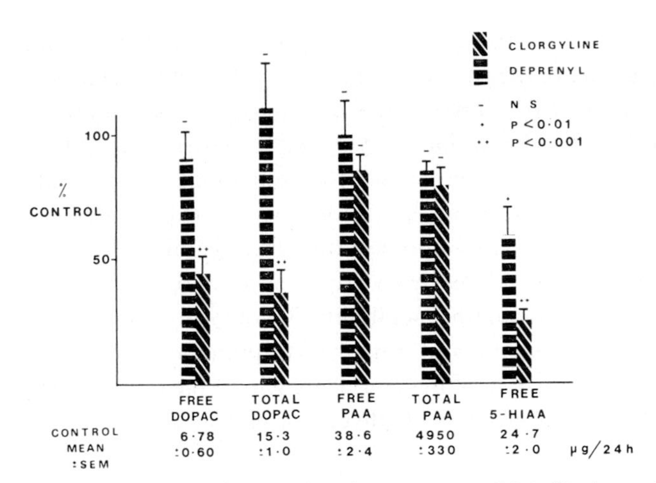

FIG. 1. Percentage decrease in urinary output of 3,4-dihydroxyphenylacetic acid (DOPAC), phenylacetic acid (PAA) and 5-hydroxyindoleacetic acid (5-HIAA) in 200 g White Wistar rats after clorgyline or deprenyl (10 mg/kg) administration, compared with control collections.

or administered *in vivo* (Neff *et al.* 1973). However, the question of whether endogenous substrates are similarly metabolized must be considered an open one.

In an attempt to elucidate this problem, we (M. Sandler, B. L. Goodwin & C. R. J. Ruthven, unpublished) have recently injected rats with the same dosage of clorgyline or deprenyl used by Neff *et al.* (1973) (these authors were able to demonstrate the separate forms in the tissues assayed after sacrifice). By measuring the major urinary metabolites in this species of the MAO A substrates, noradrenaline (4-hydroxy-3-methoxyphenylglycol) and 5-hydroxytryptamine (5-hydroxyindoleacetic acid) and the B substrate, phenylethylamine (phenylacetic acid) (Fig. 1) before and after administration of each of these drugs, any *in vivo* differential effect should have been detected. However, whereas clorgyline brought about a proportionately greater decrease in urinary 4-hydroxy-3-methoxyphenylglycol (not shown in figure) and 5-hydroxyindoleacetic acid output and, indeed, in the excretion of 3,4-dihydroxyphenylacetic acid, a major metabolite of dopamine, which is itself a substrate for both A and B forms (Neff *et al.* 1973), deprenyl did not result in the converse, a proportionately greater decrease in phenylacetic acid. Because of a possible contribution of phenylacetic acid deriving from the action of gut flora on phenylalanine residues, and perhaps from other sources, these data must be regarded as preliminary.

There are other reasons for doubting whether a simple subdivision into A and B forms is an *in vivo* reality. Although electrophoretic studies on solubilized enzyme are difficult to reconcile with the A–B classification and, indeed, with the three-dimensional structure of the enzyme as it exists on the outer membrane of the mitochondrion (Tipton *et al.*, this volume, pp. 5–16), it is worth noting that data have been obtained by this technique to suggest that there may be a separate form of MAO with a high preference for dopamine as substrate (Youdim 1973). If this form exists, it may go a long way to explaining such findings as the preferential inhibition by apomorphine of dopamine oxidation in rat brain (Di Chiara *et al.* 1974) and may even help to account for the isolated decrease in urinary output of the dopamine metabolite, 3,4-dihydroxyphenylacetic acid, compared with noradrenaline and 5-hydroxytryptamine metabolites, after administration to man of the antidepressant drug, viloxazine (M. Sandler, B. L. Goodwin, R. D. Johnson & C. R. J. Ruthven, unpublished).

MAO ACTIVITY IN DEPRESSIVE ILLNESS

Because some patients with depressive illness appear to respond favourably to MAO-inhibiting drugs, it was early assumed (Pare & Sandler 1959) that the illness might in some way derive from an overaction of MAO. However, despite considerable research effort (see Sandler *et al.* 1975, for references), the situation still remains unclear, partly, perhaps, because different investigators have used different substrates, partly because of the different tissues on which measurements have been made, and partly because of the heterogeneity of clinical material.

In 1971, Youdim *et al.* reported what they considered to be a deficit in oral tyramine conjugation in patients with dietary migraine. Monoamines, depending on their chemical nature, undergo sulphate conjugation to a greater or lesser extent when administered orally; 10–15% of oral tyramine is degraded in this way, most of the remainder being metabolized by MAO (Fig. 2) (see Sandler *et al.* 1975). Because some patients being treated with MAO inhibitors undergo severe adverse reactions when eating foods containing tyramine, whereas others can ingest them without harm, Youdim *et al.* (1971) suggested that tyramine conjugation might act as a 'safety valve' pathway, impairment of which would put the patient at risk if he were treated with an MAO inhibitor. Thus, if this interpretation had been correct, it might have been possible to develop a clinical test to distinguish such patients from others who could take the drugs with greater safety.

Sandler *et al.* (1975) therefore gathered together several groups of patients, including one group of those who had undergone an adverse reaction during MAO inhibitor therapy, and one with no history of adverse reaction, despite eating foods containing tyramine. Although the first group showed a highly

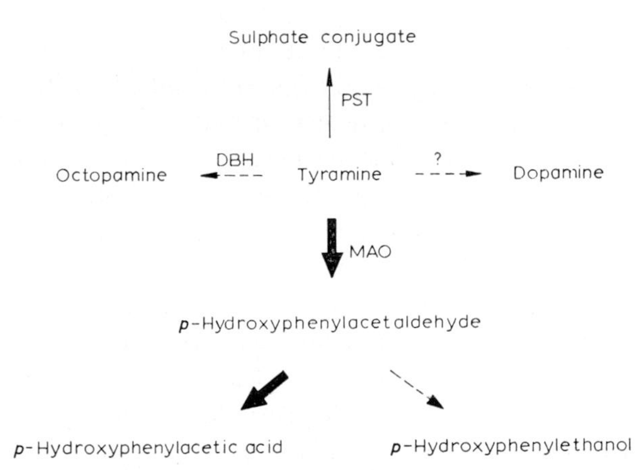

FIG. 2. Pathways of tyramine metabolism. Bold arrows show the major pathway. PST = phenolsulphotransferase; DBH = dopamine β-hydroxylase; MAO = monoamine oxidase. (Reproduced from *The Lancet*, with permission of the Editor.)

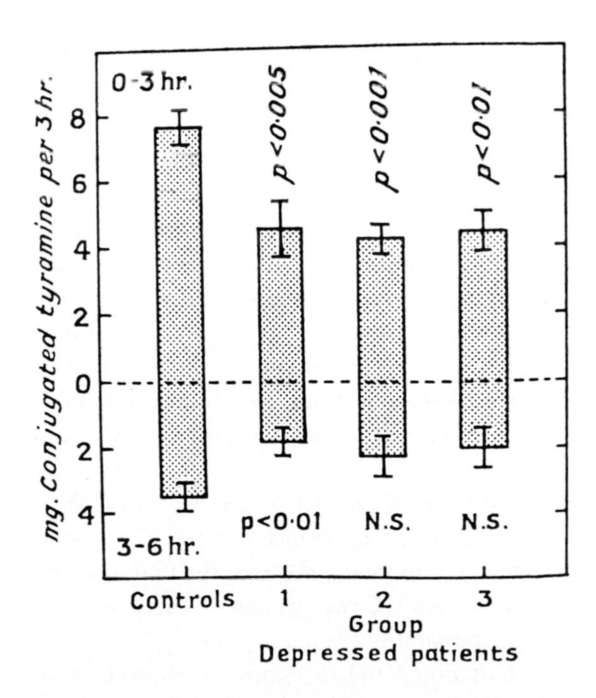

FIG. 3. Excretion of conjugated tyramine after an oral tyramine load (100 mg) by controls and depressed patients. (Reproduced from *The Lancet*, with permission of the Editor.)

significant decrease in conjugated tyramine output after an oral tyramine load, compared with controls, so did the second (Fig. 3) (Sandler *et al.* 1975). At this stage of the investigation, therefore, it seemed that there might be a sulphate conjugation deficit, existing as a manifestation of depressive illness itself rather than as a peculiarity of the subgroup which had undergone an adverse reaction. This interpretation was underscored by the presence of the biochemical difference in a group of depressed patients considered suitable for MAO inhibitor therapy, but never so treated (Fig. 3). However, it was not possible to exclude one further possibility. Youdim *et al.* (1971) thought they had ruled out an increase in MAO activity as a mechanism to explain their findings by measuring the urinary output of *p*-hydroxyphenylacetic acid, the major oxidatively deaminated metabolite of tyramine. However, on re-evaluation, it seemed unlikely that a small increase in *p*-hydroxyphenylacetic acid would have been apparent in such a large pool; the subtraction of a similar amount of tyramine equivalent from the much smaller conjugation pool, on the other hand, would be much more obvious. Such a 'back-titration' effect, if confirmed, might well prove to be a sensitive method of detecting small *in vivo* changes in MAO activity.

The most direct way of deciding between the two possibilities would have been to measure conjugating ability and MAO activity directly in gut mucosal biopsy samples. As this approach was not practicable, however, Sandler *et al.* (1975) adopted another; they tested the conjugating mechanism indirectly by administering another monoamine, isoprenaline, which is inactivated almost solely by conjugation when given by mouth (Morgan *et al.* 1969). Presumably the mechanism is a common one for all monoamines. Had the conjugation system been impaired, the urinary output 'tolerance curve' might have been expected to be flatter and take longer to fall to baseline in the group of low conjugated-tyramine excreters compared with high. In the event, the reverse tendency was present (Fig. 4). Nor was there any difference in 24 h output between the two groups (Fig. 5), although any block in conjugation might have been expected to dissipate a large proportion of the monoamine along minor pathways.

These data seem to reinstate the concept of an MAO increase as the most likely explanation for the clear-cut biochemical difference between depressives and controls—always bearing in mind the remote chance that the findings stem from some difference in gut motility between the two groups. Experiments are now proceeding to try to exclude this possibility.

Because Robinson *et al.* (1972) had noted the tendency of platelet MAO activity to increase with age, it seemed necessary to try to exclude the possibility of these findings being manifestations of disease in an older age group, rather

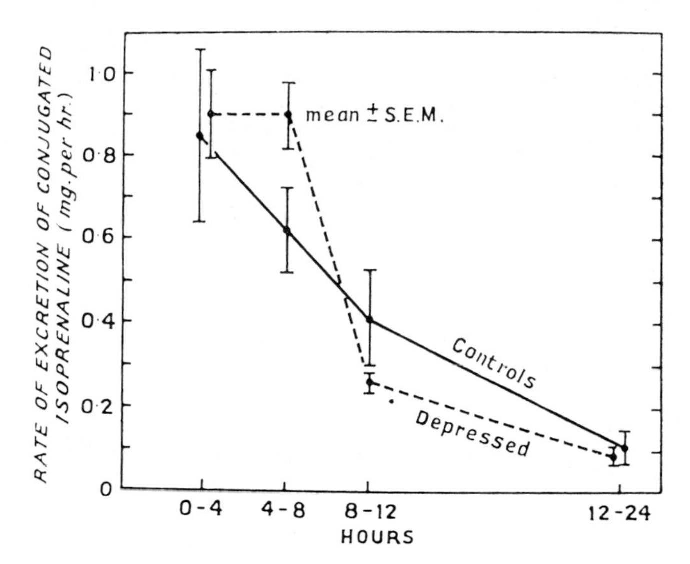

FIG. 4. Rates of excretion of conjugated isoprenaline by controls and depressed patients after sublingual isoprenaline sulphate (20 mg). (Reproduced from *The Lancet*, with permission of the Editor.)

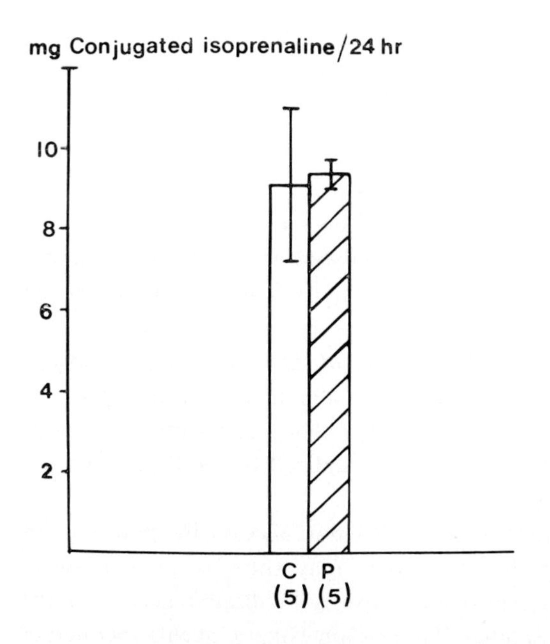

FIG. 5. Twenty-four hour excretion of conjugated isoprenaline (mean ± S.E.M.) by a control group (C) and a group of depressed patients (P) after sublingual isoprenaline sulphate (20 mg).

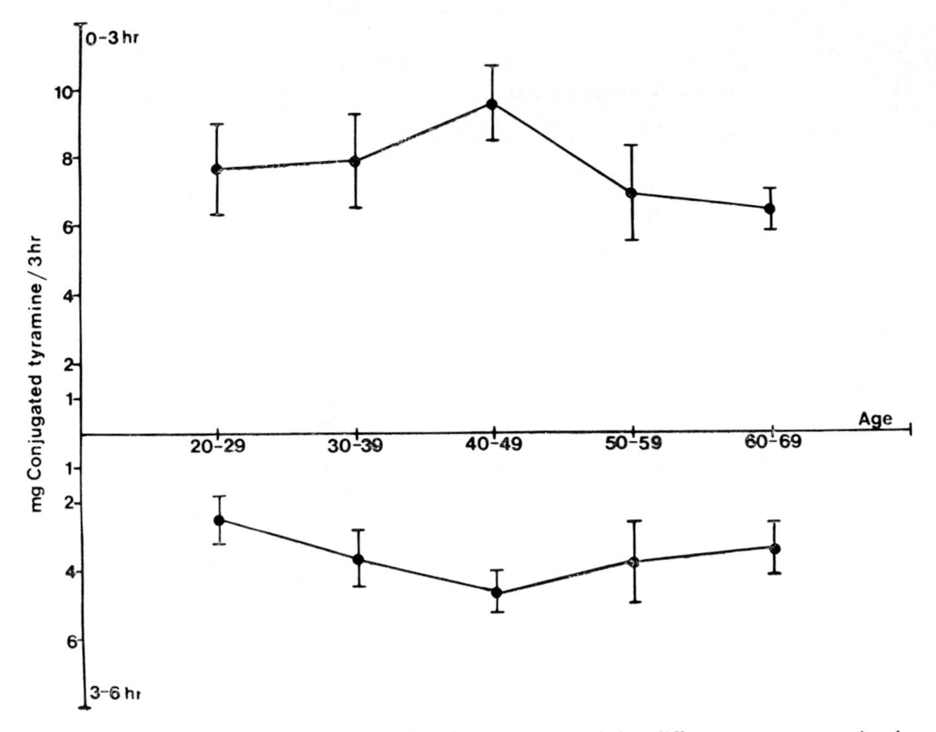

FIG. 6. Excretion of conjugated tyramine (mean ± S.E.M.) by different age groups in the control series after an oral tyramine load (100 mg).

than of the depressive illness itself. However, when the control group, which had been chosen with this point in mind, was analysed according to age, there was no evidence of any significant variation in activity from decade to decade (Fig. 6).

Although the changes detected by Sandler *et al.* (1975) were probably a manifestation of alterations in enzyme activity in the small intestine, there is suggestive evidence deriving from other systems that depression may be a generalized organic illness, and not merely one localized to the brain. Such a conclusion appears to possess certain merits.

The putative changes in small gut MAO do not necessarily point to an increase in enzyme protein as such. What they may show is an increase in *functional* activity, meaning that substrates have a facilitated *access* to the enzyme. Reserpine (Youdim & Sandler 1968) or anaesthetic agents (Schneider *et al.* 1974) may similarly bring about an increase in MAO activity. Thus, it may be appropriate to view depression not just in its narrow sense as a mental illness, but in a wider context, as a disease of membrane barriers. Other chemical systems may also be affected by such a 'leaky membrane' disease. And what is the connection with migraine? Sandler *et al.* (1975) tentatively explained the findings of Youdim *et al.* (1971), of a decrease in conjugated tyramine output in patients with dietary migraine, in terms of a possible high incidence of masked depression in the test group. However, the relationship between the two diseases obviously deserves more careful scrutiny; it should be remembered particularly that when all other therapeutic regimens have failed, MAO inhibitors have been used with success in the treatment of intractable migraine (Anthony & Lance 1969). The present sharp biochemical differences which have been demonstrated between depressed and normal subjects may even point, once more, to the MAO inhibitors being a rational form of treatment for at least some patients with depressive illness.

References

ANTHONY, M. & LANCE, J. W. (1969) Monoamine oxidase inhibition in the treatment of migraine. *Arch. Neurol. 21*, 263

CHALLACOMBE, D. N., SANDLER, M. & SOUTHGATE, J. (1971) Decreased duodenal monoamine oxidase activity in coeliac disease. *Arch. Dis. Child. 46*, 213-215

COLLINS, G. G. S. & SANDLER, M. (1971) Human blood platelet monoamine oxidase. *Biochem. Pharmacol. 20*, 289-296

DI CHIARA, G., BALAKLEEVSKY, A., PORCEDDU, M. L., TAGLIAMONTE, A. & GESSA, G. L. (1974) Inhibition by apomorphine of dopamine deamination in the rat brain. *J. Neurochem. 23*, 1105-1108

GASCOIGNE, J. E., WILLIAMS, D. & WILLIAMS, E. D. (1975) Histochemical demonstration of an additional form of rat brain monoamine oxidase. *Br. J. Pharmacol. 54*, 274*P*

HOUSLAY, M. D. & TIPTON, K. F. (1974) A kinetic evaluation of monoamine oxidase activity in rat liver mitochondrial outer membranes. *Biochem. J. 139*, 645-652

JOHNSTON, J. P. (1968) Some observations upon a new inhibitor of monoamine oxidase in brain tissue. *Biochem. Pharmacol. 17*, 1285-1297

KNOLL, J. & MAGYAR, K. (1972) Some puzzling pharmacological effects of monoamine oxidase inhibitors, in *Monoamine Oxidases—New Vistas* (Costa, E. & Sandler, M., eds.) *(Adv. Biochem. Psychopharmacol. 5)*, pp. 393-408, Raven Press, New York & North-Holland, Amsterdam

LATT, N., RIPPEY, J. J. & STACEY, R. S. (1968) Monoamine oxidase activity of platelets. *Br. J. Pharmacol. 32*, 427*P*

LEVINE, R. J., OATES, J. A., VENDSALU, A. & SJOERDSMA, A. (1962) Studies on the metabolism of aromatic amines in relation to altered thyroid function in man. *J. Clin. Endocrinol. Metab. 22*, 1242-1250

LEVINE, R. J. & SJOERDSMA, A. (1962) Monoamine oxidase activity in human tissues and intestinal biopsy specimens. *Proc. Soc. Exp. Biol. Med. 109*, 225-227

MELTZER, H. J. & STAHL, S. M. (1974) Platelet monoamine oxidase activity and substrate preferences in schizophrenic patients. *Res. Commun. Chem. Pathol. 7*, 419-431

MORGAN, C. D., RUTHVEN, C. R. J. & SANDLER, M. (1969) The quantitative assessment of isoprenaline metabolism in man. *Clin. Chim. Acta 26*, 381-386

NEFF, N. H., YANG, H.-Y. T. & GORIDIS, C. (1973) Degradation of the transmitter amines by specific types of monoamine oxidases, in *Frontiers in Catecholamine Research* (Usdin, E. & Snyder, S. H., eds.), pp. 133-137, Pergamon, New York

PAASONEN, M. K., SOLATUNTURI, E. & KIVALO, E. (1964) Monoamine oxidase activity of blood platelets and their ability to store 5-hydroxytryptamine in some mental deficiencies. *Psychopharmacologia 6*, 120-124

PARE, C. M. B. & SANDLER, M. (1959) A clinical and biochemical study of a trial of iproniazid in the treatment of depression. *J. Neurol. Neurosurg. Psychiatr. 22*, 247-251

PHILLIPS, A.W., NEWCOMBE, H. R., RUPP, F. A. & LACHAPELLE, R. (1962) Nutritional and microbial effects on liver monoamine oxidase and serotonin in the chick. *J. Nutr. 76*, 119-123

PLETSCHER, A., GEY, K. F. & BURKARD, W. P. (1966) Inhibitors of monoamine oxidase and decarboxylase of aromatic acids, in *5-Hydroxytryptamine and Related Indolealkylamines* (Erspamer, V., ed.), vol. 19, *Handbuch der experimentellen Pharmakologie*, pp. 593-735, Springer, Berlin

ROBINSON, D. S., DAVIS, J. M., NIES, A., COLBURN, R. W., DAVIS, J. N., BOURNE, H. R., BUNNEY, W. E., SHAW, D. M. & COPPEN, A. J. (1972) Ageing, monoamines and monoamine oxidase levels. *Lancet 1*, 290-291

ROBINSON, D. S., LOVENBERG, W., KEISER, H. & SJOERDSMA, A. (1968) Effect of drugs on human blood platelet and plasma amine oxidase activity *in vitro* and *in vivo*. *Biochem. Pharmacol. 17*, 109-119

SANDLER, M. & BALDOCK, E. (1963) *In vivo* monoamine oxidase activity in toxaemia of pregnancy. *J. Obstet. Gynaecol. Br. Commonw. 70*, 279-283

SANDLER, M. & COVENEY, J. (1962) Placental monoamine-oxidase activity in toxaemia of pregnancy. *Lancet 1*, 1096-1098

SANDLER, M., RUTHVEN, C. R. J. & CEASAR, P. (1967) Urinary pH and the excretion of biologically active monoamines and their acidic metabolites, in *Proc. 7th Int. Congr. Biochemistry, Tokyo*, p. 971

SANDLER, M., YOUDIM, M. B. H. & HANINGTON, E. (1974) A phenylethylamine oxidising defect in migraine. *Nature (Lond.) 250*, 335-337

SANDLER, M., BONHAM CARTER, S., CUTHBERT, M. F. & PARE, C. M. B. (1975) Is there an increase in monoamine-oxidase activity in depressive illness? *Lancet 1*, 1045-1049

SCHNEIDER, D. R., HARRIS, S. C., GARDIER, R. W., O'NEILL, J. J. & DELAUNOIS, A. L. (1974) Increased monoamine oxidase activity produced by general inhalation anesthetic agents. *Arch. Int. Pharmacodyn. Ther. 211*, 64-73

SJOERDSMA, A., GILLESPIE, L. A. & UDENFRIEND, S. (1958) A simple method for the measurement of monoamine-oxidase inhibition in man. *Lancet 2*, 159-160

SJOERDSMA, A., OATES, J. A., ZALTZMAN, P. & UDENFRIEND, S. (1959) Identification and assay of urinary tryptamine; application as an index of monoamine oxidase inhibition in man. *J. Pharmacol. Exp. Ther. 126*, 217-222

SOLATUNTURI, E. & PAASONEN, M. K. (1966) Intracellular distribution of monoamine oxidase, 5-hydroxytryptamine and histamine in blood platelets of rabbit. *Ann. Med. Exp. Biol. Fenn. 44*, 427-430

SOUTHGATE, J., GRANT, E. C., POLLARD, W. & PRYSE-DAVIES, J. (1968) Cyclical variations in endometrial monoamine oxidase: correlation of histochemical and quantitative biochemical assays. *Biochem. Pharmacol. 17*, 721-726

TIPTON, K. F., HOUSLAY, M. D. & MANTLE, J. T. (1976) This volume, pp. 5-16

TRUITT, E. B. JR, DURITZ, G. & EBERSBERGER, M. (1963) Evidence for monoamine oxidase inhibition by myristicin and nutmeg. *Proc. Soc. Exp. Biol. Med. 112*, 647-650

WURTMAN, R. J. & AXELROD, J. (1963) A sensitive and specific assay for the estimation of monoamine oxidase. *Biochem. Pharmacol. 12*, 1439-1441

YANG, H.-Y. T. & NEFF, N. H. (1973) β-Phenylethylamine: a specific substrate for type B monoamine oxidase of brain. *J. Pharmacol. Exp. Ther. 187*, 365-371

YOUDIM, M. B. H. (1973) Multiple forms of mitochondrial monoamine oxidase. *Br. Med. Bull. 29*, 120-122

YOUDIM, M. B. H., BONHAM CARTER, S., SANDLER, M., HANINGTON, E. & WILKINSON, M. (1971) Conjugation defect in tyramine-sensitive migraine. *Nature (Lond.) 230*, 127-128

YOUDIM, M. B. H. & SANDLER, M. (1968) Activation of monoamine oxidase and inhibition of aldehyde dehydrogenase by reserpine. *Eur. J.Pharmacol 4*, 105-108

YOUDIM, M. B. H., WOODS, H. F., MITCHELL, B., GRAHAME-SMITH, D. G. & CALLENDE, S. (1975) Human platelet monoamine oxidase activity in iron-deficiency anaemia. *Clin. Sci. Mol. Med. 48*, 289-285

ZELLER, E. A., BABU, B. H., CAVANAUGH, M. J. & STANICH, G. J. (1969) On the *in vivo* inhibition of human platelet monoamine oxidase (1.4.3.4.) by pargyline. *Pharmacol. Res. Commun. 1*, 20-24

Discussion

Fuller: You showed that isoprenaline was not conjugated any differently in the different groups of patients, so that is one substrate that doesn't show a difference, and tyramine is one that does. Have you tried other compounds that are metabolized by conjugation to see whether this is a general phenomenon, or is it specific in some way to tyramine?

Sandler: We haven't done that yet. We have plans to compare the ability of the body to conjugate orally administered A and B substrates.

Fuller: I am concerned that just because a structurally unrelated drug, isoprenaline, did not show a similar deficit to tyramine, you began searching for a secondary explanation, such as increased MAO activity. It would seem to me more direct, rather than looking at other MAO substrates, to stay with the tyramine structure and simply put an α-methyl group on it; you would then eliminate MAO from the picture and you could look at *p*-hydroxyamphetamine, which is what that drug is, and see if that also shows the conjugation deficit.

Sandler: That is a very useful suggestion.

Sourkes: Is orally given isoprenaline metabolized mainly through the sulphate?

Sandler: Yes (Morgan *et al.* 1969). The *O*-methylation pathway is responsible for only a small proportion (8% by C. T. Dollery's data).

Sourkes: The hippuric acid test is a time-honoured test in biochemical psychiatry; have you thought of trying that as the control for amine conjugation?

Sandler: No.

Fuller: Is viloxazine an MAO inhibitor? If so, is it reversible or irreversible?

Sandler: It is said not to be an MAO inhibitor.

Fuller: Does it inhibit the uptake of dopamine, or affect dopamine receptors? What is the explanation for its effect on dihydroxyphenylacetic acid?

Youdim: It is not an MAO inhibitor either *in vitro* or *in vivo*. We have tested it up to very high doses. It is supposed to have an effect on dopamine and noradrenaline uptake.

Sandler: I think there must be some selective inhibition, perhaps of dopamine MAO, if there is such an enzyme.

Fuller: Inhibition of dopamine uptake would explain the results just as easily.

Oreland: We collected brains from 15 patients who had committed suicide and from a corresponding control material, and then analysed the MAO activity in 13 parts of the brain with two substrates, tryptamine and β-phenylethylamine (Gottfries *et al.* 1974*a*, 1975). To our astonishment we found a highly significant decrease in MAO activity in the brains of the patients who had committed suicide. Then the real work began, in order to rule out possible errors such as the influence of age, sex, storage time, storage temperature, other kinds of brain diseases, drugs, dietary factors, and cause of death. Taking all those factors into account we still found a significantly lower activity in the brains of the suicides, which seemed to involve essentially all parts of the brains investigated. Curiously enough it also seemed to be valid for all patients in the suicide group. Of course, suicide does not necessarily mean that you have a depressed patient, so we would suggest that people with 'suicidal behaviour' have lowered MAO activities in their brains. In order to relate this to the current monoamine hypothesis we suggest that patients with 'suicidal behaviour' have low monoaminergic activities or weak monoaminergic systems, which in turn cause low induction rates of MAO, or are otherwise reflected in low MAO activities. This hypothesis is supported by the experiments on rats that I mentioned earlier (p. 221) showing that it is possible to increase MAO activity in rat brain by prolonged administration of L-dopa.

I should mention that in the suicidal group the most marked decrease in MAO activity was found in the patients with a history of alcohol abuse. To see if this was a direct effect of ethanol on the enzyme or just an expression of a close relationship between alcohol abuse and 'suicidal behaviour' in terms of monoaminergic activity in the brain we have done experiments on the chronic administration of ethanol to rats using different techniques, without any notable effect on the brain MAO activity, which thus supports the latter explanation (A. Wiberg, G. Wahlstrom & L. Oreland, in preparation).

Youdim: Have you measured the monoamine levels in these brain areas?

Oreland: I have not done that, but Drs Gottfries, Roos & Winblad (1974*b*) have started some studies on it. The preliminary finding seems to be that the amines are somewhat unstable after death so, since we know that the enzyme is rather stable, MAO activity may be a better indicator of monoaminergic activity than a direct measurement of the amines.

Blaschko: I wonder whether we don't put too much emphasis on the A and B type enzymes. This is also highlighted by Dr Oreland's observations. What I take from this symposium is the beautiful analysis that Professor Trendelenburg gave us, and also the remark in which Dr Iversen underlined that the MAO contribution of the monoaminergic neurons of the central nervous system is probably relatively small. If we assume as a working hypothesis that Dr Oreland's finding can be substantiated, we would expect significant differences to be due to aminoceptive structures in the CNS responding, rather than aminergic structures in the nervous system contributing. Am I right in this assumption?

The second point is that perhaps Dr O. Hornykiewicz might not agree with your statement that the keeping qualities of the amines in the brain are so much worse than those of MAO.

Iversen: Brain MAO may respond to flooding the system with L-dopa, but it doesn't appear to work the other way round, at least in animals, because one can destroy central aminergic pathways with 6-hydroxydopamine without causing any change in brain MAO activity.

Sandler: I am fascinated by Dr Oreland's work. I notice that you use two B type substrates, however (tryptamine and β-phenylethylamine).

Oreland: Tryptamine is an A + B substrate.

Sandler: Not according to Dr Tipton's table (p. 14). Secondly, you stressed that the suicide syndrome is not synonymous with depression; there may be some overlap but in the main they are likely to be different clinical entities. One should stress too that rats aren't men and, in this context, your data from rats must be treated circumspectly. I suppose the difference between our two approaches is that if what we were measuring was MAO *functional* activity, you were measuring the direct *in vitro* action of the enzyme itself. Perhaps our data ought to approximate one set to the other, if we studied similar clinical material. I myself am attracted by the possibility of there being an increase in MAO in depressive illness because the idea of the enzyme destroying more than its share of monoamine fits in nicely with the monoamine hypothesis of depression! I really look on depression as just one manifestation of a systemic disease. Indeed, there are many indications of physical disturbances, if we look for them, such as osteoporosis, electrolyte upsets, and so on; the actual symptom of depression might just be the tip of the iceberg, one end-organ response symptomatic of, and shared by, perhaps, a number of different physical disease complexes.

Rafaelsen: It should be emphasized that in most parts of the world, when you study a group of suicides, less than one-third have been psychotic; around 50% will have been personality problems and 25% will have been normal and stable personalities. Among the third who have been psychotic, half will have been schizophrenic and only some 10–15% will have had affective disorders. The findings of Dr Oreland should be viewed in this perspective.

Coppen: We have investigated the tyramine pressor response in depressives (Ghose *et al.* 1975). This is part of our investigation on the response of depressive patients to amitriptyline, in which we are measuring plasma levels of amitriptyline and nortriptyline, the therapeutic response, side-effects, and so on. We are also looking at various autonomic factors. One test is the tyramine pressor response in which tyramine is injected intravenously to obtain a certain rise in blood pressure (30 mmHg). We do this at baseline before the patients start therapy and then two weeks afterwards, when one finds a steady-state plasma level of the antidepressant. We found that the patients are more sensitive to tyramine than the controls; that is, they need less tyramine to produce the pressor response than do the control patients. We found no relationship with age or sex, in the number that we looked at. There could be several interpretations, one of which is a decreased activity of MAO. There could be differences in the amount of noradrenaline available for release; or there could be changes in receptor sensitivity. After two weeks we found the expected change in response when patients were on the tricyclic drug and we obtained a correlation between the nortriptyline concentration in plasma and the change in the tyramine pressor response. We are now looking at correlations between changes in the pressor response and therapeutic outcome, and this seems to be a different story. There is no positive correlation and it is beginning to look like a negative correlation. I mention this to show how murky the biogenic amine hypothesis is becoming! There was no correlation with amitriptyline and the pressor response.

References

GHOSE, K., TURNER, P. & COPPEN, A. (1975) Intravenous tyramine pressor response in depression. *Lancet 1*, 1317-1318

GOTTFRIES, C. G., ORELAND, L., WIBERG, Å. & WINBLAD, B. (1974*a*) Brain-levels of monoamine oxidase in depression. *Lancet 2*, 360-361

GOTTFRIES, C. G., ROOS, B. E. & WINBLAD, B. (1974*b*) Determination of 5-hydroxytryptamine, 5-hydroxyindoleacetic acid and homovanillic acid in brain tissue from an autopsy material. *Acta Psychiatr. Scand. 50*, 496-507

GOTTFRIES, C. G., ORELAND, L., WIBERG, Å. & WINBLAD, B. (1975) Lowered monoamine oxidase activity in brains from alcoholic suicides. *J. Neurochem. 25*, 667-674

MORGAN, C. D., RUTHVEN, C. R. J. & SANDLER, M. (1969) The quantitative assessment of isoprenaline metabolism in man. *Clin. Chim. Acta 26*, 381-386

Clinical, genetic, hormonal and drug influences on the activity of human platelet monoamine oxidase

DENNIS L. MURPHY

Section on Clinical Neuropharmacology, Laboratory of Clinical Science, NIMH, Bethesda, Maryland

Abstract Individual differences in human platelet monoamine oxidase (MAO) appear to be primarily genetically based, although hormones, nutritional states and drugs may also affect the activity of the enzyme. Several psychiatric disorders, including bipolar manic-depressive illness and chronic schizophrenia, appear to be associated with altered MAO activity. The biochemical basis for these differences in MAO may be genetic in some instances, but different factors, including sex- and age-related functions, may be contributory in other instances. The clinical data do not suggest a simple relationship between MAO activity and any psychiatric diagnostic entity, but more suggest a possible association between altered MAO activity and a vulnerability to psychopathology. This concept is supported by preliminary studies utilizing psychological tests in normal young adult populations and behavioural ratings in non-human primates which indicate relationships between both personality features and longer-term behavioural characteristics and platelet MAO activity.

Much evidence points to important contributions by the biogenic amines and other neurotransmitters to behaviour, behavioural disorders and the therapeutic effects of psychoactive drugs. Recently, it has become possible to measure in man the activity of enzymes which participate in biogenic amine degradation and synthesis, using blood cells or plasma as sources for such enzymes as monoamine oxidase, catechol *O*-methyltransferase and dopamine β-hydroxylase.

There are many potential advantages to being able to study biogenic amine metabolism in man in terms of individual differences in enzyme activities rather than relying upon the measurement of amine metabolites in urine or other body fluids, which often represent the final summation of many different metabolic processes (Murphy 1973). There are many potential pitfalls as well. Although many genetically based disorders of metabolism have been effectively studied in other areas of medicine using blood cell measurements, non-genetic factors may also affect measured enzyme activity. Our studies have attempted to evaluate the relative contributions of genetic and other factors to individual differences in human platelet monoamine oxidase activity, especially in regard to possible associations between the activity of this enzyme and behaviour.

CHARACTERISTICS OF HUMAN PLATELET MONOAMINE OXIDASE

Platelet monoamine oxidase (amine: oxygen oxidoreductase [deaminating] [flavin-containing]; EC 1.4.3.4) (MAO) is thought to be localized in mitochondria (Paasonen & Solatunturi 1965), like most other tissue monoamine oxidases. However, the small size of platelet mitochondria and the consequent problems in completely separating them from alpha granules and amine storage vesicles in platelets have rendered definitive localization studies difficult. Platelet MAO has been solubilized and purified 12-fold by Collins & Sandler (1971) who reported an estimated molecular weight of 235 000, a value at the lower end of the range for other mitochondrial MAOs.

In contrast to brain, liver and other tissues where multiple forms of MAO have been demonstrated, most evidence suggests that the platelet possesses only one molecular form of MAO. Polyacrylamide gel electrophoresis revealed a single tetrazolium-stained band (Collins & Sandler 1971). Studies with the MAO-inhibiting drugs clorgyline and deprenyl, which in other tissues have suggested the presence of two enzyme forms, MAO A and MAO B, on the basis of biphasic log inhibition plots with tyramine as substrate (Johnston 1968), have revealed only a simple sigmoid curve for the platelet enzyme (Murphy & Donnelly 1974). The 200-fold greater sensitivity of the platelet MAO to inhibition by deprenyl compared to clorgyline suggests that platelet MAO is an MAO B form. This conclusion is supported by the high activity of the platelet enzyme with the relatively specific MAO B substrates, benzylamine and β-phenethylamine, in contrast to its lesser activity with 5-hydroxytryptamine, an MAO A substrate (Robinson *et al.* 1968; Murphy & Donnelly 1974). It should be noted that apparent differences between platelet MAO activities measured with different substrates have been reported in several studies (Meltzer & Stahl 1974; Nies *et al.* 1974; Sandler *et al.* 1974*a*). Although our comparison of platelet MAO activities measured with two substrates, tryptamine and benzylamine, yielded a very high correlation coefficient ($r = 0.89$, $P < 0.001$, $n = 75$) (Murphy & Donnelly 1974), the question of whether dopamine (Sandler *et al.* 1974*a*) or octopamine and tryptamine (Meltzer & Stahl 1974) may reveal some individually different platelet MAO activities in comparison to tyramine is an intriguing possibility.

GENETIC BASIS FOR INDIVIDUAL DIFFERENCES IN PLATELET MAO ACTIVITY

Platelet MAO activity differences range over 10-fold in normal individuals. Studies comparing enzyme activity differences between monozygotic and dizygotic twins have been used to determine the magnitude of genetic contributions to the individual variations in platelet MAO. Nies *et al.* (1973) and Murphy (1973) have reported significantly higher intra-class correlation coefficients for normal monozygotic than for dizygotic twins, and these differences have been confirmed by non-parametric statistical measures as well. In addition, comparisons of sib pairs with non-related pairs matched for age and sex have also confirmed a large genetic contribution to MAO activity differences (Murphy & Donnelly 1974). Similarly high intra-class correlation coefficients have also been observed in monozygotic twins with bipolar manic-depressive disorders (Murphy *et al.* 1974) and monozygotic twins discordant for schizophrenia (Wyatt *et al.* 1973). The latter finding not only supported a major genetic contribution to the enzyme activities measured in schizophrenic patients, but also suggested that being schizophrenic itself did not appreciably alter platelet MAO activity, since the non-schizophrenic twins had MAO values closely similar to those in the schizophrenic twins.

Nies *et al.* (1973) observed a bimodal distribution pattern in platelet MAO activities of 80 normal subjects. This was, of course, of interest since it directly suggested that two enzyme forms might exist. However, our studies of 167 normals (Murphy *et al.* 1974) did not yield a bimodal distribution pattern, but rather a typical unimodal pattern. It is still possible that these differing results are due to the different substrates and assay methods used. Furthermore, the occurrence of a unimodal distribution pattern does not eliminate the possibility of more than one enzyme form existing; for example, erythrocyte acid phosphatase activities are distributed in a unimodal pattern, but electrophoretic studies have revealed five different enzyme proteins contributing in an overlapping fashion to the unimodal pattern (Harris 1971).

FACTORS INFLUENCING PLATELET MAO ACTIVITY

In addition to genetically based differences, there are a number of other factors which may affect platelet MAO activity. Of particular concern are the factors which may be specifically important to blood cell enzymes and not to enzymes in other tissues, including, for example, issues related to cell age and turnover and to the use of anticoagulant drugs in obtaining blood cell samples.

Sex. A tendency for females to have slightly higher MAO activities than males has been noted by Robinson *et al.* (1971) and has also been observed in our more recent studies with a larger number of young adult subjects, although our earlier data did not clearly identify a sex difference (Murphy & Weiss 1972). Most of the other studies of platelet MAO have either not observed sex differences or have not separated subjects by sex.

Age. Moderately increasing activity with increasing age has been observed in some studies of platelet MAO (Robinson *et al.* 1971). Our data, however, do not indicate more than a very slight increase in MAO activity with age after puberty.

Hormone, nutritional and stress effects. Menstrual cycle-related changes in platelet MAO activity occur during the human and non-human primate menstrual cycle (Belmaker *et al.* 1974; Redmond *et al.* 1975) with the highest enzyme activity occurring just before ovulation and the lowest activity during the post-ovulatory period. Peak-to-trough alterations averaged approximately 20%. Ovariectomy was also associated with elevations in platelet MAO activity (Redmond *et al.* 1975). In male rhesus monkeys, MAO activity was also observed to vary between mating and non-mating seasons, in association with marked changes in plasma testosterone levels (Redmond *et al.* 1976).

Platelet MAO activity was not altered by the ingestion of food (Belmaker *et al.* 1974). Iron-deficiency anaemia was observed to be associated with a marked reduction in platelet MAO activity, which was reversed by treatment with iron (Callender *et al.* 1974). On the basis of animal data, it would be expected that alterations in the intake of riboflavin might also affect the activity of platelet MAO. Marked clinical changes accompanied by alterations in physical activity and psychological 'stress', as in patients studied during periods of severe depression, mania and normality, do not seem to be associated with changes in MAO activity (D. L. Murphy *et al.*, in preparation).

DRUG EFFECTS ON PLATELET MAO ACTIVITY

Human platelet MAO is markedly inhibited both *in vitro* and *in vivo* by a number of drugs used clinically as antidepressant and antihypertensive agents, including tranylcypromine, phenelzine, isocarboxazid and pargyline (Robinson *et al.* 1968). In our studies, phenelzine in doses of 60 mg/day produced approximately 80% inhibition of platelet MAO with steady-state inhibition levels reached after 5–14 days of treatment (D. L. Murphy *et al.*, in preparation). Phenelzine treatment in this study was associated with changes in monoamine metabolism, including an increased urinary excretion of tryptamine and the *O*-methylated catecholamine metabolites normetanephrine and metanephrine and decreased excretion of 3-methoxy-4-hydroxyphenylglycol and vanillylmandelic acid. Other drugs given clinically, including lithium carbonate,

L-tryptophan, L-dopa, prednisone and phenothiazine antipsychotic agents, were not observed to be associated with altered platelet MAO activity. Two other reports, however, have described both increased (Bockar *et al.* 1974) and decreased (Pandey *et al.* 1975) platelet MAO activity during lithium carbonate treatment.

In vitro, a wide range of inhibitory effects was observed both with typical MAO-inhibiting drugs and with some other psychoactive drugs (Donnelly *et al.*, in preparation). Deprenyl, pargyline and tranylcypromine were among the drugs with highest potency against the platelet enzyme, with I_{50}'s of 0.02, 0.04 and 0.2 μM, respectively, when 10^{-3} M-tyramine was used as substrate. The tricyclic antidepressant drugs can also inhibit the platelet enzyme, although higher drug concentrations, 60–400 μM, were required, as Edwards & Burns (1974) have recently observed. D- and L-amphetamine and chlorpromazine were among the drugs requiring extremely high concentrations (500–790 μM) to produce platelet MAO inhibition.

ASSOCIATIONS BETWEEN CLINICAL PSYCHIATRIC DISORDERS AND PLATELET MAO ACTIVITY

Alterations in platelet MAO activity have been described in some patients with affective disorders and schizophrenia. Information available at present does not indicate that any more than a small proportion of these differences can be explained by general factors such as age or sex, or by iron deficiency, abnormalities in thyroid or testosterone states, or drug treatment (D. L. Murphy *et al.*, in preparation). While it may be that genetically based differences in MAO activity are responsible for these associations with psychiatric disorders, continued caution is necessary in reaching this conclusion, as we have repeatedly indicated (Murphy & Wyatt 1972; Murphy *et al.* 1974).

Affective disorders

While platelet MAO activity was no different in 57 patients hospitalized for depression compared to 52 normal controls, a subgroup of the depressed patients who had histories of mania and had been identified as bipolar manic-depressive individuals had significantly reduced MAO levels compared to both the controls and the non-bipolar depressed patients (Murphy & Weiss 1972). This conclusion was based on the mean of 2–5 platelet MAO measurements per patient obtained during a drug-free interval of a minimum of two weeks. In fact, only three of the bipolar patients had received phenothiazines and/or butyrophenones, and only four had received tricyclic antidepressant drugs during the preceding three months.

Somewhat contrasting data were presented in preliminary form by Nies *et al.* (1974) from a study of a large number of inpatient and outpatient depressed patients, including some bipolar patients. They observed a small but significant increase in platelet MAO activity in the depressed patients as a group. They also suggested that the higher incidence of depression in females and in older individuals might be associated with the higher platelet and brain MAO activity observed by them in females and in association with increasing age (Robinson *et al.* 1971).

Schizophrenia

Reduced platelet MAO activity has also been observed in some patients with schizophrenia. Thirty-three chronically hospitalized schizophrenic patients were reported to have markedly reduced mean MAO activities in comparison to controls of similar age and sex (Murphy & Wyatt 1972). Patients who had not received phenothiazines for an average of 30 days had equally marked differences in MAO activity as did those receiving drugs. Chronicity, severity of impairment or diagnostic subclass of schizophrenia was not correlated with MAO activity differences within this patient group.

In contrast, a study of 44 individuals hospitalized because of an acute schizophrenic episode revealed no differences in platelet MAO activity from controls (Carpenter *et al.* 1975). This difference from the results in chronic schizophrenic patients was considered of some interest because of the recently reported lack of an increased incidence of schizophrenia and schizophrenic spectrum disorders in the families of acute schizophrenic patients compared to chronic schizophrenic females (Rosenthal 1971). These patients were studied after a two-week drug-free period; most patients had been receiving antipsychotic drugs for variable but generally brief periods before hospitalization. A small number of patients with diagnoses other than acute schizophrenia or schizo-affective schizophrenia (i.e., paranoid, catatonic and chronic schizophrenia, total $n = 13$) had significantly reduced platelet MAO activity compared to the controls (Carpenter *et al.* 1975). A comparison of the six patients with the lowest MAO activity with the 15 with the highest MAO activity from the entire group of 44 patients did not reveal any significant discriminatory variables. There were some tendencies, however, for the group with low MAO activity to have more severely impaired reality testing, more paranoid and grandiose delusions, better prognostic scores and less restlessness.

Reduced platelet MAO activity in schizophrenic patients compared to controls was also observed by Meltzer & Stahl (1974) who studied four substrates, tryptamine, tyramine, *m*-iodobenzylamine and octopamine. Female patients had significant reductions in MAO with all four substrates, while in male schizophrenic patients only MAO activities measured with tyramine and *m*-iodobenzylamine were significantly different from controls. When the patients were divided into chronic and acute groups, the chronic group manifested reduced MAO activity with all four substrates, while in the acute group only tyramine and *m*-iodobenzylamine, again, were discriminatory. While the patients in this study were receiving antipsychotic drugs, no significant correlation between the dosage of phenothiazine and platelet MAO activity was observed.

Nies *et al.* (1974) studied a small number of schizophrenic patients who had experienced at least two schizophrenic episodes but who had been hospitalized for no more than four months in any one year and no more than a total of one year out of the preceding five years. They also observed differences between substrates in the platelet MAO comparisons, with the schizophrenic patients having a statistically significant reduction in MAO activity when tryptamine but not benzylamine was used as the substrate. However, it is noteworthy that the controls in this study had MAO activities with benzylamine of 22.1 $\pm$ 2.8 nmol/mg protein/h, while previously reported age-matched control values from a large number of individuals studied by the same investigators were 37 nmol/mg protein/h (Nies *et al.* 1974). This raises the question of whether an atypical control population was obtained by chance.

Friedman *et al.* (1974) also studied a mixed group of schizophrenic patients, most experiencing an acute re-exacerbation of symptoms, and did not observe any mean difference from age and sex-matched normal controls in platelet MAO activity, using tryptamine as the substrate. Schizo-affective, chronic undifferentiated and paranoid subtypes also did not show any tendency to differ in MAO activity. The female schizophrenic patients had 32% less platelet MAO activity than the female controls, although this difference was not quite statistically significant.

Shaskan & Becker (1975) reported preliminary data from anergic, depressed, schizophrenic outpatients who were observed to have no differences in platelet MAO activity compared to eight alcoholic and seven normal staff controls. Unfortunately, this study is severely limited by the few normal controls studied, and by their observations of an unusual biphasic distribution of MAO activities in all three of their groups, with no overlap between high and low activity subgroups. Their observation that platelet MAO activity remained stable over five weeks' observation, which included a drug-free baseline period and then treatment for four weeks with either chlorpromazine (100–1200 mg/day) plus imipramine (150 mg/day) or thiothixene (5–60 mg/day), is of interest in providing further information indicating that phenothiazine-related drugs in clinically used amounts do not appear to alter platelet MAO activity.

At the present time, it is not clear precisely how the apparent association between reduced platelet MAO activity and psychiatric symptomatology can best be understood. The majority of studies on this question do point towards some relationship, perhaps most clearly exemplified in the group of chronic schizophrenic patients (Meltzer & Stahl 1974; Murphy & Wyatt 1972; Wyatt *et al.* 1973) and also in the bipolar subgroup of patients with affective disorders (Murphy & Weiss 1972). The direct study of known non-genetic factors affecting platelet MAO (including age, sex, clinical state and such nutritional and hormonal elements as are reflected in plasma levels of iron, triiodothyronine and testosterone [D. L. Murphy *et al.*, in preparation]) has not yielded any evidence which can explain the reduced platelet MAO activity in these groups of patients. In addition, the study of monozygotic twins discordant for schizophrenia (Wyatt *et al.* 1973) suggests that the reduced MAO activity observed in schizophrenic patients is not a consequence of being schizophrenic (or being treated for schizophrenia) but rather may represent or correlate with some vulnerability factor to psychopathology.

The lack of understanding of the molecular basis for reduced MAO activity in those psychiatric patient groups with reduced MAO activity cautions us against attributing these differences to a genetic basis at present. Studies of kinetic parameters (Friedman *et al.* 1974; Murphy & Weiss 1972; Murphy & Wyatt 1972) and of possible dialysable inhibitors (Murphy & Weiss 1972; D. L. Murphy *et al.*, in preparation) have not revealed any differences associated with reduced MAO activity. In addition, schizophrenic patients have not been identified as having reductions in either total MAO or MAO type B activity in brain (Domino *et al.* 1973; Nies *et al.* 1974; Schwartz *et al.* 1974; Wise *et al.* 1974). Since the localization and function of MAO type B in brain and other tissues are not known, however, continued efforts to determine the basis for platelet MAO activity differences are still required.

POSSIBLE ASSOCIATION BETWEEN PLATELET MAO ACTIVITY AND GENERAL PERSONALITY AND BEHAVIOURAL FACTORS

Our studies of platelet monoamine oxidase in psychiatric patients and normals have clearly indicated that there is no simple one-to-one relationship between a particular psychiatric entity or behavioural syndrome and reduced MAO activity. Although it is still possible that there may be more similarities between the bipolar manic-depressive patient and the chronic schizophrenic patient than

Kraepelin and some subsequent psychiatric diagnosticians have suspected (Murphy 1972), there are certainly many differences in overt behaviour between these patient groups which share the most marked reductions in platelet MAO activity. Furthermore, there is a very definite overlap between patients and controls in MAO activity, with 10% of normals having platelet MAO activities as low as the mean of the chronic schizophrenic group. Thus, it is possible to have fairly marked reductions in platelet MAO activity either endogenously or in association with diseases like iron-deficiency anaemia (Callender *et al.* 1974) and migraine (Sandler *et al.* 1974*b*) without having a psychiatric disorder. This point was also evident from our study of identical twins discordant for schizophrenia, which demonstrated that the non-schizophrenic twins had MAO activities as low as the schizophrenic twins (Wyatt *et al.* 1973).

On the basis of the above information, we have suggested that reduced MAO activity might be associated with a predisposition or vulnerability to psychopathology (Wyatt *et al.* 1973; Murphy *et al.* 1974). Consequently, we have begun studies attempting to determine whether any psychological features of normal individuals might be correlated with or might be predicted from platelet MAO activity differences (M. Buchsbaum *et al.*, in preparation; D. L. Murphy *et al.*, in preparation).

Our preliminary results have suggested a greater number of significant correlations than expected on a chance basis between platelet MAO activity and scale scores from the Minnesota Multiphasic Personality Inventory (MMPI) and the Zuckerman Sensation Seeking Scale (SSS) in 95 normal young adults (D. L. Murphy *et al.*, in preparation). In addition, sex-related differences were present in the correlational patterns. A subgroup of male subjects with quite low platelet MAO activities demonstrated a definite increase in many of the MMPI and SSS scores.

Redmond & Murphy (1975) have been pursuing a similar approach, using quantitative behavioural measures obtained over several months' time in both free-ranging and corral rhesus monkeys as variables for correlational studies with platelet MAO activity. In preliminary results from this study (Redmond & Murphy 1975) the highest positive correlation with platelet MAO activity in both sexes was with 'time spent alone' ($r = 0.49$, $P < 0.001$, $n = 43$). Moderately high positive correlations were also found for inactivity ($r = 0.32$, $P < 0.05$) and passivity ($r = 0.31$, $P < 0.05$) measures. Significant negative correlations were found for some corresponding behaviours, including ambulatory movement ($r = -0.33$, $P < 0.05$), social contact ($r = -0.31$, $P < 0.05$) and play in males ($r = -0.66$, $P < 0.01$).

The conclusions from these two studies need to be regarded as tentative because they derive from correlational data. Nonetheless, the possibility that platelet MAO activity might reflect not only some proclivity to psychopathology, but also some aspects of personality function and behaviour in normal young adults and in non-human primates studied under natural conditions, is certainly of interest. While very speculative, it is possible that some 'normal' personality characteristics and behaviour could be related to the consequences of genetic and non-genetically based differences in monoamine oxidase activity as they affect the function of biogenic amines at synapses.

References

BELMAKER, R., MURPHY, D. L., WYATT, R. J. & LORIAUX, L. (1974) Human platelet monoamine oxidase changes during the menstrual cycle. *Arch. Gen. Psychiatry 31*, 553-556

BOCKAR, J., ROTH, R. & HENINGER, G. (1974) Increased human platelet monoamine oxidase during lithium carbonate therapy. *Life Sci. 15*, 2109-2118

CALLENDER, S., GRAHAME-SMITH, D. G., WOODS, H. F. & YOUDIM, M. B. H. (1974) Reduction of platelet monoamine oxidase activity in iron deficiency anaemia. *Br. J. Pharmacol. 52*, 447-448

CARPENTER, W. T., MURPHY, D. L. & WYATT, R. J. (1975) Platelet monoamine oxidase activity in acute schizophrenia. *Am. J. Psychiatry 132*, 438-441

COLLINS, G. G. S. & SANDLER, M. (1971) Human blood platelet monoamine oxidase. *Biochem. Pharmacol. 20*, 289-296

DOMINO, E. F., KRAUSE, R. R. & BOWERS, J. (1973) Various enzymes involved with putative transmitters. *Arch. Gen. Psychiatry 29*, 195-201

EDWARDS, D. J. & BURNS, M. O. (1974) Effects of tricyclic antidepressants upon platelet monoamine oxidase. *Life Sci. 15*, 2045-2058

FRIEDMAN, E., SHOPSIN, B., SATHANANTHAN, G. & GERSHON, S. (1974) Blood platelet MAO activity in psychiatric patients. *Am. J. Psychiatry 131*, 1392-1394

HARRIS, H. (1971) *The Principles of Human Biochemical Genetics*, p. 140, American Elsevier, New York

JOHNSTON, J. P. (1968) Some observations upon a new inhibitor of monoamine oxidase in brain tissue. *Biochem. Pharmacol. 17*, 1285-1297

MELTZER, H. Y. & STAHL, S. M. (1974) Platelet monoamine oxidase activity and substrate preferences in schizophrenic patients. *Res. Commun. Chem. Pathol. Pharmacol. 7*, 419-431

MURPHY, D. L. (1972) L-Dopa, behavioral activation and psychopathology, in *Neurotransmitters* (Kopin, I. J., ed.), Res. Publ. A. R. N. M. D. *50*, 472-493

MURPHY, D. L. (1973) Technical strategies for the study of catecholamines in man, in *Frontiers in Catecholamine Research* (Usdin, E. & Snyder, S., eds.), pp. 1077-1082, Pergamon Press, Oxford

MURPHY, D. L. & DONNELLY, C. H. (1974) Monoamine oxidase in man: enzyme characteristics in platelets, plasma, and other human tissues, in *Neuropsychopharmacology of Monoamines and Their Regulatory Enzymes* (Usdin, E., ed.), pp. 71-85, Raven Press, New York

MURPHY, D. L., BELMAKER, R. & WYATT, R. J. (1974) Monoamine oxidase in schizophrenia and other behavioral disorders. *J. Psychiatr. Res. 11*, 221-248

MURPHY, D. L. & WEISS, R. (1972) Reduced monoamine oxidase activity in blood platelets from bipolar depressed patients. *Am. J. Psychiatry 128*, 1351-1357

MURPHY, D. L. & WYATT, R. J. (1972) Reduced monoamine oxidase activity in blood platelets from schizophrenic patients. *Nature (Lond.) 238*, 225-226

NIES, A., ROBINSON, D. S., LAMBORN, K. R. & LAMPERT, R. P. (1973) Genetic control of platelet and plasma monoamine oxidase activity. *Arch. Gen. Psychiatry 28*, 834-838

NIES, A., ROBINSON, D. S., HARRIS, L. S. & LAMBORN, K. R. (1974) Comparison of monoamine oxidase substrate activities in twins, schizophrenics, depressives, and controls, in *Neuropsychopharmacology of Monoamines and Their Regulatory Enzymes* (Usdin, E., ed.), pp. 59-70, Raven Press, New York

PAASONEN, M. K. & SOLATUNTURI, E. (1965) Monoamine oxidase in mammalian blood platelets. *Ann. Med. Exp. Biol. Fenn. 43*, 98-100

PANDEY, G. N., DORUS, E. B., DEKIRMENJIAN, H. & DAVIS, M. M. (1975) Effect of lithium treatment on blood COMT and platelet MAO in normal human subjects. *Fed. Proc. 34*, 778

REDMOND, D. E., JR & MURPHY, D. L. (1975) Behavioral correlates of platelet monoamine oxidase (MAO) activity in Rhesus monkeys. *Psychosom. Med. 37*, 80

REDMOND, D. E., MURPHY, D. L., BAULU, J., ZIEGLER, M. G. & LAKE, C. R. (1975) Menstrual cycle and ovarian hormone effects on plasma and platelet monoamine oxidase (MAO) and plasma dopamine-beta-hydroxylase (DBH) activities in the Rhesus monkey. *Psychosom. Med. 37*, 417

REDMOND, D. E., BAULU, J., MURPHY, D. L., LORIAUX, D. L., ZIEGLER, M. G. & LAKE, C. R. (1976) The effects of testosterone on plasma and platelet monoamine oxidase (MAO) and plasma dopamine-B-hydroxylase activities in the male Rhesus monkey. *Psychosom. Med.* in press

ROBINSON, D. S., LOVENBERG, W., KEISER, H. & SJOERDSMA, A. (1968) Effects of drugs on human blood platelet and plasma amine oxidase activity *in vitro* and *in vivo*. *Biochem. Pharmacol. 17*, 109-119

ROBINSON, D. S., DAVIS, J. M., NIES, A., RAVARIS, C. L. & SYLWESTER, D. (1971) Relation of sex and aging to monoamine oxidase activity of human brain, plasma, and platelets. *Arch. Gen. Psychiatry 24*, 536-539

ROSENTHAL, D. (1971) Two adoption studies of heredity in the schizophrenic disorders, in *The Origin of Schizophrenia* (Bleuler, M. & Angst, J., eds.), pp. 21-34, Verlag Hans Huber, Bern

SANDLER, M., CARTER, S. B., GOODWIN, B. L., RUTHVEN, C. R. J., YOUDIM, M. B. H., HANINGTON, E., CUTHBERT, M. F. & PARE, C. M. B. (1974*a*) Multiple forms of monoamine oxidase: some *in vivo* correlations, in *Neuropsychopharmacology of Monoamines and Their Regulatory Enzymes* (Usdin, E., ed.), pp. 3-10, Raven Press, New York

SANDLER, M., YOUDIM, M. B. H. & HANINGTON, E. (1974*b*) A phenylethylamine oxidizing defect in migraine. *Nature (Lond.) 250*, 335-337

SCHWARTZ, M. A., WYATT, R. J., YANG, H.-Y. T. & NEFF, N. H. (1974) Multiple forms of brain monoamine oxidase in schizophrenic and normal individuals. *Arch. Gen. Psychiatry 31*, 557-560

SHASKAN, E. G. & BECKER, R. E. (1975) Platelet monoamine oxidase in schizophrenics. *Nature (Lond.) 253*, 659-660

WISE, C. D., BADEN, M. H. & STEIN, L. (1974) Postmortem measurements of enzymes in human brain: evidence of a central noradrenergic deficit in schizophrenia. *J. Psychiatr. Res. 11*, 221-248

WYATT, R. J., MURPHY, D. L., BELMAKER, R., COHEN, S., DONNELLY, C. H. & POLLIN, W. (1973) Reduced monoamine oxidase activity in platelets: a possible genetic marker for vulnerability to schizophrenia. *Science (Wash. D.C.) 179*, 916-918

[for discussion of this paper, see pp. 363-369]

An investigation of platelet monoamine oxidase activity in schizophrenia and schizo-affective psychosis

I. BROCKINGTON, T. J. CROW, EVE C. JOHNSTONE and F. OWEN

Division of Psychiatry, Clinical Research Centre, Northwick Park Hospital, Harrow

Abstract Platelet monoamine oxidase (MAO) activity was investigated in a series of 56 male patients, age range 25 to 80 years, suffering from chronic schizophrenia. All of them required long-term hospitalization and none of them had been treated with neuroleptic drugs in the previous six months. Platelet MAO was assayed by the radiometric method of Robinson using [^{14}C]tyramine and [^{14}C]tryptamine as substrates. Age-matched controls were selected from a population of male patients attending for routine medical examination. Platelet MAO activity was also assessed in 55 male and female patients who had suffered from schizo-affective illnesses diagnosed by standardized Present State Examination criteria.

Comparison of the platelet MAO activities of the patients with chronic schizophrenia with the controls revealed no significant differences. Patients were rated on various clinical characteristics and subdivided according to the presence of positive or negative symptoms. There was no significant difference between the mean MAO levels in the two groups or between each separate group and the control group. The values of platelet MAO were compared in a group of patients who were given a six-month course of neuroleptic medication and in a group who remained drug-free. There was a significant trend towards an increase in MAO levels after medication.

Platelet MAO activities in the group of patients with schizo-affective illnesses were significantly higher than in the controls, but this difference could be accounted for by the higher levels in female patients.

It is concluded that low platelet MAO activity is not a necessary feature of chronic schizophrenic illnesses, and is unlikely to be a genetic marker for the disease.

Investigations of the possible role of disturbances of monoamines in psychiatric illness have been plagued by the difficulty of obtaining good indices of monoamine function in the central nervous system. For the reason that monoamine oxidase (MAO) activity in platelets might in some way reflect that in the central nervous system, investigations of this parameter have been carefully pursued in psychiatric disease. The report of Murphy & Weiss (1972) that MAO activity is somewhat reduced in depressed patients with bipolar affective illness, and the subsequent report (Murphy & Wyatt 1972) of a greater reduction in young patients (mean age 27.9 years) suffering from schizophrenic illnesses, aroused considerable interest. The findings with respect to schizophrenia were replicated in substance by Meltzer & Stahl (1974), Nies *et al.* (1974) and Owen *et al.* (1974). In a study of twins, where one of the pair suffered from schizophrenic illness, Wyatt *et al.* (1973) found the lowest levels in the psychotic twin, but levels which were lower in the non-psychotic twin than in their control group. They concluded that low platelet MAO activity might be a genetic marker for the disease. On the other hand, Shaskan & Becker (1975), in a small series of schizophrenic out-patients, and Friedman *et al.* (1974), in a group of young hospitalized patients, found no differences in platelet MAO between their groups of patients suffering from schizophrenic illnesses and the controls. Carpenter *et al.* (1975) found normal levels in 44 patients with acute schizophrenic illnesses and suggested that low MAO levels might be characteristic of the group of patients who progress to chronic schizophrenia.

In view of these differences and the interest of the problem, we have investigated platelet MAO in two somewhat unusual populations—a group of patients with long-standing schizophrenic illnesses untreated with neuroleptic medication, and a carefully selected group of patients suffering from schizo-affective illnesses.

METHODS

Subjects

Fifty-six patients suffering from schizophrenia (age range 25 to 80 years; mean age 56.6 $\pm$ 12.3 years) were selected from the long-stay wards of an area mental hospital on the grounds that they had not been treated with neuroleptic medication in the previous six months. Most had been drug-free for many years as a consequence of a general emphasis on milieu methods of treatment and care. Diagnostic selection was based on the criteria proposed by Feighner *et al.* (1972) for identifying a core group of patients with undoubted schizophrenic illnesses, and an assessment of current mental state was made using the rating scales of Goldberg. Seventy age-matched controls (mean age 47.9 $\pm$ 11.2 years) were selected from a population of patients attending for an insurance medical examination.

In the second investigation 55 patients (mean age 43.2 $\pm$ 16.5 years) suffering from schizo-affective illnesses were selected on the grounds that the clinical

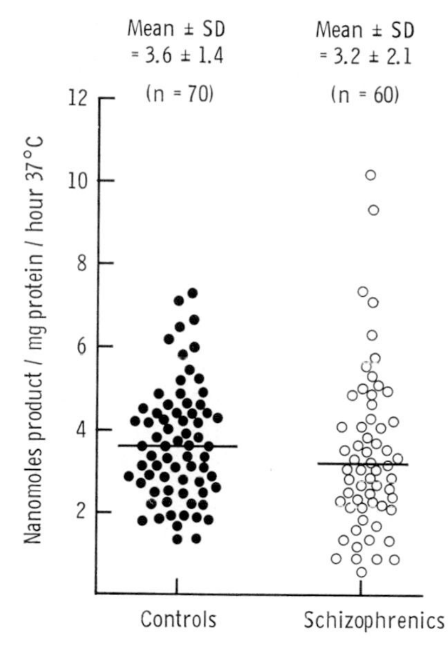

FIG. 1. Platelet monoamine oxidase activity of controls and schizophrenics, when tryptamine was used as substrate.

picture met the criteria for *both* a schizophrenic or paranoid psychosis *and* a manic or depressive psychosis, according to symptoms and signs as defined in the glossary of the Present State Examination, 9th edition (Wing *et al.* 1974).

Biochemical techniques

Twenty ml of venous blood were collected into ACD for each subject. The platelets were separated by differential centrifugation and incubated at pH 7.2 for one hour at 37 °C with radioactive substrate ([^{14}C]tyramine or [^{14}C] tryptamine). The radioactive products were solvent extracted at pH 1 and quantified by scintillation counting. Results are expressed as nanomoles of product formed per mg of platelet protein per hour. This technique is closely similar to that of Robinson *et al.* (1968). Serum iron and iron-binding capacities were measured in 38 of the patients in the first group.

RESULTS

Unmedicated schizophrenic patients

Mean platelet MAO activities did not differ between patients and controls when tryptamine (Fig. 1) was used as substrate. The values with tyramine

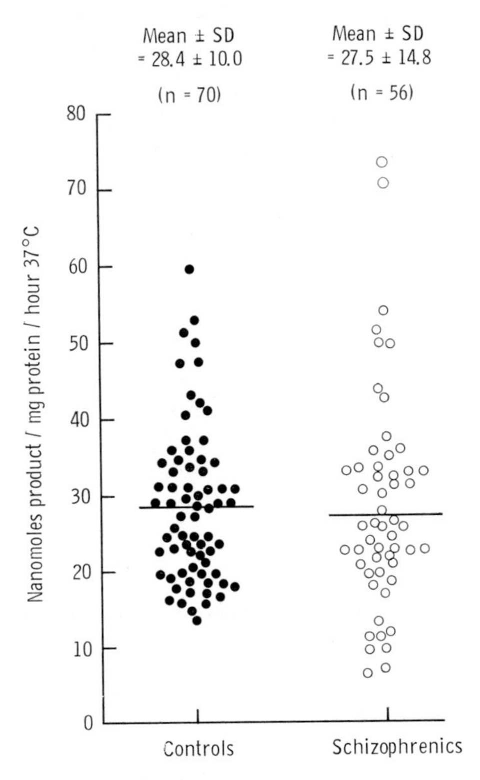

FIG. 2. Platelet monoamine oxidase activity of controls and schizophrenics, when tyramine was used as substrate.

were similar (Fig. 2); however, the range of values in the patients was wide, and with each substrate there were eight patients whose values fell below, and two above, the range of control values. When the patient group was subdivided according to the presence of positive symptoms (thought disorder, hallucinations or delusions) or negative symptoms (flattening of affect, mutism) there was no significant difference between the mean levels in the two groups, or between each separate group and the control group (Fig. 3 shows the results for tyramine; the findings with tryptamine were similar). In 38 patients serum iron was found to be 18.5 $\pm$ 5.8 μmol/litre and iron-binding capacity 64.5 $\pm$ 10.3 μmol/litre; within this group there was no relationship between either of these parameters and platelet MAO activity. Neuroleptic medication was subsequently begun for eight patients (seven on depot injections of flupenthixol and one subject on fluphenazine decanoate 25 mg, three times weekly). The values of platelet MAO activities in these patients after a mean of six months

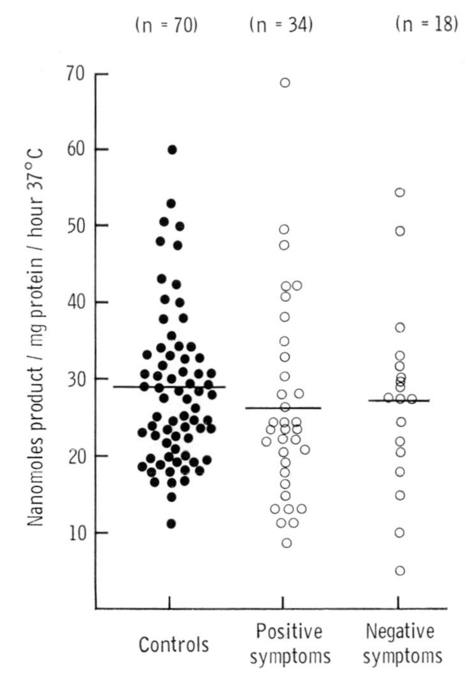

FIG. 3. Relationship between platelet monoamine oxidase activity (tyramine as substrate) and psychiatric state.

on medication were compared with those in a group of nine patients who remained free of medication for a similar period (Fig. 4). There is considerable variation with time in the unmedicated group, but a trend towards an increase in MAO levels after medication ($P < 0.01$; paired t-test). Analysis by age revealed no change with increasing age in either the patients or control groups (Fig. 5).

Schizo-affective patients

In the 55 patients with schizo-affective illnesses the scatter of values was considerably greater than in either the control group or in patients with chronic schizophrenia, and the mean was significantly greater ($P < 0.001$, Fig. 6). However, this greater mean activity was due to the higher platelet MAO activities found in the female patients of the schizo-affective group ($P < 0.05$, Fig. 7). The platelet MAO activity of the male schizo-affectives (mean $\pm$ S.D, 33.9 $\pm$ 18.1 nanomoles product/mg protein/hour) did not differ significantly from either the control or the schizophrenic groups. The

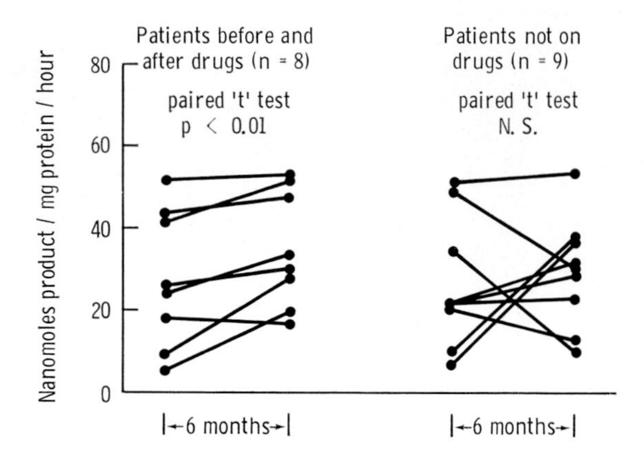

FIG. 4. Effects of neuroleptics on platelet monoamine oxidase activity of schizophrenics.

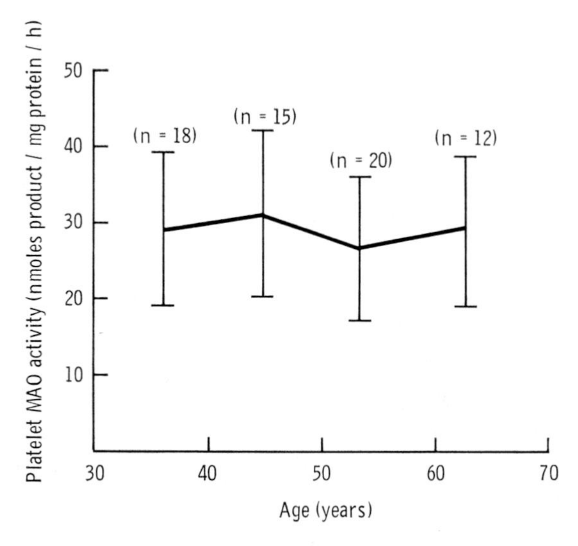

FIG. 5. Relationship between platelet monoamine oxidase activity (tyramine as substrate) and age.

finding of increased platelet MAO in females compared with males is in agreement with the work of Robinson *et al.* (1971). An analysis of the schizo-affective group by the predominant type of affective symptomatology—that is, 'schizo-depression' versus 'schizomania'—revealed no significant differences in platelet MAO activities between the two sub-groups (Fig. 8) although there was some evidence of reduced platelet MAO activity in male depressives compared with male manics ($P < 0.05$).

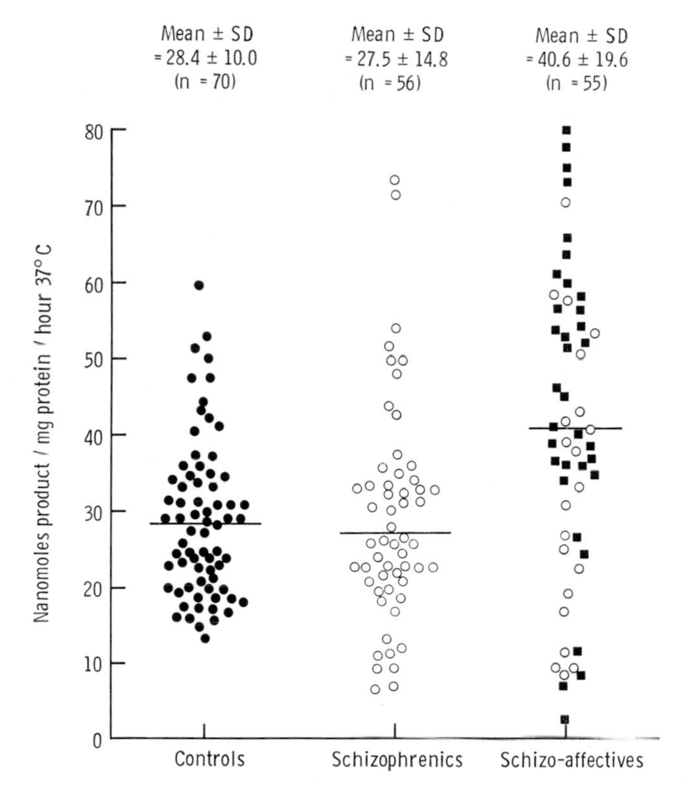

FIG. 6. Platelet monoamine oxidase activity of controls, schizophrenics and schizo-affectives (tyramine as substrate; ○, males; ■, females).

DISCUSSION

The findings in the group of patients with schizophrenia are in contrast with those of some workers (Murphy & Wyatt 1972; Meltzer & Stahl 1974; Nies *et al.* 1974; Owen *et al.* 1974) but in agreement with two more recent studies (Friedman *et al.* 1974; Shaskan & Becker 1975). Therefore we have a situation in which some workers find highly significant differences between their control and patient populations and others do not. The factors which may account for these differences require careful examination.

Our patients were not treated with neuroleptic drugs, an unusual situation with chronic schizophrenic patients, and indeed they were selected on this basis. Meltzer & Stahl report most of their patients as being on phenothiazines, but Murphy & Wyatt found platelet MAO to be as low in patients from whom medication had been withdrawn for a mean of 30 days as in those on medication. Shaskan & Becker give no details of their patients' medication prior to a two-

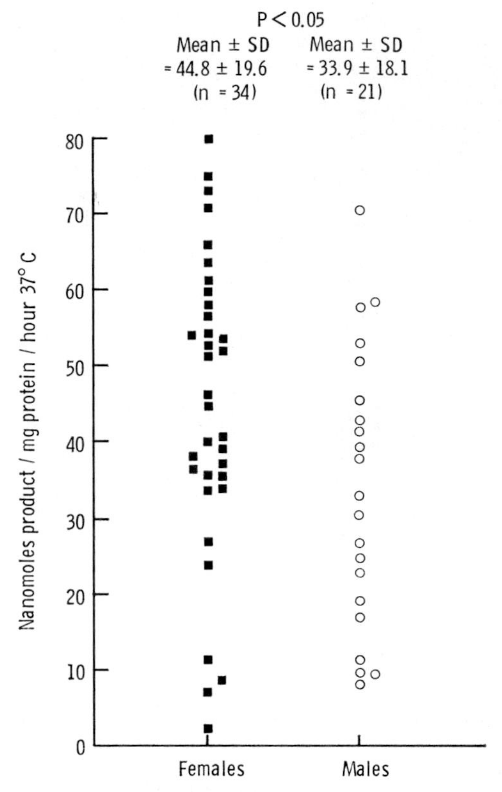

FIG. 7. Platelet monoamine oxidase activity of female and male schizo-affectives (○, males, age 42.2 ± 16.4 years; ■, females, age 43.8 ± 16.8 years).

week drug-free period. Long-term medication therefore may be a possible difference between the present and previous studies, but our limited study of the effects of six months' treatment with neuroleptics does not suggest that such treatment lowers platelet MAO activity. On the other hand, one must consider that by studying a drug-free population we have selected an atypical sample. This is a possibility, although we think that the absence of medication was determined more by the therapeutic environment than by the patients' state; but we have not yet studied a comparable drug-treated population.

Murphy & Wyatt (1975) suggest that low platelet MAO activity may be a characteristic not of acute schizophrenic illnesses, but of illnesses which progress to chronicity. Our findings do not support such a hypothesis; nor, in so far as we have studied a representative sample of the total schizophrenic population, do they support the concept of low platelet MAO activity as a genetic marker for the disease.

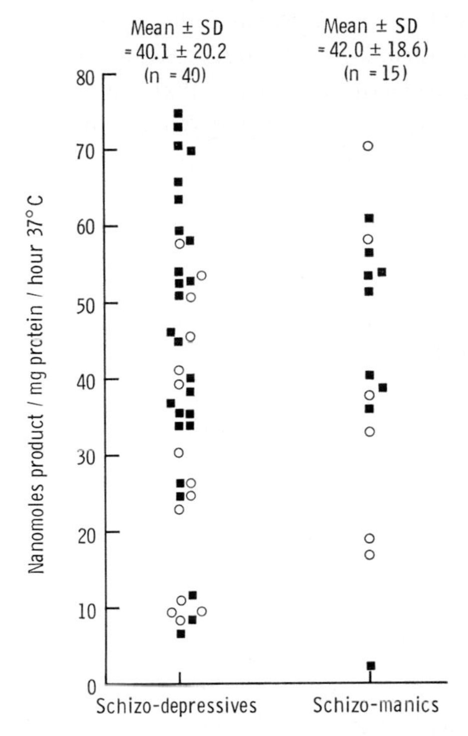

FIG. 8. Comparison of platelet MAO activity of schizo-depressives and schizo-manics (○, males; ■, females).

A second obvious difference between our own and previous studies is age. The mean age of our series was 56.6 ± 12.3 years; that of Murphy & Wyatt's was 27.9 ± 1.8 and of Meltzer & Stahl's, 29 ± 5.4 years. Again, however, within our own series we found no significant changes in platelet MAO activity with age and no differences between the schizophrenics and controls in the younger age groups.

Other factors known to affect platelet MAO activity include sex (Robinson *et al.* 1971), phase of the menstrual cycle (Belmaker *et al.* 1974) and iron deficiency (Youdim *et al.* 1975). Low levels have been reported in patients suffering from migraine (Sandler *et al.* 1974). These variables are unlikely to account for the differences between our present and some previous findings. It seems likely that there are further determinants of platelet MAO activity yet to be identified.

Our findings in a small group of schizophrenic patients untreated with drugs suggest that in psychiatric patients at least platelet MAO activity varies with time. This point is emphasized particularly in a single case study (Cookson *et al.* 1975) in which very low platelet MAO levels were seen at the height of an acute psychosis but the levels returned to normal with recovery. At the present time, therefore, we are unable to go beyond the conclusion that the factors leading to the wide reported variations in platelet MAO in psychiatric disease have yet to be determined.

References

BELMAKER, R. H., MURPHY, D. L. & WYATT, R. J. (1974) Human platelet monoamine oxidase changes during the menstrual cycle. *Arch. Gen. Psychiatry 31*, 553-556

CARPENTER, W. T., MURPHY, D. L. & WYATT, R. J. (1975) Platelet monoamine oxidase activity in acute schizophrenia. *Am. J. Psychiatry 132*, 438-441

COOKSON, I. B., OWEN, F. & RIDGES, A. P. (1975) Platelet monoamine oxidase activity during the course of a schizophreniform psychosis. *Psychol. Med. 5*, 314-316

FEIGHNER, J. P., ROBINS, E., GUZE, S. B., WOODRUFF, R. A., WINOKUR, G. & MUNOZ, R. (1972) Diagnostic criteria for use in psychiatric research. *Arch. Gen. Psychiatry 26*, 57-63

FRIEDMAN, E., SHOPSIN, B., SATHANANTHAN, G. & GERSHON, S. (1974) Blood platelet monoamine oxidase activity in psychiatric patients. *Am. J. Psychiatry 131*, 1392-1394

MELTZER, H. Y. & STAHL, S. M. (1974) Platelet monoamine oxidase activity and substrate preferences in schizophrenic patients. *Res. Commun. Chem. Pathol. Pharmacol. 7*, 419-431

MURPHY, D. L. & WEISS, R. (1972) Reduced monoamine oxidase activity in blood platelets. from bipolar depressed patients. *Am. J. Psychiatry 128*, 1351-1357

MURPHY, D. L. & WYATT, R. J. (1972) Reduced monoamine oxidase activity in blood platelets from schizophrenic patients. *Nature (Lond.) 238*, 225-226

MURPHY, D. L. & WYATT, R. J. (1975) Platelet monoamine oxidase in schizophrenics. *Nature (Lond.) 253*, 659-660

NIES, A., ROBINSON, D. S., HARRIS, L. S. & LAMBORN, K. R. (1974) Comparison of MAO substrate activities in twins, schizophrenics, depressives and controls. *Psychopharmacol. Bull. 10*, 10-11

OWEN, F., COOKSON, I. B. & RIDGES, A. P. (1974) Platelet monoamine oxidase activity—a genetic marker for schizophrenia? *Acta Genet. Med. Gemellol. (Roma) 23*, 371-376

ROBINSON, D. S., DAVIS, J. M., NIES, A., RAVARIS, C. L. & SYLWESTER, D. (1971) Relation of sex and aging to monoamine oxidase activity of human brain, plasma, and platelets. *Arch. Gen. Psychiatry 24*, 536-539

ROBINSON, D. S., LOVENBERG, W., KEISER, H. & SJOERDSMA, A. (1968) Effect of drugs on human blood platelet and plasma amine oxidase activity *in vitro* and *in vivo*. *Biochem. Pharmacol. 17*, 109-119

SANDLER, M., YOUDIM, M. B. H. & HANINGTON, E. (1974) A phenylethylamine oxidising defect in migraine. *Nature (Lond.) 250*, 335-337

SHASKAN, E. G. & BECKER, R. E. (1975) Platelet monoamine oxidase in schizophrenics. *Nature (Lond.) 253*, 659-660

WING, J. K., COOPER, J. E. & SARTORIUS, N. (1974) *The Measurement and Classification of Psychiatric Symptoms*, Cambridge University Press, London

WYATT, R. J., MURPHY, D. L., BELMAKER, R., COHEN, S., DONNELLY, C. H. & POLLIN, W. (1973) Reduced monoamine oxidase activity in platelets: a possible genetic marker for vulnerability to schizophrenia. *Science (Wash. D.C.) 179*, 916-918

YOUDIM, M. B. H., WOODS, H. F., MITCHELL, B., GRAHAME-SMITH, D. G. & CALLENDER, S. (1975) Human platelet monoamine oxidase activity in iron-deficiency anaemia. *Clin. Sci. Mol. Med. 48*, 289-295

Discussion of the two preceding papers

Youdim: There are a number of other enzymes in the platelet, including succinate dehydrogenase. You have shown that this enzyme did not change in its activity in schizophrenia, Dr Murphy. Since this enzyme is in the inner membrane and monoamine oxidase in the outer membrane, is the effect specific for the outer mitochondrial membrane enzymes?

Murphy: We would have liked to have measured another outer mitochondrial membrane enzyme, but we did not find enough activity to permit the use of other marker enzymes with the small amount of tissue we had.

Roth: I noticed that monoamine oxidase levels appeared to be correlated with the amount of activity, and I wonder what the effect would be of partialling out the contribution of activity in this heterogeneous group of patients? Another variable that would differ in severity between such groups would be anxiety.

Murphy: In terms of activity, I imagine you are thinking of the characteristics of the individual over long periods of time. We have not seen changes in platelet MAO in the same individuals studied while acutely manic as against depressed, or acutely psychotic as against improved. The correlation between physical activity and MAO levels was something that was found in the study of the rhesus monkey, in that the activity dimension was one that correlated inversely with platelet MAO. Monkeys with lower MAO had more activity, more social contact, and, in males, more play activity. These are only correlations, and this is the weakness of the approach, but it is interesting. So far, in our studies of psychiatric patients we have no information on correlations of MAO with anxiety.

Roth: Floridly psychotic patients may be much more active in unobtrusive ways than neurotics; seemingly immobile patients often prove to have a great deal of muscular activity.

Crow: In our studies we found no correlation of platelet MAO activities with anxiety, with our somewhat crude rating scale methods.

van Praag: Dr Murphy, is anything known about the relationship between platelet MAO and MAO in the CNS? Secondly, what is the functional significance of a 40% or 50% reduction in peripheral MAO; did you measure acid monoamine metabolites in urine, with or without loading with monoamines? Do you know anything about the c.s.f. values of monoamine metabolites in connection with decreased MAO activity? Can you measure MAO in human c.s.f.?

Murphy: The studies made so far have shown no differences in brain MAO in any of the chronic schizophrenic patient groups, including the study in which Dr Neff was involved (Schwartz *et al.* 1974) which looked at A and B substrates and showed no differences in brain MAO with phenylethylamine or 5-hydroxytryptamine as substrates. Nobody has yet reported platelet MAO and brain MAO activities measured in the same individual, although Dr Oreland and also our group are examining this issue in humans and animals now. Until we know that, and we know more about who the different patient groups are, I don't think we can readily cross-relate the two sets of findings from platelet and brain.

In regard to urinary and c.s.f. metabolites, a couple of things have been looked at. For example, three-hour urinary MHPG excretion in normal subjects did not correlate with platelet MAO (Belmaker *et al.* 1975) but, given the fact that platelet MAO is a B type enzyme, and catecholamine metabolism is only partially sensitive to MAO inhibition, that result is not surprising. Most of the metabolites that are measurable in c.s.f. (5HIAA, HVA) are also A type or mixed MAO metabolites, so until we have measures of, for example, phenylethylamine metabolites, we may not be able to easily relate differences in platelet enzyme activity to levels of metabolites in the c.s.f.

The detection of a small amount of MAO activity in c.s.f. has been reported, but I would presume that it is much more likely to be the soluble, plasma-type amine oxidase rather than mitochondrial MAO.

van Praag: On the MMPI, what is your feeling about the correlation between what clinicians tend to diagnose in behavioural disorders, and the outcome of the MMPI profile? In my opinion, the correlation is rather unconvincing.

Murphy: The question of a convincing relationship with clinical state is one reason why we were interested in using the MMPI. We found marked differences between unipolar and bipolar depressed patients in their MMPI characteristics (Donnelly & Murphy 1973). The patients rated as quite severely depressed among the bipolar group on our behavioural scales show relatively few differences from normals on the MMPI, while unipolar depressed patients have quite markedly pathological MMPI profiles. The point is that the MMPI has many more dimensions than a psychiatric diagnosis by which individual differences can be detected, and there are many reports of differences between various psychiatric groups and their responses to psychiatric treatment on MMPI measures. We thought it might be an interesting preliminary way of looking at some personality features in normals to use the MMPI for a correlational study of individual differences in platelet MAO activity.

Coppen: What is the current consensus of opinion on the clinical effects of MAO inhibitors in schizophrenia? I recall that MAO inhibitors given alone were rather harmful in schizophrenia.

Murphy: The earlier literature suggested that they produced activation in chronic schizophrenic patients, which in a few patients was seen as therapeutic, but in others as detrimental because of increased agitation and psychosis. There are a few more recent case reports suggesting that giving MAO inhibitors together with phenothiazines may be of benefit.

Youdim: The question of diet was mentioned. We have obtained some interesting results in rats by accident. Recently, we were studying strain differences in MAO and phenylalanine hydroxylation in groups of highly inbred Wistar rats from a number of laboratories. They were put on the same diet for three weeks before enzyme activities were measured. Our preliminary studies showed significant differences between the groups of rats. When these experiments were repeated with another group of the same rats we could not find the same differences. It was brought to our attention that the diet had been changed. We are re-investigating this problem further.

Kety: Were there any gross differences between the two diets, for example in protein content or type?

Youdim: I am not sure.

Murphy: On the issue of diet, we studied patient groups in two hospitals, the Clinical Center and Saint Elizabeths Hospital, and found low MAO activity in patients with bipolar depressions and chronic schizophrenia, but normal activity in the acute schizophrenic group and in unipolar patients in the same Clinical Center wards, so we can't say that it is a dietary factor unless the patient groups on the same wards are eating very different things. That is not inconceivable, but not very likely, in my view, having been close to those patients.

Kety: Were any of your chronic schizophrenic patients with low platelet MAO values in the Clinical Center?

Murphy: Yes. This was a subgroup of the group we initially reported on. Seven or eight patients were on a Clinical Center ward and were included in that group, and they had low platelet MAO activity.

Kety: Had they been living there for some time?

Murphy: Yes. They weren't included in the subsequent study.

Green: Have you looked at racial differences, since the population of Saint Elizabeths Hospital is almost entirely black?

Murphy: Only in a limited way; we have no evidence, among either the normals or the patient groups, of racial differences in platelet MAO levels.

Kety: Dr Murphy, how do you account for the difference between your result and Dr Crow's?

Murphy: I don't know! I would like to know more about your first group, Dr Crow, and also about possible population differences between the two groups of patients. Were they in different hospitals?

Crow: The first group I mentioned was a group of schizophrenic patients studied in Liverpool by Dr F. Owen and Dr I. B. Cookson. In our present series the first group was a chronic inpatient group in Shenley Hospital in North-West London. The second group, of 'schizo-affective' patients, was a mixture of inpatients and outpatients attending the Maudsley Hospital, London.

van Praag: Dr Crow, you rather tended to generalize about drugs in schizophrenia. I think you were referring to neuroleptics. I don't think one is justified in generalizing in biochemical terms about neuroleptics, because they are different in their actions, certainly on catecholamine metabolism (van Praag 1975; van Praag *et al.* 1975; van Praag & Korf 1975).

Crow: This point is well taken. In fact, we ascertained that the patients had not been receiving neuroleptic drugs for the previous six months and mostly had not been on drugs at all, because of a feeling in these particular wards that drugs should be avoided where possible. In our study of the effects of drugs on platelet MAO the drugs given were mostly flupenthixol, with one patient receiving fluphenazine.

Murphy: I think one has to ask about the fact that this large number of people were being maintained without drugs; but perhaps there were good reasons why they were? The fact that some were put on drugs in your study suggests that they probably might have been treated with drugs under other circumstances, so I don't think that explains the difference.

Crow: This is obviously a valid point. Have we selected a particular population who did not need drug therapy because they were less severely ill? Our impression was that the fact that they were off drugs was much more a function of the particular wards they were on than of clinical severity. The evidence that neuroleptic drugs are beneficial in long-stay inpatients is somewhat equivocal (see e.g. Letemendia & Harris 1967) and there was a general emphasis in these wards on the therapeutic community approach. The patients who subsequently went on neuroleptic medication did so partly because of a change in the consultant in charge, and partly because of the introduction of a therapeutic trial involving some of the patients. Our impression was that the total group of patients included many who were as disordered as drug-treated patients elsewhere in the same institution, but we cannot exclude the possibility that there is a bias in our study against the inclusion of the more severely ill patient.

On the other hand, our diagnostic criteria (the Feighner criteria and the Wing P. S. E.) have been stringent and we have little doubt that the patients in both series would be considered to be suffering from schizophrenic illnesses by all psychiatric observers. Yet their platelet MAO activities are the same as those of the controls.

Murphy: I am somewhat impressed by the general consistency of the results in the other studies. I don't know whether this is an initial halo effect; I was prepared to see a more mixed bag of results. Psychiatric patient populations are, as we heard earlier, diagnosed in different ways in different centres. There are a lot of issues in the clinical study of psychiatric patient populations that will always plague attempts to establish biological correlates. That is why we think that the approach in terms of normal behaviour suggests more convincingly that there are some relationships between the MAO activities we can measure in platelets and behaviour.

Kety: You have done studies on the families of schizophrenics. What were the findings there?

Murphy: Essentially we find, as would be expected from our twin data, that platelet MAO levels definitely are similar within families. We found some families with multiple psychiatric disorders and low platelet MAO activity, and diagnoses of all sorts, ranging from psychotic depression to schizo-affective states, in different family members (Murphy *et al.* 1974). These studies are still in progress and we have not defined the mode of inheritance of platelet MAO.

Kety: In the families where the platelet MAO levels are low, are they low in a general sense, or are there a few individuals with low values while the others are normal?

Murphy: MAO activities tended to be low in all individuals in such families. It is again a mix, and so far there hasn't been any simple relationship between MAO and clinical diagnoses within families. Our concept certainly fits with the comments that Sir Martin Roth made in his paper. Our view of how we might think about reduced MAO is very much in terms of a risk or vulnerability factor, something that might be part of the background of the individual, which might then, in relation to various life circumstances, make that individual more liable to psychiatric disorders of various types. Dr Kety has already expressed views similar to this in discussions of his adoption studies (Kety *et al.* 1971). We feel that a 'contributory but not determining' vulnerability factor is the most reasonable hypothesis for the apparent association between platelet MAO activity and psychiatric disorders.

Oreland: We have been thinking along similar lines, in trying to interpret our results on MAO activity in the brains of suicides (p. 338). There might be families or individuals who have a genetically weak monoaminergic system reflected in low MAO activities, perhaps in both platelets and brain, and one expression of such a weak system might be 'suicidal behaviour' or other psychiatric disease.

Sandler: You yourself, Dr Kety, have drawn attention over the years to the possible importance of diet and gut flora in assessing institutionalized patients. I mentioned earlier that germ-free chicks have a higher hepatic MAO (p. 98). Some such influence may explain the findings on platelet MAO in an institutionalized schizophrenic population; this is certainly one thing that ought to be ruled out. Another possibility is that some people take things in their diet—and 'the pill' provides a precedent here (Southgate *et al.* 1970)—which cause an increase in MAO activity.

Kety: One could think of many possibilities to account for the difference in the chronic schizophrenics and in Dr Crow's Liverpool study, as opposed to outpatients. Patients in hospital often have low dietary proteins, inadequacies of vitamins, hepatic dysfunction, and so on. But what is difficult to explain on any of those bases is the finding in monozygotic discordant twins, who were not institutionalized and were not on drugs.

Roth: If vulnerability to schizophrenia is the associated characteristic, one probably deals with a continuously distributed trait and, if so, differences in platelet MAO activity between different groups would be quantitative and affected by sampling. Dr Kety mentioned genetic differences, in the light of the differences in concordance between monozygotic and dizygotic twins in respect of platelet MAO activity. Kallmann & Reisner (1943) published data in relation to tuberculosis showing a marked discrepancy in concordance rate between monozygotic and dizygotic twins. It is interesting to place the figures alongside those for schizophrenia, because the similarities are striking. But Kallmann was presumably studying a form of susceptibility subject to continuous variation.

Kety: The genetic factors in the host response to tuberculosis are clear. Lurie was able to breed rabbits which either would show a fulminating pneumonia and galloping consumption, such as the Negro shows, or which were practically immune to tuberculosis and, if they developed the disease with the same high degree of exposure, would show small fibro-caseous lesions. There are clear genetic vulnerability factors in tuberculosis.

Roth: Yes, but that sharp difference was achieved through selective breeding. In man susceptibility to tuberculosis infection is universal in some measure and variation in respect of it is presumably continuous.

Neff: Dr Murphy, there may be other explanations for the different results obtained that could be related to the method used to measure MAO. Some investigators have been using similar methods but different substrate concentrations, in some cases apparently less than saturating concentrations of substrate and in other cases optimal concentrations of substrate. Perhaps many of the differences, and the inability to show differences, are due to the method used to measure the enzyme rather than to the patient population.

Murphy: As far as my reading of the studies goes, everyone has used saturating levels of substrate.

Fuller: Certainly the method of Wurtman & Axelrod (1963) used a lower than saturating substrate concentration. Dr Crow, what tryptamine concentration did you use?

Crow: Dr Neff's point would not explain our results. These two investigations, one done in Liverpool and one in London, were done by the same biochemist (Dr F. Owen) and have produced widely discrepant results, using exactly the same techniques and with tyramine as substrate at saturation concentrations. So I think there is a difference which cannot be attributed to differences in the technique. It must be something to do with the populations, or their environment, or the treatment.

The concentration of tryptamine was 33 μM (i.e. non-saturating, as in Dr Murphy's report). Therefore we used a saturating concentration of tyramine and a non-saturating concentration of tryptamine, but the results with the two substrates correlated highly ($r = 0.89$ for patients, 0.84 for controls).

References

BELMAKER, R., BECKMANN, H., GOODWIN, F., MURPHY, D., POLLIN, W., BUCHSBAUM, M., WYATT, R., CIARANELLO, R. & LAMPRECHT, F. (1975) Relationships between platelet and plasma monoamine oxidase, plasma dopamine-b-hydroxylase, and urinary 3-methoxy-4-hydroxy-phenylglycol. *Life Sci. 16*, 273-280.

DONNELLY, E. F. & MURPHY, D. L. (1973) Primary affective disorder: MMPI differences between unipolar and bipolar depressed subjects. *J. Clin. Psychol. 29*, 303-306

MURPHY, D. L., BELMAKER, P. & WYATT, R. J. (1974) Monoamine oxidase in schizophrenia and other behavioral disorders. *J. Psychiatr. Res. 11*, 221-247

KALLMANN, F. R. & REISNER, D. (1943) *Am. Rev. Tuberc. Pulm. Dis. 47*, 549

KETY, S. S., ROSENTHAL, D., WENDER, P. & SCHULSINGER, F. (1971) Mental illness in the biological and adoptive families of adopted schizophrenics. *Am. J. Psychiatry 128*, 87-94

LETEMENDIA, F. J. J. & HARRIS, A. D. (1967) Chlorpromazine and the untreated chronic schizophrenic: a long-term trial. *Br. J. Psychiatr. 113*, 950-958

SCHWARTZ, M. A., WYATT, R. J., YANG, H.-Y. T. & NEFF, N. H. (1974) Multiple forms of brain monoamine oxidase in schizophrenic and normal individuals. *Arch. Gen. Psychiatry 31*, 557-560

SOUTHGATE, J., COLLINS, G. G. S., PRYSE-DAVIES, J. & SANDLER, M. (1970) Effect of contraceptive steroids on monoamine oxidase activity. *J. Clin. Pathol. Suppl.* (Assoc. Clin. Pathol.), *3*, 43-48

VAN PRAAG, H. M. (1975) Neuroleptics as a guideline to biological research in psychotic disorders. *Compr. Psychiatry 16*, 7-22

VAN PRAAG, H. M. & KORF, J. (1975) Neuroleptics, catecholamines and psychotic disorders. A study of their interrelation. *Am. J. Psychiatry 132*, 593-597

VAN PRAAG, H. M., DOLS, L. C. W. & SCHUT, T. (1975) Biochemical versus psychopathological action profile of neuroleptics: A comparative study of chlorpromazine and oxypertine in acute psychotic disorders. *Compr. Psychiatry 16*, 255-263

WURTMAN, R. J. & AXELROD, J. (1963) A sensitive and specific assay for the estimation of monoamine oxidase. *Biochem. Pharmacol. 12*, 1439-1441

General discussion

FUTURE INVESTIGATIONS

Maître: If I may start with a short general comment and a simple question, for people working in the laboratory, whether with rats or mice or monkeys, it is essential to know the exact feeling and judgement of the clinicians about the substances we use routinely as reference materials. We need this feedback from the clinic and are highly dependent on it. This is another reason why we appreciate this type of meeting, in which an extensive exchange is possible between observations made in the clinic and those made in animal experiments —observations originating from the administration of the same substances, or at least the same type of substance. This leads me to a question to our clinical colleagues: if we exclude tyramine potentiation (the cheese reaction), are there any therapeutic effects or side-effects of MAO inhibitors that are not shared by classical tricyclic antidepressants?

van Praag: This is a difficult question. The number of investigations comparing tricyclics and MAO inhibitors is small, and I think the MAO inhibitor story ended several years ago, apart from a possible upsurge of the use of phenelzine in the treatment of neurotic depression or depression with a fair amount of anxiety. My personal feeling is that one cannot generalize about MAO inhibitors. I collected a lot of experience with iproniazid and I think it was a good antidepressant, especially for endogenous depression. We have had other experience with other MAO inhibitors like isocarboxazid and phenelzine, and we found them rather weak in the treatment of endogenous depression (van Praag 1962). So it is difficult at the moment to say that MAO inhibitors as a group are similar to or different from the tricyclics in this or that respect.

While I am in favour of introducing more biological parameters for diagnosing depression and evaluating the results of treatment than has been done so far and trying to differentiate subgroups characterized mainly by biochemical criteria, it could well be that there are subgroups of depression which react more favourably to certain kinds of MAO inhibitors than to tricyclics; but so far I can't give you an answer. It could be an interesting field for research to compare tricyclics with MAO inhibitors, if possible more selective MAO inhibitors, and to relate the therapeutic results not only to psychopathological criteria but also to biochemical criteria (van Praag 1974, 1976).

Murphy: There is one physiological effect that distinguishes the MAO inhibitors. I don't think any other known drug is able to eliminate REM (rapid eye movement) sleep 100%, as both Wyatt *et al.* (1969) and Akindele *et al.* (1970) described. Tricyclics can diminish REM sleep but not reduce it to zero over a sustained period of time. Incidentally, the paper by Akindele *et al.* is one study where altered behaviour, including euphoria, was also observed after MAO inhibition in a few non-psychiatric patients.

van Praag: This is a peculiar thing, if it is right. In my department there have been some investigations of sleep psychology—the subjective experience of the quality of sleep the following morning. You really eliminate REM sleep with MAO inhibitors such as tranylcypromine, but the feeling of having slept well or badly isn't changed; this is a peculiar experience. The amount of REM sleep evidently has not very much to do with the subjective feeling of having slept badly or well.

Pare: So far as I know, the muscular jactitation and hyper-reflexia only occur with amine oxidase inhibitors, although admittedly in quite large doses. I don't recall seeing this with the tricyclics.

Rafaelsen: As one who constantly moves from the laboratory to the clinical ward and back again, I am acutely aware of the problems that are the basis for Dr Maître's question. When we go back from this symposium to our own countries, I am afraid that the clinicians will say that MAO is in a bad state and the biochemists will say that depression is even worse than before! What we might need is a new type of model. The models we have are either psychiatric, and they will not solve any of our problems in biological psychiatry, or they are psychopharmacological. Even if some of them, like the amine model, have been persuasive, basically they are static models which do not tell us why some patients go into depression and out again. The models have not been directly translatable from the one frame of reference to the other. They are always in terms of one or the other. As to the MAO inhibitors, I feel, like Dr Pare and many others, that they work on a subgroup of depressions, but this subgroup is difficult to characterize. But it might perhaps help us in understanding the heterogeneity behind depressive syndromes. I have also been impressed by the sensitivity of dosage. When you treat bipolar patients with an

MAO inhibitor you find the optimum dose for a particular patient. After some months he will become depressed again if you reduce the dose by 5 or 10 mg; or if you increase it by another 5 mg he develops manic symptoms. This delicate balance is important. It may be that some sort of clinical *and* biochemical instability in depression has been changed, by the MAO inhibitor, to give clinical well-being and a new sort of instability, which is also delicately balanced.

Another important aspect when we construct models of affective disorders is that they should be able to explain why some patients get only melancholia (depression); those are called unipolar, and that is perhaps the major advance in trying to make a subdivision that has come from clinically oriented research in the last 10 years. Others are bipolar and such patients get both depression and mania—we don't know why, nor why some only get a few episodes during a life-time and others seem to cycle continuously between depression and mania.

The model I propose (Rafaelsen 1974), which has very little basis in actual findings, is inspired by several sources; one of these is the studies made on amino acid transport defects. I suggest that to concentrate on the diseased part of the patient's life may not be the most productive. The special mode of reaction, which is of course a special vulnerability, is one which I think must be genetically determined, and means that a certain subgroup of the population will react much more readily with depression or mania than the rest. The psychosis is connected with disease and treatment; the potential for a particular reaction is connected with prophylaxis (prevention), and in the long run that will be more important.

We have to accept that at the base of this mode of reaction there may be heterogeneity or homogeneity, but the former is the most probable, in my view. Conversely, this basic reaction potential may lead to homogeneity in the manifestation, if we see depression as a final common pathway of many different metabolic aberrations.

Scriver (1972) worked with iminoglycinurias (a group of rather harmless amino acid transport defects) and found, from kinetic evidence and from loading tests, that he had to invoke two different transport systems, one handling a *group* of amino acids and the other handling *individual* amino acids. So one could have an iminoglycinuria affecting one amino acid, say proline, and in another case both proline and hydroxyproline were affected. These different pumps, as we may call them—facilitating transport systems of various sorts—have been described for several amino acids, both the *specific* pump and the *group* pump. The latter is characterized by high capacity, low affinity, and group specificity. The individual pump has low capacity, high affinity and substrate specificity.

PUMP	UNIDIRECTIONAL	BIDIRECTIONAL
GROUP	1. □	3. □
SPECIFIC	2. □□□□	4. □□□□
TYPE AFFECTIVE DISORDER	UNIPOLAR	BIPOLAR

FIG. 1 (Rafaelsen). Model for the affective disorders, which suggests that all types of affective disorder are pump deficiencies concerning normal physiological substances (morphological evidence of membrane disorders being unobtainable at present). The figure illustrates the four main groups of deficiencies (see text).

We can attempt to translate this scheme into the affective disorders. To be provocative, I shall suggest that all types of affective disorders are pump deficiencies, concerning normal physiological substances; no morphological changes have been demonstrated and none can be demonstrated until membrane abnormalities of this subtlety are within the reach of the electron microscopist. There are four main groups of deficiencies (Fig. 1). Two are concerned with pumps which are unidirectional, which under all circumstances transport from one side of the membrane to another. The others are bidirectional and such pumps also exist; for lactate, fatty acids, and so on. Those individuals who are affected by unidirectional pump deficiencies are unipolar; they can only get one type of swing and nearly always this is a depressive. Those with bidirectional pump deficiencies will be bipolar patients where the deficiency can work one way or another. So we have a unidirectional group pump and a unidirectional specific pump; and the same for the bidirectional pumps.

The consequences of transport defects of a unidirectional group pump (Fig. 2) consist only of melancholias (or, rarely, manias) as a result of transport competition with other substances, often influenced by external and internal environmental factors. They are often explained as due to nutritional, climatic, psychological or social factors.

Fig. 3 illustrates a deficiency of a specific unidirectional pump. Here we have only melancholias again (or rare mania) as a result of transport insufficiency in relation to one substance. This is much less influenced by external or internal factors and will often be unexplained ('endogenous').

The next example (Fig. 4) is the bidirectional group pump. Deficiency results in both melancholias and manias, as a result of transport competition

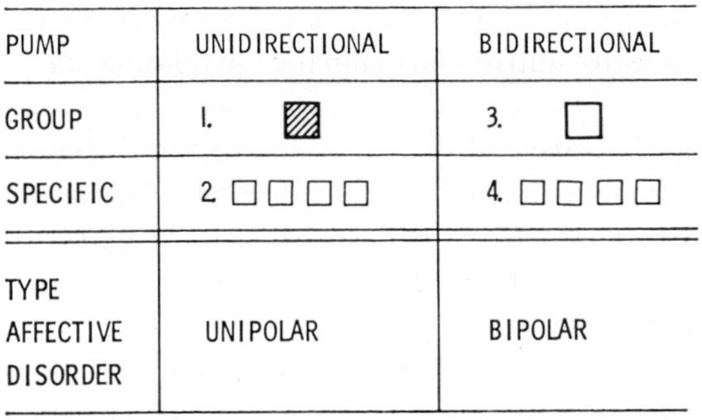

PUMP	UNIDIRECTIONAL	BIDIRECTIONAL
GROUP	1. ▨	3. □
SPECIFIC	2. □ □ □ □	4. □ □ □ □
TYPE AFFECTIVE DISORDER	UNIPOLAR	BIPOLAR

FIG. 2 (Rafaelsen). Deficiency of *general* pump, *unidirectional* type: only melancholias (or in rare instances only manias) as a result of transport competition with other substances, often influenced by external or internal environmental factors. Will often be explained as due to nutritional, climatic, psychological or social factors.

PUMP	UNIDIRECTIONAL	BIDIRECTIONAL
GROUP	1. □	3. □
SPECIFIC	2. □ ▨ □ □	4. □ □ □ □
TYPE AFFECTIVE DISORDER	UNIPOLAR	BIPOLAR

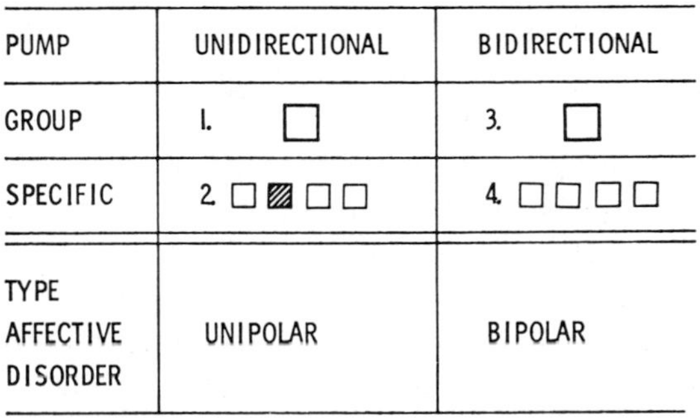

FIG. 3 (Rafaelsen). Deficiency of *specialized* pump, *unidirectional* type: only melancholia (or in rare instances only manias) as a result of transport insufficiency concerning one substance. Lesser influence by external or internal factors. Will often be unexplained ('endogenous').

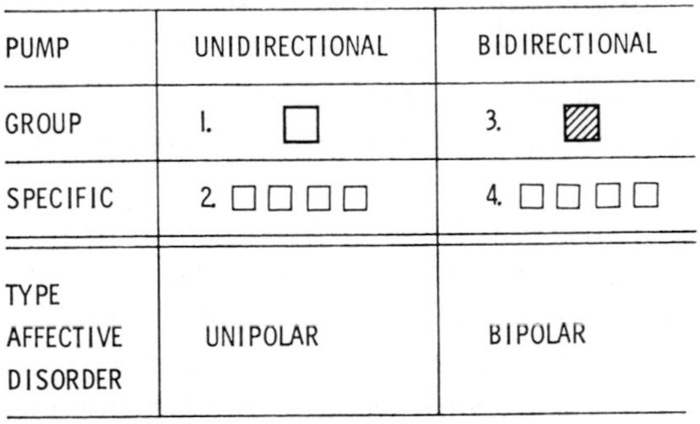

PUMP	UNIDIRECTIONAL	BIDIRECTIONAL
GROUP	1. □	3. ▨
SPECIFIC	2. □ □ □ □	4. □ □ □ □
TYPE AFFECTIVE DISORDER	UNIPOLAR	BIPOLAR

FIG. 4 (Rafaelsen). Deficiency of *general* pump, *alternating* type: both melancholia and manias as a result of transport competition with other substances, often influenced by external or internal environmental factors. Will often be explained as due to nutritional, climatic, psychological or social factors.

with other substances, often influenced by external and internal environmental factors and often explained as due to nutritional, climatic, psychological or social factors.

The last is the specific pump for one substance, a bidirectional pump, causing both melancholias and manias, with less influence by external or internal factors because this transport insufficiency chronically affects one substance; therefore stability cannot be obtained and this leads to the rare cyclic type of manic-melancholic disorder (Fig. 5).

The advantage of this type of model is that it should facilitate experiments in which one challenges the organism, loading it with substances of particular interest. It should thereby be easier to validate or falsify such a hypothesis.

Pare: What exactly do you mean by a 'group' abnormality? Is this a generalized transport abnormality, for instance to explain our findings on the uptake of 5HT into platelets (see p. 294) or Dr Coppen's findings with the tyramine pressor effect (p. 339), which could perhaps be explained on the basis of uptake of tyramine into noradrenaline sites, or your own work on erythrocytes, Dr Rafaelsen? There is also earlier work on sodium transport across the blood–brain barrier. Is that what you mean by an abnormality of group transport?

Rafaelsen: The only case where we know of group transport systems for sure is the amino acids. In several tissues group pumps catering for three or four amino acids are known. We have no idea whether a hypothesized group pump deficiency will be only in the blood–brain barrier, in the synaptosomal membrane, in the erythrocytes or in the gastric mucosa. An example taken from another field where this type of thinking has been productive is Oldendorf's (1973) explanation of the abnormality in phenylketonuria. He demonstrated that it is the group pump concerned with phenylalanine transport which is deficient, thereby showing that it was not phenylalanine which is toxic to the brain, but that the huge amount of it going into the brain reduces the capacity to transport the two or three other amino acids catered for by the same transport system. The shortage of these amino acids is damaging to the brain. This is a good example of competition of substances for transport via a common system.

van Praag: This is an attractive model, and certainly a broader concept than the monoamine model; but I don't agree that the monoamine model is static. If one accepts the monoamine hypothesis as stating that a relative or absolute deficiency of monoamines occurs at the receptor sites in certain types of depression, and if you show that an imbalance between receptor function and available transmitter exists, this is as dynamic or static as what you propose. I don't see any fundamental difference in dynamics between the two. What is more static about speaking about imbalances between receptors and available

PUMP	UNIDIRECTIONAL	BIDIRECTIONAL
GROUP	1. □	3. □
SPECIFIC	2. □□□□	4. □▨□□
TYPE AFFECTIVE DISORDER	UNIPOLAR	BIPOLAR

FIG. 5 (Rafaelsen). Deficiency of *specialized* pump, *alternating* type: both melancholias and manias as a result of transport insufficiency chronically affecting one substance. Lesser influence of external or internal factors. Stability cannot be obtained. Leads to a cyclic type of manic–melancholic disorder.

transmitter than about transport abnormalities in the central nervous system?

Rafaelsen: It depends how you view them, certainly. I have had difficulties on the monoamine theory in understanding why some people are more susceptible than others and why, among those who break down, some do it repeatedly and others do not.

van Praag: That is difficult to explain on either model.

Rafaelsen: The hypothesis of a specific pump of the bidirectional type will explain why stability in affected individuals can never be obtained (Fig. 5). There must always be a plus or a minus in relation to the normal state. It is difficult to see how this could be explained by the monoamine theory. But this is not the most important point. The value of such models is whether they enable us to do better experiments.

Fuller: What do you think about the possibility that these hypothetical pumps might be the neuronal membrane pumps that are responsible for the re-uptake of noradrenaline, 5HT or dopamine or some other monoamine? That might bring together the two models.

Rafaelsen: Certainly. I don't know whether it is the blood–brain barrier, or the neuronal cell body, or its mitochondrial membrane, but one hopes that the amine theory could be integrated into the model I have presented.

Sandler: We have speculated that in various disease states such as depression there may be a state of increased membrane permeability towards amines, which thus gain abnormal access to mitochondrial MAO. In other words, the disease may be a manifestation of a general defect (see my paper, p. 335). This concept seems to fit in well with your hypothesis.

Sourkes: You gave the example of phenylketonuria and Dr Charles Scriver's data. These deal with competitive phenomena with presumably normal transport mechanisms, whereas monoamine theories of depression postulate a deficient mechanism for providing an amine; so there is a difference. Theoretically phenylketonuria could be got around if you gave the patient enough tryptophan, tyrosine and branched-chain amino acids to compensate for those deficiencies, *if* the deficiencies really have something to do with the disease.

Rafaelsen: The complete normalization of the abnormal functions and the lack of any histological findings in affective disorders suggest to me that the reduced capacity of normal pumps rather than more specific abnormalities will be the underlying defects.

Sandler: Dr Coppen mentioned his interesting findings where he gave tyramine intravenously and found an increased sensitivity in the depressive group (Ghose *et al.* 1975; and see p. 339). As a possible interpretation he suggested a decrease of MAO activity, but in fact when you give many monoamines intravenously, although this has never been done with tyramine, they are almost completely inactivated by pulmonary uptake. Thus, if the fate of intravenous 5-hydroxytryptamine (Alabaster & Bakhle 1970; Junod 1972) can be used as a precedent, tyramine given by this route is likely to be taken up first by the lungs before being dealt with at leisure by MAO. So I don't think that Dr Coppen's experience, although very interesting, is necessarily relevant to the MAO status in depressive illness.

I am still very puzzled by our own findings, which I dwelt on in my paper, in relation to migraine and depression (see p. 330). Having demonstrated a platelet MAO B deficit in migrainous subjects one might have expected not to find an apparent increase of activity, as gauged by the output of conjugated tyramine, in the dietary migraine group of patients. It may be that the group of dietary migraine sufferers we investigated were masked depressives. It was, perhaps, a loaded group, self-selected in answer to a questionnaire sent out by my colleague Dr E. Hanington. One interesting point about migraine, however, is that in certain intractable cases, when all else fails, Jim Lance in Australia (Anthony & Lance 1969) gives an MAO inhibitor and claims it to be by far the most effective treatment for this kind of patient.

Oreland: But how do you explain the difference Dr Coppen found? If it is as you say, he should not obtain any tyramine response.

Sandler: I would explain it by a physical difference in pulmonary uptake in the two groups, perhaps not necessarily primarily to do with MAO.

Coppen: I think that the marked pressor response that we found in both controls and depressives shows that intravenous tyramine is not by any means completely inactivated by the lungs. I have already listed possible reasons for the differences in sensitivity to tyramine. Perhaps we should add pulmonary uptake as another one, although at the moment I would put it at the bottom of the list of possible explanations.

Trendelenburg: Is it safe to extrapolate from 5HT in the lung to the sympathomimetic amines? In the old days, when people perfused hind legs, the preparation worked only when the lungs were included in the circulation; now we know that the lung took care of the circulating 5HT, and the vessels did not constrict. I don't think the lungs play a corresponding role for sympathomimetic amines.

Sandler: We know that catecholamines are not taken up by lung. Many agents are inactivated, however, some on first pass through the lungs, others only in the liver. I agree that one cannot necessarily extrapolate from one monoamine to another.

Trendelenburg: Is it possible that tyramine goes through the lung, while 5HT gets stuck there?

Sandler: You may be right. MAO may yet prove to be implicated. We need more information.

Murphy: I would like to raise the issue of increased monoamine oxidase in relation to behaviour. It is generally held that MAO is present in tissues in excess. It has been suggested that there may be some connection between the observations of higher MAO activity with increasing age, and the higher incidence of depression in older individuals, but I'm not sure how a mechanism for such an association might be formulated. If, as is generally held, there is great excess of MAO activity in tissues under normal circumstances, it would seem difficult for a further increase in MAO activity to be associated with any functional changes. Secondly, if there is, in fact, an increased effective MAO level within nerve endings, one might expect it to reduce the free intraneuronal amine concentration. From what is known, it would seem that as free amine concentrations are thought to regulate amine synthesis, an effective chronic increase in MAO activity would be expected to produce an increase in synthesis in response to the lesser amount of free amine, which would tend, at least according to the traditional biogenic amine theory of depression, to have an antidepressant effect rather than a depressant effect. So there is a problem here, although this view takes into account only one of the mechanisms by which MAO might alter amine function and amine metabolism.

I want to briefly mention some of the other biochemical sequelae of MAO inhibition, by way of asking what this symposium might conclude to be the most important functional effects of MAO inhibition, particularly in relation to neuronal function. Some of the more obvious ways in which reduced MAO levels might have biological effects which in turn might have behavioural effects are shown in Table 1. For the major neurotransmitter amines—nor-

TABLE 1 (Murphy)

Some examples of the biological effects of monoamine oxidase-inhibiting drugs relevant to neurotransmitter function and behaviour

I. Increased accumulation of neurotransmitter amines (noradrenaline, dopamine, 5-hydroxytryptamine)
   - A. Unequal effects on different amines
   - B. Different MAO inhibitors produce different effects

II. Accumulation of other amines (tyramine, octopamine, phenethylamine, tryptamine)—'false transmitters'
   - A. Endogenous amine depletion
   - B. Release of other amines having lesser effects at synapses
   - C. Inhibition of endogenous amine synthesis
   - D. Increased likelihood of occurrence and/or potentiation of (*a*) methylated amine metabolites, (*b*) dimethyltryptamine, (*c*) 6-hydroxydopamine, 5,6-dihydroxytryptamine

III. Decreased formation of aldehydes from biogenic amines

IV. Decreased synthesis of biogenic amines via feedback mechanisms

adrenaline, dopamine and so forth—increased accumulation has certainly been well documented. But there may be unequal effects of the different MAO-inhibiting agents on different amines, as is shown by the selective inhibitor story. The other big issue with MAO inhibition is the accumulation of less well-studied amines such as tyramine, octopamine, phenylethylamine, and tryptamine, which may function as false transmitters, whose presence may be associated with amine depletion, secondary inhibition of endogenous amine synthesis, and metabolic pathway shifts. Also, if tryptamine is accumulated, there are enzymes known to convert tryptamine into dimethyltryptamine and related substances with behavioural effects. The evidence that aldehydes themselves may have some sedative effects or interactions with other neurochemical mechanisms is an additional point. That is only a brief summary, but I think these possibilities (which are reviewed elsewhere [Murphy *et al.* 1974]) should be brought out.

Green: We certainly have to consider the effect of MAO inhibition on the action of amines that are normally found only peripherally. When 5-methoxytryptamine was injected peripherally into rats it was not found in brain 30 min later and there were no behavioural changes. When the rats had been pretreated with an MAO inhibitor, concentrations of 5-methoxytryptamine in the brain rose and there were gross behavioural changes (Green *et al.* 1975). One should perhaps add to Dr Murphy's list the fact that if you are giving a monoamine oxidase inhibitor, compounds that might not normally produce centrally mediated effects, because they are broken down rapidly by MAO on entering the brain, will no longer be metabolized.

Sandler: I go along with that. Dr Murphy, we presumably agree that when one gives an MAO inhibitor, if it works at all it does so by virtue of its MAO inhibitor reaction?

Murphy: I think we have to say that that is still an open question. Some MAO inhibitors certainly have other effects, such as on the uptake and release of amines, as well as other non-amine effects.

Sandler: If Dr Oreland is right, and if you give an MAO inhibitor to inhibit further something that you think is inhibited already, how do you interpret any therapeutic action of MAO inhibitors in terms of their MAO-inhibiting activity?

Oreland: That would be no problem, as far as I can see. If our suggestion is correct, one has a low MAO activity because of a low monoaminergic activity, and one way of increasing that activity would be to inhibit the enzyme. There is no contradiction there.

Kety: There is some support for Dr Oreland's hypothesis. In chronic electroconvulsive shock in animals there has been controversy about whether there is an increase in noradrenaline turnover and noradrenaline levels in the brain, and even more with 5HT and dopamine, but one fairly consistent observation is that of an increase in MAO activity in the brain after repeated electroconvulsive shocks (Pryor 1974).

Youdim: I have recalculated the values in this work and found no significant difference between the values. There was no significant increase in brain MAO.

Kety: As I recall, Pryor found MAO to be increased even for a few weeks after the last shock.

Green: I have just completed a study on 5HT metabolism in repeatedly shocked rats and found no evidence of increased turnover or synthesis of 5HT after 10 days of shocks (Evans *et al.* 1976). After one shock there were transient changes in 5HT metabolism but not after repeated shocks.

Neff: As I understand it from the literature, the monoamine theory of depression states that the patient is depressed because he has insufficient catecholamines at receptor sites. From Dr Murphy's list (Table 1, p. 380), I have the impression that almost every mechanism proposed for the mode of action of MAO inhibitors limits the concentration of catecholamines at receptors. For example, false transmitters would compete for the receptor site with the normal transmitter, and if they reduce synthesis there is less available for interaction with receptors, and so on.

Murphy: You are left with a cell that *contains* more transmitter but perhaps less effectively available transmitter. That may fit with the fact that MAO inhibitors are not very good antidepressants.

Neff: Maybe the MAO inhibitors work because they cut back on the synthesis of the catecholamines?

Murphy: Yes. In fact, that's the way the tricyclic story has been developing, that perhaps the receptor-blocking effect that develops later may explain some of the antidepressant effects of the tricyclic drugs. Several phenothiazines like chlorpromazine and thioridazine have been shown to be as effective in some trials as the tricyclics; there are a lot of data suggesting that drugs that don't inhibit the uptake of amines (such as iprindole) nearly as well as a standard tricyclic may be equally effective, perhaps because they also have receptor-blocking effects. The hypothesis that antidepressant drugs act by increasing effective amine levels may need to be inverted, and this is one of the current points of interest.

Gorkin: Dr Murphy said that biogenic aldehydes have to be considered. If so, we can't disregard the aldehyde reductases and aldehyde dehydrogenases. We then also have to consider the amount of pyridine nucleotides in tissues, so the situation becomes very complex. I believe Dr Tipton has relevant information on the activity of aldehyde dehydrogenases, in which he demonstrated how they interact with MAO metabolism in general (Tipton & Turner 1974)?

Tipton: Certainly the 'biogenic aldehydes' must be considered, since there is a great deal of evidence that they have physiological properties that are distinct from those of their parent amines. In addition, high steady-state levels of these aldehydes will cause product inhibition of monoamine oxidase, thus slowing down the rate of amine degradation (Turner *et al.* 1974). It is not clear whether the metabolism of these aldehydes will be greatly affected by fluctuations in the steady-state levels of the oxidized and reduced pyridine nucleotides in the brain. The available data indicate that both aldehyde dehydrogenase and the aldehyde reductases will be saturated with the correct coenzymes for aldehyde metabolism under normal conditions and the amount of inhibition by the product coenzyme will be small (Tipton & Turner 1974). A large flux through this pathway may disturb the coenzymes' redox balance sufficiently to alter this situation, but at present there are insufficient data to know whether this is likely or not. Another effect that must be considered is the effect of coenzyme utilization by this pathway on the activities of other enzymes that use pyridine nucleotides and here, at least at low amine concentrations, there should be differences that are dependent on the amines involved, since the metabolism of amines that lack a β-hydroxyl group preferentially follows the oxidative pathway using NAD^+ as the coenzyme, whereas the metabolism of amines that contain this group involves the reductive pathway with NADPH as the coenzyme to a much greater extent.

Sourkes: If one has to take into account the aldehyde dehydrogenase and aldehyde reductase one will have produced some reduced pyridine nucleotide, or will use some up, depending on which enzyme predominates. If the aldehyde goes on mainly to the acid, there will be a net increase in reduced pyridine nucleotide which might then get back to the tyrosine hydroxylase co-factor, or the tryptophan hydroxylase co-factor, if it needs one, and help it out. On the other hand, in a noradrenaline-containing fibre reduced pyridine nucleotide will be used up, to convert the aldehyde to the glycol, and you may be detracting from the biosynthesis of acidic metabolites; while we cannot prove this, we know there is a net formation of glycol from noradrenaline and a net formation of the acids from dopamine and 5HT. This is therefore the only factor I would add to Dr Murphy's list, in line with the suggestion that there will be different influences depending upon which monoaminergic fibre is being considered.

Sharman: Dr Neff made the point (p. 381) that if you reduce the synthesis of the amine, you are spoiling the monoamine concept. There is nothing wrong with Dr Murphy's concepts. Pharmacologically the effectiveness of the system will be related to the concentration of the active substance in the region of the receptor and the time that it stays there. If you have a system, such as we see in the brain, forming metabolic products from a transmitter substance, this is probably a reflection of the wastage. Provided you have not changed any of the other physical parameters in the intact system in the region of the nerve ending, the rate of synthesis of the transmitter will give some index of how fast the transmitter is being released into the synaptic cleft and taken up again. Once you have put in a drug to upset part of this system, the rate of synthesis need bear no relation to the effect of the transmitter at its receptor and, in fact, if you can make the system more efficient so that less transmitter is wasted, it will be a lot easier for enough transmitter to get onto the receptor.

Fuller: What Dr Sharman says is important, and is illustrated particularly well by uptake inhibitors, which decrease the rate of turnover of monoamines and presumably do so by virtue of an increase in their concentration at the receptors. The principle seems to apply generally in the case of monoamines, but the point can be made particularly with reference to 5HT. The turnover of 5HT in the presynaptic nerve ending can be influenced by agents that influence the action of 5HT at the receptor. For example, a receptor-blocking drug, methiothepin, accelerates the synthesis of 5HT. A receptor agonist, quipazine, decreases its turnover. An uptake inhibitor, which also increases the stimulation of the receptor by increasing the 5HT concentration at it, decreases the turnover of 5HT. But with MAO inhibitors this doesn't seem to happen in the case of 5HT.

We have some illustrative data with the uptake inhibitor Lilly 110140. This

TABLE 1 (Fuller)

Effects of Lilly 110140 on the turnover of 5-hydroxytryptamine (5HT)

(1) *Turnover measured by the rate of 5-hydroxyindoleacetic acid (5HIAA) increase after giving probenecid to block the efflux of 5HIAA*

	Rate of 5HIAA increase ($ng\ g^{-1} h^{-1}$)
Control	176
110140-treated	66

(2) *Turnover measured by the rate of 5HT increase after MAO inhibition*

	Rate of 5HT increase ($ng\ g^{-1} h^{-1}$)
Control	157
110140-treated	165

drug reduces the turnover of brain 5HT as measured by a variety of experimental approaches. It decreases the rate of decline of 5HT after the administration of *p*-chlorophenylalanine to inhibit synthesis; it decreases the rate of accumulation of 5HTP after decarboxylase inhibition; it reduces the rate of synthesis of labelled 5HT from labelled tryptophan; and it decreases 5HIAA levels in brain. The drug also decreases the firing of single neural units in the midbrain raphe, as studied by Dr J. A. Clemens. All these data agree that the uptake inhibitors lead to a decrease in turnover of 5HT. The results in Table 1 show two ways of measuring turnover.

The experiment with probenecid shows clearly that 110140 decreased turnover. The fact that after MAO inhibition the rate of 5HT increase was linear and occurred at the same rate as 5HIAA accumulation after probenecid suggests that MAO inhibition in itself had not altered the synthesis of 5HT. The uptake inhibitor did not decrease the rate of 5HT accumulation. This finding was at first puzzling because it seemed to be at variance with the other ways of measuring turnover. However, Meek & Fuxe (1971) have studied another inhibitor of 5HT uptake, clomipramine, and two other MAO inhibitors, nialamide and pargyline, with identical results. They concluded that the normal control of 5HT synthesis is lost after MAO inhibition. The mechanism of this control of turnover is not understood; it might be due to a presynaptic receptor which senses the level of stimulation of the postsynaptic receptor, or it might be through a trans-synaptic feedback loop. But according to their data and ours it seems that after MAO inhibition this type of control may be lost.

Trendelenburg: When we discuss turnover and rate of synthesis in relation to function (that is, the release of transmitter), we must remember that turnover and synthesis consist of two different components. On the one hand, turnover is determined by the rate at which the transmitter is released by exocytosis (and in this case a 'response' accompanies the event); on the other hand, turnover is also determined by intraneuronal metabolism of the transmitter *before* it leaves the nerve ending (and in this case no 'response' can be expected). Any increase in intraneuronal MAO activity should lower the axoplasmic level of noradrenaline and thereby increase its rate of synthesis. However, this increase in rate of synthesis need not lead to any increase in the rate of release by exocytosis, since increased synthesis may well lead to increased leakage from storage vesicles into the axoplasm and increased deamination of amine that has never left the cell. Thus, when we talk about 'responses of effector organs', we should consider only that component of turnover that is concerned with exocytosis of transmitter.

van Praag: If we are trying to pull the diverse data together, we should remember all the evidence showing that it is difficult to correlate biological data to classical syndromal and nosological entities. We have heard during the symposium several examples of MAO disturbances that can be found in depression and in schizophrenia. The same is true of studies of the c.s.f.; you can find monoamine disturbances there in schizophrenia, depression, Parkinson's disease and so on. We studied dopamine in extrapyramidal disorder, in psychosis and in depression, and we think that the disturbances in dopamine metabolism are non-specific from the point of view of nosology, and of syndromes, but may be specific from the point of view of symptoms, that are specific disorders in psychological or motor functions (van Praag & Korf 1971*a*, *b*; van Praag *et al.* 1975). Thus dopamine is related, in our experience, to activity states (both hyperactivity and hypoactivity). Perhaps disturbances in 5HT are much more related to mood regulation as such, irrespective of the syndrome and of the nosological entity in which the mood abnormality occurs. So I think the future for biological psychiatry will be more in the direction of close cooperation with experimental psychology in sorting out specific dysfunctions in the field of psychology, rather than in trying to correlate specific biological dysfunctions with complex patterns of abnormal behaviour, like depression or psychosis or anxiety states. Thus the finding that in a heterogeneous group of suicides (heterogeneous with regard to psychopathology and personality structure) there was one specific metabolic dysfunction could be the consequence of a relationship between biological factors and more or less specific psychological dysfunctions. I am therefore a believer in specificity, but on a symptomatological level.

STANDARDIZATION OF ASSAYS

Tipton: A great deal of effort is currently being directed towards measurement of the monoamine oxidase activity in organs or cells taken from patients who may be receiving various forms of therapy. Clearly the situation with respect to variation between individual patients and differences in the details of their treatment is one that can lead to a great deal of variability and in order to lessen the possibility of other variables Professor Singer, Dr Youdim and I would like to make a plea for some standardization of the MAO activity determinations.

In considering assay methods, several points should be borne in mind:

(1) If one is working with cell material it is necessary to eliminate any effects of permeability barriers, that could possibly become rate limiting, by making sure that the cells are broken.

(2) The rate of the reaction should be determined at a stage where product formation is proceeding linearly with time. Many workers take a single determination of the amount of product formed (or substrate used) after a fixed time as a measure of the enzyme activity. Clearly, results from such determinations will be meaningless unless the reaction is proceeding linearly up to this time. Curvature of reaction time-courses can be due to one or more of several common causes (Dixon & Webb 1964) and although it cannot be completely eliminated it is convenient to adjust the conditions so that linearity is maintained for a reasonable length of time. One common cause of early departure from linearity in monoamine oxidase assays is the use of too small a substrate concentration. If the substrate concentration is not much greater than the K_m value, depletion of substrate during the assay will cause the enzyme to become progressively less saturated and hence the velocity will fall off. The obvious remedy is to use very high concentrations of substrate so that the enzyme remains saturated throughout the period of assay. A substrate concentration of 10 times the K_m value will normally achieve this, but such high substrate concentrations may not always be practicable.

(3) Controls must be used to show that the measured change is due to an enzymic reaction, and also proportionality between the velocity of the reaction and the amount of enzyme-containing material added should be established. At least a 10-fold range of enzyme concentration, which covers the concentration range to be used in other experiments, should be used, and the straight line obtained should pass through zero activity at zero enzyme concentration. Such a result will also establish the linearity of the reaction time-course under the conditions used.

(4) It is important to use pure substrates, since impurities could act as inhibitors of the enzyme and may also behave like the products of the reaction in the separation procedures used in radioactive assay methods, giving rise to high blanks. If a radioactive assay is to be used it should ideally be compared

with the results obtained with another assay method (e.g. oxygen consumption or spectrophotometric determination of aldehyde production) in order to establish its validity.

(5) The comparison of results obtained by different workers would be facilitated if assays were done under defined conditions. A temperature of 30 °C is generally accepted as the standard temperature for enzyme work, and I would consider a pH value of 7.2 to be close to what we believe may be the physiological pH at which monoamine oxidase works.

Singer: My own particular worry concerns the entire approach of using radioactive MAO substrates for assay. In theory, it would be best to correlate these with an absolute assay like spectrophotometric or polarographic assays but, for practical purposes, someone who is going to use platelets from hundreds of patients is obviously not going to do this. So I would suggest that a recommended assay should include reliable data on what sort of aldehyde-trapping agent should be included in the assay, lest the labelled aldehyde that arises from the action of MAO recombines with one of the thousands of proteins present in the biological material and thus vitiates the validity of the entire assay. This does happen, even with pure preparations of MAO, so I don't see why it might not also happen with crude biological material.

Neff: It should also be remembered that the platelet contains high concentrations of amines which could compete with the added substrates, which may in turn lead to unusual results.

Gorkin: I entirely agree with Dr Tipton's recommendations. However, many workers in this field have found that monoamine oxidases are easily inhibited by excess of substrate, so if we add a concentration of a substrate which is 5–10 times the K_m value, we may get substrate inhibition. In every case one should titrate the concentration of substrate which would be optimal, otherwise it may be that in the normal sample we are working with the correct concentration of a substrate but in the pathologically altered sample we have 10 times the K_m value and get substrate inhibition.

Tipton: I agree that one should check for high substrate inhibition, but in our experience the inhibition of this type that is often encountered with monoamine oxidase results from impurities in the substrate (Houslay & Tipton 1973) and may thus be avoided by using pure substrate preparations. We generally convert free amines to their hydrochlorides and then recrystallize.

[See Appendices 1 and 2, pp. 393-406, for recommended standard procedures.]

References

AKINDELE, M. O., EVANS, J. I. & OSWALD, I. (1970) Monoamine oxidase inhibitors, sleep and mood. *Electroencephalogr. Clin. Neurophysiol. 29*, 47-55

ALABASTER, V. A. & BAKHLE, Y. S. (1970) Removal of 5-hydroxytryptamine in the pulmonary circulation of rat isolated lungs. *Br. J. Pharmacol. 40*, 468-482

ANTHONY, M. & LANCE, J. W. (1969) Monoamine oxidase inhibition in the treatment of migraine. *Arch. Neurol. 21*, 263

DIXON, M. & WEBB, E. C. (1964) *Enzymes*, pp. 8-10, Longman, London

EVANS, J. P. M., GRAHAME-SMITH, D. G., GREEN, A. R. & TORDOFF, A. F. C. (1976) Electroconvulsive shock increases the behavioural responses of rats to brain 5-hydroxytryptamine accumulation and central nervous system stimulant drugs. *Br. J. Pharmacol. 56*, 193-199

GHOSE, K., TURNER, P. & COPPEN, A. (1975) Intravenous tyramine pressor response in depression. *Lancet 1*, 1317-1318

GREEN, A. R., HUGHES, J. P. & TORDOFF, A. F. C. (1975) The concentration of 5-methoxytryptamine in rat brain and its effects on behaviour following its peripheral injection. *Neuropharmacology 14*, 601-606

HOUSLAY, M. D. & TIPTON, K. F. (1973) The reaction pathway of membrane-bound rat liver mitochondrial monoamine oxidase. *Biochem. J. 135*, 735-750

JUNOD, A. F. (1972) Uptake, metabolism and efflux of ^{14}C-5-hydroxytryptamine in isolated perfused rat lungs. *J. Pharmacol. Exp. Ther. 183*, 341

MEEK, J. L. & FUXE, K. (1971) Serotonin accumulation after monoamine oxidase inhibition. *Biochem. Pharmacol. 20*, 693-706

MURPHY, D. L., BELMAKER, R. & WYATT, R. J. (1974) Monoamine oxidase in schizophrenia and other behavioral disorders. *J. Psychiatr. Res. 11*, 221-247

OLDENDORF, W. H. (1973) Saturation of blood brain barrier transport of amino acids in phenylketonuria. *Arch. Neurol. 28*, 45-48

PRYOR, G. T. (1974) Effect of repeated ECS on brain weight and brain enzymes, in *Psychobiology of Electro Convulsive Therapy* (Fink, M., Kety, S. & McGaugh, J., eds.), Winston, Washington D.C.

RAFAELSEN, O. J. (1974) Manic-depressive psychosis or manic-melancholic mode. *Dan. Med. Bull. 21*, 81-87

SCRIVER, C. R. (1972) Familial iminoglycinuria, in *The Metabolic Basis of Inherited Disease* (Stanbury, J. B., Wyngaarden, J. B. & Fredrickson, D. S., eds.), 3rd edn, pp. 1520-1535, McGraw-Hill, New York

TIPTON, K. F. & TURNER, A. J. (1974) Computer reconstruction of tyramine breakdown in brain. *Biochem. Pharmacol. 23*, 1906-1910

TURNER, A. J., ILLINGWORTH, J. A. & TIPTON, K. F. (1974) Simulation of biogenic amine metabolism in the brain. *Biochem. J. 144*, 353-360

VAN PRAAG, H. M. (1962) Een kritisch onderzoek naar de betekenis van monoamineoxidaseremming als therapeutisch principe bij de behandeling van depressies. [A critical study of the significance of MAO inhibition as a therapeutic principle in the treatment of depressions], Dissertation, Utrecht

VAN PRAAG, H. M. (1974) Towards a biochemical typology of depressions? *Pharmacopsychiatry 7*, 281-292

VAN PRAAG, H. M. (1976) Monoamines and affective disorders, in *Current Developments in Psychopharmacology*, vol. 3, Spectrum, New York, in press

VAN PRAAG, H. M. & KORF, J. (1971*a*) Retarded depression and the dopamine metabolism. *Psychopharmacologia 19*, 199-203

VAN PRAAG, H. M. & KORF, J. (1971*b*) Endogenous depressions with and without disturbances in the 5-hydroxytryptamine metabolism: a biochemical classification? *Psychopharmacologia 19*, 148-152

VAN PRAAG, H. M., KORF, J., LAKKE, J. P. W. F. & SCHUT, T. (1975) Dopamine metabolism in depressions, psychoses, and Parkinson's disease: the problem of the specificity of biological variables in behaviour disorders. *Psychol. Med. 5*, 138-146

WYATT, R. J., KUPFER, D. J., SCOTT, J., ROBINSON, D. S. & SNYDER, F. (1969) Longitudinal studies of the effect of monoamine oxidase inhibitors on sleep in man. *Psychopharmacologia 15*, 236-243

Conclusion

S. S. KETY

Department of Psychiatry, Harvard Medical School, Boston, Massachusetts

In bringing the symposium to a close, I shall not attempt a summary of the large body of information presented, but instead may speculate with regard to future research. First, I hope that the recommendations for standard techniques for the assay of monoamine oxidase in clinical studies, prepared by Dr Tipton and Dr Youdim (see Appendices, pp. 393-406), will be adopted, because it may well be that some of the controversy in the literature has arisen because people have been using different techniques.

I am sure the biochemists don't need any suggestions from me but, clearly, if one could obtain a pure enzyme, one could use this in the development of an immunofluorescent label for MAO. This would be a remarkable contribution to the problem of the localization of MAO in tissues, especially in the brain. I am thinking of the recent beautiful demonstration of dopamine β-hydroxylase within noradrenergic pathways in the brain by means of an immunofluorescent technique (Hartman *et al.* 1972), and I wonder whether it is not possible to develop a technique like that, which should resolve many of our questions about what forms of the enzyme occur in the brain and where they occur specifically. Harvey Shein felt that one way of approaching the question clinically, in addition to using platelets, would be by culturing certain available cells from patients and from controls. He had made plans for culturing fibroblasts and glia from schizophrenic patients, normal controls, and patients with various forms of affective disorder. His premature death prevented him from carrying out these studies and I am glad to hear that similar plans are being put into action in other laboratories. There are precedents in the demonstration of genetically determined enzyme defects in fibroblasts grown in culture which are ready for application to the study of putative genetic markers in the major psychoses.

At another level, the kind of elegant clinical phenomenological and statistical research that Sir Martin Roth presented is one means of further describing and dissecting apart the very heterogeneous syndromes that we call affective disorder, or schizophrenia. Dr Sourkes posed the question that if the state of clinical psychiatry is such that we are still in no real agreement about what these subgroups may be, what is the poor biochemist to do? My response would be that perhaps the biochemist can come into the fray and contribute to the further delineation of these subgroups. It may be that ultimately one will be able to characterize them on the basis of biochemical characteristics in addition to clinical, psychological and behavioural attributes. Some progress seems to have been made in that direction in terms of some of the catecholamine metabolites in the urine. There are three groups in the United States who appear to be able to separate certain types of patients on the basis of 1-(4-hydroxy-3-methoxyphenyl)-ethan-1,2-diol (MHPG) excretion (Maas 1975) and if Sir Martin's patient groups were examined from that point of view, I should not be surprised if that provided another discriminant.

We have been impressed with the findings on the platelets of schizophrenics. It would certainly be satisfying if Murphy & Wyatt's findings were correct, because they tie in very nicely with two current hypotheses about schizophrenia, both the dopamine and the transmethylation hypotheses. It is interesting that at this meeting several contributors, including Dr Maître, Dr Neff, and others, have indicated that MAO seems to act very clearly on dopamine, so that a diminution in MAO in certain regions of the brain in schizophrenia might be compatible with an overactivity of dopaminergic synapses. It is also possible that this could be compatible with the transmethylation hypothesis of schizophrenia. I continue to find it interesting that we all have in our brains an enzyme that can methylate tryptamine to dimethyltryptamine (Saavedra *et al.* 1973), and therefore that we all have the potentiality of making the hallucinogen in our brains. It is possible that one thing that differentiates the schizophrenic from the non-schizophrenic is his ability to detoxify that compound, but the disagreement of Crow's observations with those of Wyatt & Murphy remains to be explained.

Two new studies on MAO in platelets may be of interest. One of these is by J. Schildkraut, who found a subgroup of schizophrenic patients with auditory hallucinations and delusions (personal communication). In this subgroup there was a significant reduction in MAO in the platelets as opposed to the other schizophrenics in his group. This is based on an examination of 32 patients, evenly divided between those with auditory hallucinations and delusions and those without. The difference was significant at the 0.001 level. Another is by Zeller *et al.* (1975), in a large series of patients. A highly significant reduction in MAO in platelets, in both males and females, was found in the schizophrenic groups. This was in response to three different substrates, but the substrate-activity profile was quite different for the enzyme in the platelets of schizophrenics compared to that in normals. The MAO from schizophrenics was significantly less active on tyramine and on *m*-iodobenzylamine than on *p*-methoxybenzylamine. They suggest a qualitative difference in the MAO in the schizophrenic. Perhaps these studies will provide a starting point for further investigations into MAO in normal and disease states.

References

HARTMAN, B. K., ZIDE, D. & UDENFRIEND, S. (1972) The use of dopamine β-hydroxylase as a marker for the central noradrenergic nervous system in rat brain. *Proc. Natl. Acad. Sci. U.S.A. 69*, 2722-2726

MAAS, J. (1975) in *The Psychobiology of Depression* (Mendels, J., ed.), pp. 1-6, Spectrum Publications, New York

SAAVEDRA, J. M., COYLE, J. T. & AXELROD, J. (1973) The distribution and properties of the nonspecific *N*-methyltransferase in brain. *J. Neurochem. 20*, 743-752

ZELLER, E. A., BOSHES, S., DAVIS, J. M. & THORNER, M. (1975) Molecular aberration in platelet monoamine oxidase in schizophrenia. *Lancet 1*, 1385

Appendix 1

Assay of monoamine oxidase

K. F. TIPTON* and M. B. H. YOUDIM†

**Department of Biochemistry, University of Cambridge and †Medical Research Council Unit and University Department of Clinical Pharmacology, Radcliffe Infirmary, Oxford*

INTRODUCTION

Many different methods have been used for the assay of monoamine oxidase activity (see Tipton 1975; Youdim 1975 for reviews) but this appendix will be restricted to assay methods using radioactively labelled substrates, since the high sensitivity of such methods has led to their wide use with crude tissue preparations that contain relatively low activities.

Principles of the methods

The most commonly used methods employ ^{14}C-labelled substrates although ^{3}H-labelled substrates have been used in some assay methods (McCaman *et al.* 1965; Jarrott 1971). The enzyme is incubated with the radioactive substrate and after fixed time intervals the reaction is stopped and the aldehyde product formed is separated from the unchanged amine and is determined by liquid scintillation counting. The aldehyde formed by the action of monoamine oxidase may undergo further transformations during the assay: oxidation to the corresponding acid will tend to occur in the presence of oxygen and this process will be accelerated in crude preparations that contain aldehyde dehydrogenase and NAD^+; in addition the aldehyde may be reduced to the corresponding alcohol if aldehyde reductase and NADPH are present in the sample. Thus any separation method employed must be capable of separating the aldehyde, acid and alcohol if the results are to be valid.

Two separation methods are in wide use. The products may be separated by passage of the mixture through a negatively charged ion exchange resin which will adsorb the positively charged amine but allow the products to pass through unretarded. Alternatively the products may be extracted into an organic solvent at low pH values, where they will be uncharged, whilst the positively charged amine remains in the aqueous phase. Both these methods have their strong protagonists but when used correctly they are each capable of yielding satisfactory results, and we will describe them both in detail.

Absolute values for enzyme activities

Although both the methods to be described are capable of yielding highly reproducible results, the values obtained are not easy to relate to the true specific activity of the enzyme with any given substrate. A number of studies have shown that the proportion of the total product separated varies with the substrate used and this appears to be due, at least in part, to strong binding of the aldehyde product to proteins in the enzyme preparation. Thus one would expect the yield to be also dependent on the nature of the enzyme preparation used. Although we list values for the recoveries of the products derived from different amines in this appendix, these values should not be assumed to be directly applicable with different enzyme preparations. Whilst the results obtained directly from the radiochemical assays may be satisfactory for many types of study involving a single substrate and enzyme preparation, they will not provide data that can be used for comparisons between very different enzyme preparations or different substrates unless the efficiency of separation is calculated for each substrate and preparation. This may be done directly by using a known amount of product added to the assay mixture (including the enzyme preparation) or indirectly by correlating the activity determined in the radiochemical assay with the results given by an assay method that gives absolute values—such as the polarographic determination of oxygen consumption (Tipton 1967) or the spectrophotometric determination of aldehyde production (Houslay & Tipton 1973*a*)—in order to provide a 'correction factor'.

General points concerning assay technique

One of us has already discussed a number of the points that must be borne in mind when assaying monoamine oxidase (Tipton, in the General Discussion at this symposium, p. 385) and we will not repeat these at great length here. The following points should, however, be considered.

(a) Substrate concentration. The linearity of any assay will, of course, depend on the presence of an adequate substrate concentration. Ideally the substrate concentration should be much greater than the K_m values so that changes in substrate concentration during the assay do not result in a decrease in the saturation of the enzyme, but the possibility of high substrate inhibition must

be checked. In the case of oxygen, which is the second substrate for monoamine oxidase, the K_m value exhibited by the enzyme is relatively high, being similar to the value for the oxygen solubility in air-saturated water at 30 °C (see e.g. Tipton 1975). Fortunately, the oxygen concentration in the assay may be maintained if the mixture is shaken vigorously and many workers have observed that such a procedure prolongs the linearity of the assay.

(b) Substrate purity. Impurities in the substrates may act as inhibitors of enzymes as well as causing errors in the calculation of the substrate concentration. In the case of monoamine oxidase, contamination of the amine substrates with the aldehyde product causes the appearance of high substrate inhibition as well as altering the K_m value (Houslay & Tipton 1973*b*). Amine substrates that are liquid as the free base may be purified by distillation or by conversion to their hydrochlorides followed by recrystallization, and those supplied as crystalline solids may require recrystallization. Most radioactive amines are supplied as crystalline salts at a high degree of purity, but recrystallization may be necessary with some samples. The extent of contamination with the aldehyde may be determined using aldehyde dehydrogenase, or the amount of labelled material that can be separated from the amine by ion exchange chromatography or organic solvent extraction may be used to gauge whether radioactive amines require further purification.

(c) Linearity of the assay. Since it is often convenient to assay samples by determining the amount of product formed in replicate samples after a fixed period of time, it is essential to establish that the reaction proceeds linearly for this time under all the conditions that are to be used in subsequent experiments. In addition it is necessary to show that the reaction velocity is proportional to the concentration of the enzyme preparation added.

Preparation of substrate solutions for radioactive assays

The radioactive amine is dissolved in 0.01 M-HCl and stored frozen at −20 °C in small aliquots. The solution for the assays may be prepared by adding sufficient of the radioactive amine and water to a freshly prepared solution of the unlabelled amine to give a final concentration of 5 mM and a specific radioactivity of about 50 μCi/mmol. Clearly the sensitivity of the assay method may be altered by either increasing or decreasing the specific radioactivity of the substrate. The substrate concentration recommended here is such that the addition of 100 μl of substrate to a reaction mixture with a total volume of 500 μl will give a final substrate concentration of 1 mM, which is considerably greater than the K_m values shown by monoamine oxidase for most substrates (see Houslay & Tipton 1974).

ASSAY PROCEDURES

I *The ion exchange method*

Materials required

0.1M-sodium phosphate buffer (pH 7.4)
Radioactive substrate solution (see above)
Amberlite CG 50 (H) Type I (100–200 mesh) ion exchange resin (from Rohm and Hass Inc. or BDH Ltd) prepared as described below
Pasteur pipettes (10 × 0.5 cm)
Constant-temperature water-bath with shaker
Glass beads (2.5–3.5 mm diameter)
Scintillation vials
Bray's solution (see below) or Insta Gel (Packard Inc., Illinois)

Preparation of the ion exchange resin. The method described is that of Pisano (1960). 250 g of the Amberlite are stirred in 750 ml of distilled water for 10 minutes and allowed to settle for about 30 minutes and the supernatant is removed by aspiration. This washing procedure is repeated three more times after which the supernatant liquid should be clear. The resin is then suspended in 750 ml of distilled water and 500 ml of 10.0M-NaOH are added over a period of 15 minutes, during which time the suspension is continuously stirred. The stirring is continued for a further two hours, after which the suspension is allowed to settle for 30 minutes and the supernatant is removed by aspiration. The resin is then washed five times (by stirring and decantation) each with 750 ml of distilled water. The resin, which is now in the sodium form, is converted to the acid form by the addition of 250 ml of 6M-HCl to a suspension in 750 ml of water and stirring for 30 minutes. After settling and aspiration of the supernatant the resin is washed with 750 ml of distilled water five times. The resin is then converted back to the sodium form by treatment with NaOH as described above and washed eight times with 750 ml of distilled water. The resin is then suspended in 750 ml of distilled water and glacial acetic acid is added with continuous stirring until the pH remains steady for 30 minutes at approximately 6.2–6.3 (about 25 ml of the acetic acid will be required). The resin may then be stored at 4 °C until required. The pH of the suspension should remain constant during storage but a check should be made each time before use and the pH readjusted to 6.2–6.3 with HCl or NaOH if necessary.

Preparation of Amberlite columns. One glass bead is inserted into each of the Pasteur pipettes and the columns are filled with the prepared Amberlite resin to a height of 2.5–3.0 cm when fully settled. The columns are each washed with 2.0 ml of distilled water.

Testing the Amberlite resin. Since the assay relies on the ability of the Amberlite column to retain all the unchanged amine, a column of the prepared resin should be checked for this ability, which is a measure of successful preparation of the resin. 0.1 ml of a suitable ^{14}C-labelled amine substrate is applied to the top of the column which is above a scintillation vial.Two ml of distilled water are added to the top of the column and allowed to pass through into the scintillation vial. Ten ml of a suitable scintillation mixture (see below) are then added and the mixture is counted in a scintillation counter. The column should retain 100% of pure radioactive amine and preparations of the Amberlite that do not do this should not be used for assays. Columns that have been used for testing in this way should not be used in subsequent assays.

Bray's solution (Bray 1960). 60 g naphthalene, 4 g PPO (2,5-diphenyloxazole), 0.2 g POPOP (1,4-di-2-(5-phenyloxazolyl)benzene), 100 ml methanol and 20 ml ethylene glycol are made up to 1 litre with *p*-dioxane.

Assay procedure. The method is taken from Robinson *et al.* (1968) as modified by Southgate & Collins (1969).

0.3 ml of 0.1M-phosphate buffer and 0.1 ml of the enzyme preparation are placed in test-tubes and equilibrated with shaking at 30 °C. The reaction is initiated by the addition of 0.1 ml of the substrate. After the reaction has been allowed to proceed for a fixed time the tubes are transferred to a bath containing an ice–salt mixture to stop the reaction by rapid cooling. The contents of each test-tube are then transferred to the top of one of the Amberlite columns which is positioned above a scintillation vial. After the reaction mixture has run into the column the test-tube is rinsed with 2 ml of distilled water which is then passed through the column. Ten ml of Bray's solution are then added to each of the scintillation vials and the contents are mixed and counted for radioactivity.

Blanks may be prepared either by using a sample of the enzyme that has been inactivated by boiling or by adding the native preparation to a buffer–substrate mixture after it has been cooled in the salt–ice mixture and immediately before it is applied to one of the Amberlite columns.

Special points and limitations. The presence of large amounts of protein or a

TABLE 1

Recoveries of amine metabolites from Amberlite CG-50 columns

Compound	*Percentage recovery* Pure compound	*Percentage recovery* Compound plus MAO	Reference
Indoleacetaldehyde	95–100	57	Southgate & Collins (1969)
Indoleacetic acid	95–100	76	Southgate & Collins (1969)
Tryptophol	100	100	Southgate & Collins (1969)
p-Hydroxyphenylacetaldehyde	85–93	65	M. B. H. Youdim (unpublished)
p-Hydroxyphenylacetic acid	89–95	77	M. B. H. Youdim (unpublished)
Dihydroxyphenylacetic acid	95–100	53	M. B. H. Youdim (unpublished)
5-Hydroxyindoleacetic acid	90–100	78	M. B. H. Youdim (unpublished)

Solutions of the compounds were applied to the columns and eluted as described in the text. In the samples containing monoamine oxidase (MAO) the samples were made 1–3 mg/ml with a preparation of the enzyme that had been solubilized and concentrated by ammonium sulphate precipitation (Youdim & Sourkes 1966).

high ionic strength may cause elution of radioactive amine from the column and thus it is essential to limit the concentration of the enzyme solution to 3 mg/ml or less and to use a buffer concentration of 100 mM or less. Under these conditions the reaction has been found to proceed linearly for thirty minutes in many cases. A number of substrates are suitable for assay by this method and amongst the biogenic amines 5-hydroxytryptamine and noradrenaline (or adrenaline) are the only substrates that do not yield satisfactory results. This may be due to the aldehydes derived from these amines forming relatively strong complexes with the proteins in the enzyme preparation (see Southgate & Collins 1969; Alivisatos & Ungar 1968).

In the absence of any added protein the aldehyde, acid and alcohol products can be quantitatively separated from the parent amine by this method (Southgate & Collins 1969; Jain *et al.* 1973), but unfortunately the presence of preparations of monoamine oxidase affects the separation (see Table 1). Although the available data are not exhaustive, it appears from the results shown in Table 1 that the presence of protein causes a decrease in the recoveries of the aldehyde and acid products with, perhaps, little change in the recovery of the alcohol. The different recoveries of the products could lead to errors in the estimation of velocities, particularly if the activities of preparations contaminated with aldehyde dehydrogenase or aldehyde reductase (plus their appropriate coenzymes) are compared with samples that are not so contaminated. The addition of aldehyde dehydrogenase and NAD^+ to the enzyme preparation has been used to ensure that the aldehyde produced is rapidly converted to the corresponding acid (Lovenberg *et al.* 1962; Hidaka *et al.* 1967), but it will be noted from Table 1 that this would not be expected to yield 100% recoveries of product from the column, since there appears to be some binding of the acid metabolite to protein in the enzyme preparation. The degree of binding of the acid metabolite would be expected to depend on the enzyme preparation used and thus introduce errors when the activities of different enzyme preparations are being compared. The use of aldehyde reductase plus NADPH to convert the aldehyde to the alcohol would seem to be a more satisfactory procedure, but the K_m values of this enzyme are such that relatively high concentrations would be needed to ensure rapid removal of the aldehyde (see Turner & Tipton 1972), and, to our knowledge, no one has yet attempted the radioactive assay under these conditions. An alternative approach used by Jain *et al.* (1973) is to trap the aldehyde formed by the presence of semicarbazide (20 mM) in the reaction mixture, and the results obtained by these workers suggest that this may well be the most satisfactory method. We do not, however, recommend the addition of cyanide as recommended by these workers since this compound acts as a reversible inhibitor of monoamine oxidase (Houslay & Tipton 1973*b*) and can also affect the sensitivity of the enzyme to some irreversible inhibitors (Davison 1958).

II *The solvent extraction method*

Materials required

0.1M-sodium phosphate buffer (pH 7.2)
2.0M-citric acid
Radioactive substrate solution (see above)
Extraction solvent (toluene or benzene: ethyl acetate—1:1, vol:vol) containing 0.6% PPO (2,5-diphenyloxazole)
McCartney bottles (10 × 1.5 cm screw-cap culture tubes) with plastic screw-caps
Bench centrifuge (capable of accepting McCartney bottles)
Constant-temperature water-bath with shaker
Scintillation vials
Vortex mixer (desirable but not essential)
Ethanol plus dry ice or a freezer set at about −20 °C

Procedure. The method is essentially that of Otsuka & Kobayashi (1964).

Each McCartney bottle containing 0.3 ml phosphate buffer and 0.1 ml of enzyme is incubated at 30 °C in the shaking water bath for five minutes to allow the contents to come to that temperature. 0.1 ml of the radioactive substrate is then added to initiate the reaction and the mixtures are shaken for the desired time. The reaction is then terminated by the rapid addition of 0.5 ml of the citric acid solution.

Ten ml of the extraction solvent containing PPO are then added and each McCartney bottle is capped and shaken vigorously (either by hand or using a vortex mixer) for 30 seconds and the bottles are then centrifuged briefly (e.g. 2000 r.p.m. for 20 seconds) in the bench centrifuge to separate the two layers. The organic layer may then be separated from the aqueous layer by 'freezing out'. The bottles may either be placed in a deep freezer until the aqueous layer is frozen, or in a vessel containing 96 % ethanol, which has been cooled by the addition of an excess of dry ice, for one minute. Either of these procedures will freeze the aqueous layer and allow the organic layer to be poured off into a scintillation vial for counting. Blanks may be prepared by adding enzyme to the assay mixture after the addition of citrate or by using a sample of the enzyme that has been inactivated by boiling.

Choice of extraction solvent. Many different extraction solvents have been used with this assay (see e.g.: Otsuka & Kobayashi 1964; McCaman *et al.* 1965; Wurtman & Axelrod 1963; Southgate & Collins 1969; Jain *et al.* 1973). Unfortunately no single solvent is suitable for all amine substrates and, as can be seen from Table 2, toluene, which is suitable for most amines, is poor at extracting the products derived from 5-hydroxytryptamine although benzene: ethyl acetate is suitable in this case.*

Special points and limitations. In general this method is not greatly affected by high ionic strengths although it is important that the buffering power of the assay medium does not prevent the citrate from lowering pH sufficiently to protonate the unchanged amine and any of the acid product that may be formed (see Otsuka & Kobayashi 1964). The time-course of the reaction has been shown to be linear for at least 15 minutes under the conditions of the standard assay and at enzyme concentrations (rat liver mitochondria) of up to 2.5 mg/ml in the assay mixture.

The efficiency of extraction of the products derived from any given amine

* Benzene is extremely toxic and T. J. Mantle (unpublished observations) has recently shown that toluene: ethyl acetate (1:1, vol:vol) is suitable for extracting the products of 5-hydroxytryptamine oxidation.

TABLE 2

Recoveries of amine metabolites from a single extraction with an organic solvent

Products from the oxidation of:	*Percentage extracted*			*Reference*
	Anisole	*Toluene*	*Benzene: ethyl acetate*	
Tyramine	70	35	–	Otsuka & Kobayashi (1964)
	54	25	65	T. J. Mantle, N. J. Garrett & K. F. Tipton (unpublished)
Tryptamine	95	>95	58	
5-Hydroxytryptamine	45	6	91	
Benzylamine	85	93	71	
Dopamine	16	2	18	

The extraction efficiencies are quoted as percentages of the total product recovered. Reactions and extractions were carried out as described in the text. In the experiments of Otsuka & Kobayashi a preparation of hog kidney monoamine oxidase at a concentration of 0.112 mg N per assay was used and the total volume of the assay mixture was four times that described in the text. In the other experiments a preparation of rat liver mitochondrial outer membranes at a concentration of about 1.0 mg per assay was used.

will depend on the partition coefficient between the two phases and 100% recovery would not be expected in a single extraction (see Otsuka & Kobayashi 1964; Jain *et al.* 1973). Thus this method relies on the ability of the extraction to separate a fixed proportion of the products reproducibly. The situation regarding differences in the extractabilities of the different products that may be formed is far from clear; Southgate & Collins (1969) reported that indoleacetic acid was more readily extracted into toluene than indoleacetaldehyde but Jain *et al.* (1973) found that the aldehyde was more readily extracted into the same solvent. Whatever the reason for this discrepancy it would seem that such differences do not significantly affect the assay method, since neither these workers nor Wurtman & Axelrod (1963) could detect any significant change in the amount of product extracted from tryptamine oxidation in assay mixtures containing aldehyde dehydrogenase. Similarly the presence of aldehyde dehydrogenase has been found to have no effect on the activities determined by this method when tyramine (Otsuka & Kobayashi 1964) or 5-hydroxytryptamine (T. J. Mantle, N. J. Garrett & K. F. Tipton, unpublished work) is used as the substrate. Binding of the products of the reaction to proteins in the enzyme preparation may affect their extractability (Southgate & Collins 1969) and the addition of high concentrations (100 mg/ml) of bovine serum albumin has been found to decrease the amount of product that can be extracted in a single extraction after the oxidation of benzylamine or 5-hydroxytryptamine

(T. J. Mantle, N. J. Garrett & K. F. Tipton, unpublished work). This binding would appear to be reversible, since successive extractions have been shown to remove the products quantitatively under the normal assay conditions (Otsuka & Kobayashi 1964; Jain *et al.* 1973). Jain *et al.* (1973) recommend the addition of semicarbazide (20 mM) to assay mixtures to trap the aldehyde formed and their results suggest that this may be the most satisfactory procedure. As stated earlier, we do not recommend the inclusion of cyanide in the assay mixture.

As can be seen from Table 2, this method is suitable for most amines but no suitable extraction solvent has yet been found for dopamine. In addition, we know of no solvent that is suitable for determining activity with noradrenaline as the substrate.

INTERPRETATION OF RESULTS

The counts registered by the scintillation counter must be corrected for quenching, so that the results may be expressed as disintegrations per minute (d.p.m.). Methods for determining quench corrections have been reviewed by Dyer (1974) and Peng (1970). The amount of product formed can then be calculated from the formula

$$\frac{(\text{Total d.p.m. of product formed} \sim \text{total d.p.m. of blank}) \times K}{\text{Specific activity of substrate}}$$

where the specific activity of the substrate is expressed in d.p.m. per nmole and

$$K = \frac{100}{\text{percentage recovery of product}}$$

which, as previously mentioned, must be determined separately. Application of this formula will allow the activity of the enzyme to be expressed in nmoles of product formed per min per mg of enzyme preparation protein.

References

ALIVISATOS, S. G. A. & UNGAR, F. (1968) Incorporation of radioactivity from labelled serotonin and tryptamine into acid insoluble material from subcellular fractions of brain. I. The nature of the substrate. *Biochemistry 7*, 285-292

BRAY, G. A. (1960) A simple efficient liquid scintillator for counting aqueous solutions in a liquid scintillation counter. *Anal. Biochem. 1*, 279-285

DAVISON, A. N. (1958) Physiological role of monoamine oxidase. *Physiol. Rev. 38*, 729-747

DYER, A. (1974) *An Introduction to Liquid Scintillation Counting*, pp. 57-74, Heyden & Sons, London

HIDAKA, H., NAGATSU, T. & YAGI, K. (1967) Micro-determination of monoamine oxidase using serotonin as substrate. *J. Biochem. 62*, 621-623

HOUSLAY, M. D. & TIPTON, K. F. (1973*a*) The nature of the electrophoretically separable multiple forms of rat liver monoamine oxidase. *Biochem. J. 135*, 173-186

HOUSLAY, M. D. & TIPTON, K. F. (1973*b*) The reaction pathway of membrane-bound rat liver mitochondrial monoamine oxidase. *Biochem. J. 135*, 735-750

HOUSLAY, M. D. & TIPTON, K. F. (1974) A kinetic evaluation of monoamine oxidase activity in rat liver mitochondrial outer membranes. *Biochem. J. 139*, 645-652

JAIN, M., SANDS, F. & VON KORFF, R.W. (1973) Monoamine oxidase activity measurements using radioactive substrates. *Anal. Biochem. 52*, 542-554

JARROTT, B. (1971) Occurrence and properties of monoamine oxidase in adrenergic neurones. *J. Neurochem. 18*, 7-16

LOVENBERG, W., LEVINE, R. J. & SJOERDSMA, A. (1962) A sensitive assay of monoamine oxidase activity *in vitro:* application to heart and sympathetic ganglia. *J. Pharmacol. Exp. Ther. 135*, 7-10

MCCAMAN, R. E., MCCAMAN, M. W., HUNT, J. M. & SMITH, M. S. (1965) Microdetermination of monoamine oxidase and 5-hydroxytryptophan decarboxylase activities in nervous tissues. *J. Neurochem. 12*, 15-23

OTSUKA, S. & KOBAYASHI, Y. (1964) A radioisotopic assay for monoamine oxidase determinations in human plasma. *Biochem. Pharmacol. 13*, 995-1006

PENG, C. T. (1970) A review of methods of quench correction in liquid scintillation counting, in *The Current Status of Liquid Scintillation Counting* (Bransome, E. D., ed.), pp. 283-292, Grune & Stratton, New York

PISANO, J. (1960) A simple analysis for normetanephrine and metanephrine in urine. *Clin. Chim. Acta 5*, 406-412

ROBINSON, D. S., LOVENBERG, W., KEISER, H. & SJOERDSMA, A. (1968) Effects of drugs on human blood platelet and plasma amine oxidase activity *in vitro* and *in vivo*. *Biochem. Pharmacol. 17*, 109-119

SOUTHGATE, J. & COLLINS, G. G. S. (1969) The estimation of monoamine oxidase using ^{14}C-labelled substrates. *Biochem. Pharmacol. 18*, 2285-2287

TIPTON, K. F. (1967) The submitochondrial localization of monoamine oxidase in rat liver and brain. *Biochim. Biophys. Acta 135*, 910-920

TIPTON, K. F. (1975) Monoamine oxidase, in *Handbook of Physiology*, Section 7: *Endocrinology*, vol. 6: *Adrenal Gland* (Blaschko, H. K. F. & Smith, A. D., eds.), pp. 667-691, American Physiological Society, Washington, D.C.

TURNER, A. J. & TIPTON, K. F. (1972) The characterization of two reduced nicotinamide–adenine dinucleotide phosphate-linked aldehyde reductases from pig brain. *Biochem. J. 130*, 765-772

WURTMAN, R. J. & AXELROD, J. (1963) A sensitive and specific assay for the estimation of monoamine oxidase. *Biochem. Pharmacol. 12*, 1439-1441

YOUDIM, M. B. H. (1975) Assay and purification of brain monoamine oxidase, in *Research Methods in Neurochemistry* (Marks, N. & Rodnight, R., eds.), pp. 167-208, Plenum Press, New York

YOUDIM, M. B. H. & SOURKES, T. L. (1966) Properties of purified, soluble monoamine oxidase. *Can. J. Biochem. 44*, 1397-1400

Appendix 2

Preparation of human platelets

M. B. H. YOUDIM

Medical Research Council Unit and University Department of Clinical Pharmacology, Radcliffe Infirmary, Oxford

A great deal of attention has recently been focused on the monoamine oxidase activities in human platelets as a possible index of mental abnormalities (see e.g. the contributions by Murphy, pp. 341-351, and by Crow, pp. 353-369, to this symposium). Variability in the results of such studies may arise from failure to make pure preparations or to satisfactorily disrupt the platelets to allow free access of substrate to the enzyme, and this appendix outlines a method for the preparation and disruption of human platelets which has proved to be reproducible. It should be emphasized that the preparation of platelets depends *inter alia* on the density of the original blood sample and the method outlined below will not necessarily be applicable to preparations from other species.

Twenty ml samples of blood are collected in plastic centrifuge tubes containing 2 ml of 0.129M-sodium citrate. The tubes are centrifuged at 200 *g* for 10 minutes and the supernatant (platelet-rich plasma) is carefully poured into clean plastic centrifuge tubes and centrifuged at 2000 *g* for 10 minutes. The supernatant is discarded and the sedimented platelets are washed three times with 1.0 ml aliquots of 0.3M-sucrose by suspension and resedimentation at 2000 *g* for 10 minutes. The platelets are finally resuspended in 1.0 ml of 0.3M-sucrose and disrupted by sonication for 20 seconds at a low energy (Dawe Soniprobe; with microtip; 2 kHz, 3mA). All these procedures are carried out at 4 °C. The platelet preparation obtained in this way may be stored for up to one month at −20 °C without appreciable loss in monoamine oxidase activity (Youdim *et al.* 1975).

Platelet monoamine oxidase from a number of sources has been shown to be predominantly the B species in terms of inhibitor sensitivity and substrate specificity (see e.g. Sandler *et al.* 1974; Murphy & Donnelly 1974). Thus the most suitable substrates, for obtaining good activities, would include benzylamine or 2-phenethylamine (Murphy & Donnelly 1974; Houslay & Tipton 1974).

References

HOUSLAY, M. D. & TIPTON, K. F. (1974) A kinetic evaluation of monoamine oxidase activity in rat liver mitochondrial outer membranes. *Biochem. J. 139*, 645-652

MURPHY, D. L. & DONNELLY, C. H. (1974) Monoamine oxidase in man: enzyme characteristics in platelets, plasma and other human tissues. *Adv. Biochem. Psychopharmacol. 12*, 71-85

SANDLER, M., BONHAM CARTER, S., GOODWIN, B. L., RUTHVEN, C. R. J., YOUDIM, M. B. H., HANINGTON, E., CUTHBERT, M. F. & PARE, C. M. B. (1974) Multiple forms of monoamine oxidase: some *in vivo* correlations. *Adv. Biochem. Psychopharmacol. 12*, 3-10

YOUDIM, M. B. H., WOODS, H. F., MITCHELL, B., GRAHAME-SMITH, D. G. & CALLENDER, S. (1975) Human platelet monoamine oxidase activity in iron-deficiency anaemia. *Clin. Sci. Mol. Med. 48*, 289-295

Index of contributors

Entries in **bold** *type indicate papers; other entries are contributions to discussions*

Indexes compiled by William Hill

Subject index